# ADVANCES IN MEDICINAL CHEMISTRY

*Volume 1* • 1992

# ADVANCES IN
# MEDICINAL CHEMISTRY

*A Research Annual*

*Editors:* BRUCE E. MARYANOFF
*Medicinal Chemistry Department*
*R.W. Johnson Pharmaceutical*
*Research Institute*

CYNTHIA A. MARYANOFF
*Chemical Development Department*
*R.W. Johnson Pharmaceutical*
*Research Institute*

VOLUME 1 • 1992

 JAI PRESS INC.

*Greenwich, Connecticut*          *London, England*

RS400
A424
vol. 1
1992

# CONTENTS

# LIST OF CONTRIBUTORS

Harry A. Albrecht

Hoffmann-La Roche, Inc.
Nutley, New Jersey

James G. Christenson

Hoffmann-La Roche, Inc.
Nutley, New Jersey

Cynthia M. Cribbs

Department of Chemistry
Stanford University
Palo Alto, California

Israel Hanin

Department of Pharmacology and
    Experimental Therapeutics
Loyola University Stritch
    School of Medicine
Maywood, Illinois

Alan P. Kozikowski*

Departments of Chemistry and Behavioral
    Neuroscience
University of Pittsburgh
Pittsburgh, Pennsylvania

Allen Krantz

Syntex Research
Mississauga, Ontario
Canada

Douglas A. Livingston†

The Upjohn Company
Kalamazoo, Michigan

Joseph M. Muchowski

Syntex Research
Palo Alto, California

---

*Now with: Neurochemistry Research, Mayo Clinic Jacksonville, Jacksonville, Florida
†Now with: LaJolla Pharmaceutical Company, San Diego, California

*Ichiro Shinkai*                                    Merck Sharp & Dohme
                                                         Research Laboratories
                                                     Rahway, New Jersey


*Nolan H. Sigal*                                    Merck Sharp & Dohme
                                                         Research Laboratories
                                                     Rahway, New Jersey


*X. C. Tang*                                         Department of Pharmacology and
                                                         Experimental Therapeutics
                                                     Loyola University Stritch
                                                         School of Medicine
                                                     Maywood, Illinois


*Edda Thiels*                                        Department of Behavioral Neuroscience
                                                     University of Pittsburgh
                                                     Pittsburgh, Pennsylvania


*Paul A. Wender*                                    Department of Chemistry
                                                     Stanford University
                                                     Palo Alto, California

# INTRODUCTION TO THE SERIES:
# AN EDITOR'S FOREWORD

The JAI series in chemistry has come of age over the past several years. Each of the volumes already published contain timely chapters by leading exponents in the field who have placed their own contributions in a perspective that provides insight to their long-term research goals. Each contribution focuses on the individual author's own work as well as the studies of others that address related problems. The series is intended to provide the reader with in-depth accounts of important principles as well as insight into the nuances and subtleties of a given area of chemistry. The wide coverage of material should be of interest to graduate students, postdoctoral fellows, industrial chemists and those teaching specialized topics to graduate students. We hope that we will continue to provide you with a sense of stimulation and enjoyment of the various sub-disciplines of chemistry.

Department of Chemistry                                          Albert Padwa
Emory University                                              *Consulting Editor*
Atlanta, Georgia

# PREFACE

Although medicinal chemistry, in the modern sense, has been practiced since about the turn of the century, its status as a scientific discipline really blossomed just after World War II. This was sparked largely by major successes, such as those associated with sulfa drugs, opiate analgesics, penicillin, and hydrocortisone. Around 1960, the field took a leap forward in terms of scientific respectability with the advent of the *Journal of Medicinal Chemistry* (1958; the current journal name is used here) and the *Annual Reports in Medicinal Chemistry* (1965). Medical research has burgeoned over the past three decades and activity in medicinal chemistry has followed proportionately. Now, the discipline is pursued by a variety of practitioners, people who call themselves organic chemists, bioorganic chemists, biochemists, pharmacologists, inorganic chemists, physical chemists, theoretical chemists, and computational chemists . . . and, of course, medicinal chemists. Medicinal chemistry is particularly exciting because it demands a multidisciplinary approach to achieve meaningful goals. Also, it frequently deals with sophisticated molecules, interesting biochemical mechanisms, and underexplored physiological processes.

Given the present hotbed of activity in medicinal chemistry and allied fields throughout the world, we are inaugurating a new book series called *"Advances in Medicinal Chemistry."* This series will bring together in one forum accounts of drug research and drug development, involving many of the diverse fields mentioned above. This forum will collect important chemical work from industrial, academic, and government laboratories. The emphasis will be on synthetic chemistry, chemical development of drugs, structure-activity studies, enzyme- and receptor-based drug design, concepts of de novo drug design, and chemical approaches to elucidating biochemical processes of relevance to drug discovery. As one of the hallmarks of this series, we have encouraged authors to use an informal narrative style, and have afforded them broad latitude for telling their story. The authors even have license to describe their efforts as they really occurred, with all of the behind-the-scenes background and realism (if they so dare).

Welcome to this first volume of *Advances in Medicinal Chemistry*, which is comprised of seven chapters contributed by noted specialists. These chapters span a wide range of topics, from Alzheimer's disease to "SNAP" chemistry. Sit back and enjoy.

The volume opens with an account of the quest in Dr. Wender's research group for a tumor-promoter pharmacophore. He and Dr. Cribbs describe the importance of protein kinase C (PKC) by virtue of its role in cellular signal transduction and in the regulation of cell proliferation. The authors coherently review the structural basis for PKC activation and inhibition, and highlight its importance to carcinogenesis. In the end, their work offers a logical explanation for sterically dissimilar families of potent tumor promoters and a structural template for simplified compounds that exhibit high affinities for the PKC receptor.

The second chapter reflects on the need for effective and safe immunosuppressants in the era of transplantation therapy, and relates this to cyclosporin A and FK-506. Drs. Shinkai and Sigal discuss the first enantioselective total synthesis of FK-506, a 23-membered macrolide discovered in 1985 in fermentation broths, and review the synthetic endeavors of other laboratories. They also provide information on the cellular and biochemical actions of FK-506. This area has recently been very hot, as it deals with a newly identified class of enzymes called peptidyl-prolyl *cis-trans* isomerases, which appear to play a critical role in lymphocyte signal transduction.

Ketorolac is the result of a mission at Syntex to develop and market a new, highly efficacious, nonopiate analgesic. Dr. Muchowski delivers a stirring account of the development track for ketorolac, from a chemical perspective. Besides yielding an important drug, work on this project led to novel contributions to pyrrole chemistry. This product is now on the market in several countries for the treatment of moderate-to-severe pain.

The primary focus of the fourth chapter, by Dr. Livingston, is the development of a method for the practical utilization of Upjohn's "sitosterol pile." He discusses various synthetic attempts to produce valuable steroidal materials, which culminated in a novel silicon-based homologation of cyanohydrins. This efficient and versatile technology is able to transform the readily available 17-ketones into commercially significant corticosteroids, 17α-hydroxy progesterones, and progesterones.

Dr. Kozikowski and co-workers bring huperzine A, a polycyclic alkaloid first identified in China, to center stage. This compound has memory-enhancing (nootropic) properties attributed to its ability to inhibit acetylcholinesterase; thus, it is potentially useful for treating Alzheimer's disease. They describe a concise total synthesis of huperzine A (racemic and enantiomeric forms), the preparation of analogues, and biological studies.

The β-lactam ring of a cephalosporin with antibacterial activity is generally cleaved by bacterial enzymes, often liberating the 3'-substituent. When the eliminated group is, itself, an antibacterial agent, the drug can exert a dual mechanism of action. Mr. Albrecht and Dr. Christenson relate the application of this novel idea to finding a drug that combines the activity profile of a traditional cephalosporin with that of a quinolone-type agent.

The final chapter by Dr. Krantz deals with the intriguing concept of quiescent affinity labels. Dipeptidyl acyloxymethyl ketones, of general structure Z-[AA']-[AA]-CH$_2$OC(O)Ar, serve as selective, irreversible inhibitors of the cysteine proteinase cathepsin B. The key to these molecules is the weak acyloxy leaving group, which is displaced by a nucleophilic group present in the enzyme's active site.

Our goal is for this book, and this series as a whole, to serve the scientific community in a manner that complements other reviews and books in medicinal chemistry. It should be clear from many of the chapters in Volume 1 that we wish to emphasize synthetic chemistry and the crucial role that it serves in various facets of medicinal chemical research. We also want to promote a liberal style that confers some personality and candor to the accounts. In future volumes, we will continue to present research stories that are not necessarily mainstream. And, if we do have a topic that is overexposed in the current secondary literature, we will endeavor to have it presented from a fresh perspective, which emanates more from an author's individual research experiences than from the intent of reviewing a field. If you have an idea for a chapter of this type, dealing with medicinal chemistry from a discovery and/or development viewpoint, please consider sending us a proposal.

Finally, we thank the contributing authors for their fine chapters, Dr. Albert Padwa for recruiting us over drinks at a symposium in Blacksburg, Virginia, and the R.W. Johnson Pharmaceutical Research Institute for supporting our editorial efforts.

R.W. Johnson Pharmaceutical                                    Bruce E. Maryanoff
    Research Institute                                         Cynthia A. Maryanoff
Spring House, Pennsylvania 19477                                    *Series Editors*

# COMPUTER ASSISTED MOLECULAR DESIGN RELATED TO THE PROTEIN KINASE C RECEPTOR

Paul A. Wender and Cynthia M. Cribbs

Advances in Medicinal Chemistry,
Volume 1, pages 1–53.
Copyright © 1992 by JAI Press Inc.
All rights of reproduction in any form reserved.
ISBN: 1-55938-170-1

## I.  INTRODUCTION

Protein kinase C (PKC) was first discovered in 1977 as a proteolytically activatable enzyme in bovine cerebellum and rat brain tissues.[1,2] Since then, eight isozymes have been identified in species as diverse as nematodes,[3] fruit flies,[4] and sea urchins,[4] as well as various vertebrates;[5] in mammals, PKC has been observed in every tissue type with the exception of red blood cells.[4,6,7] PKC's importance in signal transduction and in the regulation of cell proliferation has made it one of the most aggressively pursued subjects of chemical and biological research in the last decade. Indeed, there has been a virtual explosion of information as evidenced by the rapid rise in the number of publications on this subject appearing each year in journals covered by the Chemical Abstracts Service (Figure 1).

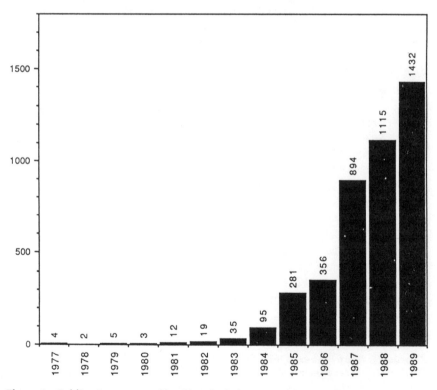

**Figure 1.**  Publications  covered by *Chemical Abstracts* each year referring to "protein kinase C" (PKC).

Contributing significantly to the expansion of interest in PKC was the finding in the early 1980s that PKC is the receptor for tumor-promoting phorbol esters.[8,9] Phorbol myristate acetate (PMA; also referred to as 12-*O*-tetradecanoylphorbol-13-acetate, TPA), the most potent of the tumor promoters, is isolated from the seeds of the leafy shrub *Croton tiglium* L., a member of the *Euphorbiaceae* family of plants indigenous to southeast Asia.[10,11] The phorbol esters have themselves been a target of research since the discovery a half-century ago that these plant-derived compounds enhance the effect of carcinogens while possessing no innate carcinogenicity.

The intent of this article is to provide an overview of the structural basis for PKC activation and inhibition. The physical properties of PKC will be summarized along with the role of PKC in signal transduction, ion channeling, and tumor promotion. Finally, the structural features of naturally occurring PKC agonists and antagonists will be surveyed, and a pharmacophore model for PKC activation will be presented, along with the *de novo* design and synthesis of PKC activators based on the model.

## II. PHYSICAL PROPERTIES OF PKC

The first reports of the purification and characterization of a cyclic nucleotide-independent protein kinase came from Nishizuka's laboratories in 1977.[1,2] Initially, the enzyme was found as a proteolytically activated protein kinase which, independent of cyclic nucleotides, was capable of phosphorylating histone and protamine. It was first thought to be produced from a proenzyme by proteolytic degradation catalyzed by a calcium-dependent protease that was present in the same tissue. The catalytic fragment was initially referred to as protein kinase M. It is now understood that the "proenzyme" is, in fact, the species referred to as protein kinase C (PKC). PKC is normally present in the cytosol of cells in an inactive form; however, in the presence of micromolar concentrations of $Ca^{2+}$, the enzyme is reversibly associated with phospholipids of the cell membrane where it is activated by endogenous factors.[12] Proteolysis can occur between the regulatory and catalytic domains to give a fragment which is active in the absence of the endogenous factors.[13-15] This proteolysis is more facile for membrane-bound PKC than for cytosolic PKC.[16]

Once the initial identification of PKC was made, the enzyme was found in a variety of species and cell types.[3-7] The complete primary structure of PKC was determined independently by several groups in the mid-1980s.[17-24] PKC was purified to approximately 70% purity from bovine brain. Then, through the use of oligonucleotide probes based on partial amino acid sequences, complementary DNA (cDNA) clones were derived from bovine brain cDNA libraries, thus allowing for the determination of the complete bovine brain PKC amino acid sequence.[17] One of the clones, designated PKCα, expressed a protein which corresponded identically to the isolated protein sequence. Two closely related cDNA sequences,

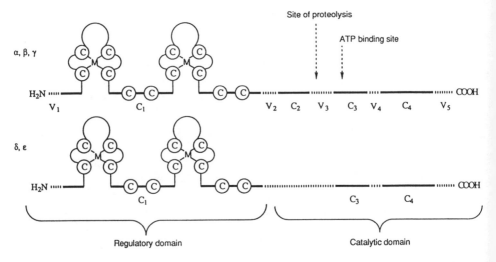

**Figure 2.** Structural features of the two major classes of PKC isozymes as determined by their sequence homology. C = cysteine, M = metal. (See reference 27 for further description.)

termed PKCβ and PKCγ, were found to correlate to two highly homologous proteins.[18] To date, eight isozymes of PKC have been identified that are encoded by cDNA clones α, β, γ, δ, ε, and ζ, all located on different chromosomes. By alternative splicing, the cDNA corresponding to the β isozyme is expressed as two different isozymes, βI and βII, which differ only in a region of approximately 50 amino acid residues at the carboxy terminus, and even there show a high degree of homology.[25] Two different cDNAs, encoding the sequences of ε and ε' isozymes, are nearly identical except for a small region at the 5' terminus and may also arise from alternate splicing of a single mRNA transcript.[26]

Figure 2 depicts the conserved and variable regions in the sequences of the various PKC isozymes, which are separated into two classes based on the degree of homology.[27] The α, βI, βII and γ isozymes are the most closely related, each having four conserved regions $C_1$ through $C_4$, and five variable regions $V_1$ through $V_5$.[28] The regulatory domain at the amino terminus is proteolytically cleaved from the catalytic domain in the third variable region.[13–15] The highest degree of homology is shared by the two β isozymes. The second class of PKC isozymes, containing the δ, ε, ε', and ζ isozymes, lacks the second conserved region.[26] Isozyme types do not vary significantly between different animals. For example, bovine PKCβ is closer to human PKCβ, with an amino acid sequence conservation of 98.4%, than it is to bovine PKCα.[18] The high degree of conservation across animal species is unusual among enzyme systems studied to date but is consistent with the multitude of functions of PKC. In related families of enzymes, divergent

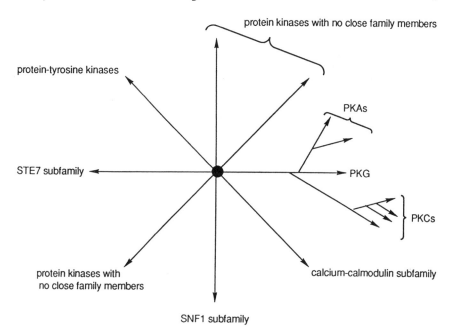

protein kinases with no close family members

protein-tyrosine kinases

PKAs

STE7 subfamily

PKG

PKCs

protein kinases with
no close family members

calcium-calmodulin subfamily

SNF1 subfamily

**Figure 3.** Depiction of a phylogeny tree of the catalytic domains of protein kinases whose amino acid sequences were reported by November, 1987. (See reference 33 for the actual tree and details concerning its generation.)

sequences play one of two roles: for some, the variable regions are merely spacers to facilitate protein folding,[29] while in others, they are thought to confer specificity for ligand binding.[30–32] In the case of PKC, the latter role is thought to be operative because there is a high degree of conservation within the variable regions of each isozyme across species.[18]

The eight isozymes that make up the PKC family are part of a larger group of protein kinases comprising nearly 100 members.[33] The catalytic domains of the protein kinases range from 250 to 300 residues, or about 30 kDa, and have certain key residues in highly conserved regions which characterize the amino acid residues as representing protein kinases.[34,35] In general, the catalytic domain lies near the carboxy terminus while the regulatory domain lies near the amino terminus. Protein kinases fall into two major classes with respect to substrate specificity, phosphorylating the hydroxyl group of either serine/threonine residues (as does PKC, for example[36]) or tyrosine residues. Hank et al. prepared a phylogenic tree of protein kinase catalytic domains whose sequences were available as of November 1987, depicted in Figure 3.[33]

Examination of the amino acid sequences in the region of the phosphorylation site of substrate proteins of PKC has shown a requirement for basic residue determinants, with potent synthetic substrates containing basic residues on either side of the phosphoryl acceptor. In the first conserved region of each isozyme, there is a pseudosubstrate region composed of 18 amino acid residues. This roughly corresponds to sequences found in many of the substrate proteins in terms of a juxtaposition of basic residues, with the exception that the pseudosubstrate region possesses no serine or threonine residue to act as a phosphoryl acceptor. The pseudosubstrate region is thought to occupy the substrate binding site in the absence of PKC activators. Consistent with this analysis, a synthetic peptide corresponding to the pseudosubstrate sequence (Arg$^{19}$/Phe-Ala-Arg-Lys-Gly-Ala$^{25}$-Leu-Arg-Glu-Lys-Asn-Val$^{31}$) is a potent inhibitor of PKC with a 0.15 micromolar inhibition constant. Substitution of Ala$^{25}$ with a serine transforms this pseudosubstrate into a potent substrate.[37]

With the exception of PKCε' and PKCζ,[25] by far the smallest of the eight isozymes, each of the PKC isozymes contains in the first conserved region, $C_1$, a tandem repeat of a cysteine-rich sequence $C-X_2-C-X_{13-14}-C-X_2-C-X_7-C$ where X represents any amino acid.[38] Such a tandem repeat is characteristic of a zinc-finger motif, which has been identified in DNA binding proteins involved with transcriptional regulation and in metalloproteins.[39,40] There is no indication at this time, however, that PKC is involved in DNA interactions or in binding to metals (other than divalent calcium). It has been demonstrated with various deletion and point mutants of the γ isozyme of PKC that the cysteine-rich region is essential for binding of the phorbol esters to the enzyme, but it is not known whether the region interacts directly with the phorbol esters or influences their binding allosterically.[38] However, the conservation of the region suggests that the PKC family may play some as-yet undefined role in control of gene expression. One point of interest is that the $C_2$ region, lacking in the δ, ε, and ζ isozymes, appears to play some role in calcium sensitivity since a polypeptide fragment containing only $C_1$ could bind PKC activators in the absence of $Ca^{2+}$, while a fragment containing $C_1$ and $C_2$ required $Ca^{2+}$ in order for binding to occur.[38] In support of this hypothesis, it has been observed that the δ, ε, and ζ isozymes show measurable catalytic activity in the absence of $Ca^{2+}$.[25]

Table 1 lists the physical properties of PKC isozymes isolated from mammalian tissues,[41] while Table 2 lists the distribution of α, β, and γ isozymes throughout various organs.[42,43] The distribution of the δ, ε, and ζ isozymes has not yet been determined. Because there are differences among the enzymatic properties of the PKC isozymes, and because their locations within cells and in different cell types vary, there is a heterogeneity of response to PKC activation.[44] For instance, it is known that the endogenous activators of PKC are diacylglycerols. The response of the isozymes, however, is variable: PKCγ is activated to a lesser extent than PKCβ by diacylglycerols and phospholipids but to a greater extent than PKCβ by 25–100 micromolar concentrations of free arachidonic acid in the absence of $Ca^{2+}$.[45] PKCα has properties similar to PKCγ but it is activated by free arachidonic acid only when the $Ca^{2+}$ concentration is increased, and maximal activity is

**Table 1.** Physical Properties of the Eight Isozymes of PKC Isolated from Mammalian Tissues

| Isozyme | α | βI | βII | γ | δ | ε | ζ |
|---|---|---|---|---|---|---|---|
| Amino Acid Residues | 672 | 671 | 673 | 697 | 673 | 737 | 592 |
| Calculated Molecular Weight | 76,799 | 76,790 | 76,933 | 78,366 | 77,517 | 83,474 | 67,740 |
| Chromosome Location (Human) | 17 | 16 | 16 | 19 | ? | ? | ? |
| Activators (In addition to PS + DAG + Ca$^{2+}$) | AA + Ca$^{2+}$ | | | AA | | | |
| Tissue Expression | Universal | Some tissues and cells | Many tissues and cells | Brain and spinal cord | Many tissues | Brain only | Many tissues |

7

***Table 2.*** Distribution throughout Organs of the α, βI, βII, and γ Isozymes of PKC

| Isozyme: | Soluble Kinase Activity (Percent of Total) | | |
|---|---|---|---|
| | α | βI + βII | γ |
| Whole brain | 24.8 | 49.1 | 26.1 |
| Cerebellum | 14.4 | 33.5 | 52.1 |
| Cerebral cortex | 17.5 | 62.7 | 19.8 |
| Hippocamus | 34.8 | 40.5 | 24.9 |
| Hypothalamus | 37.7 | 44.1 | 18.3 |
| Spinal cord | 47.1 | 49.7 | 3.2 |
| Sciatic nerve | 92.9 | 7.1 | |
| Heart | 80.3 | 19.7 | |
| Kidney | 81.7 | 18.3 | |
| Liver | 68.7 | 31.3 | |
| Lung | 72.6 | 27.4 | |
| Spleen | 32.5 | 67.5 | |
| Testis | 48.3 | 51.7 | |

obtained with a free arachidonic acid concentration of 400 micromolar. Another difference arises from the rate of depletion of the isozymes from the cell following activation and translocation to the cell membrane, apparently due to different rates of proteolysis in the variable region $V_3$.[46–50] A hindrance in the study of the different responses of PKC isozymes to $Ca^{2+}$, diacylglycerol, and phospholipid activation has been the in vitro dependence on the phosphate acceptor. Proteins phosphorylated in vivo are only a subset of those phosphorylated in cell-free systems,[51] a reflection of the fact that there is a diverse distribution of PKC isozymes among various cell types and even in intracellular compartments. Additionally, dependence on $Ca^{2+}$ and phospholipids varies depending on the phosphate acceptor; in the extreme, protamine is phosphorylated in the absence of both $Ca^{2+}$ and phospholipid. Clearly, the question concerning in vivo isozyme distributions and selective agonism or antagonism of the isozymes will need to be addressed before a therapeutic agent is likely to be useful.[52–56]

## III. ROLE OF PKC IN PHYSIOLOGICAL PROCESSES: PKC AND SIGNAL TRANSDUCTION

The first report of phosphatidylinositol turnover upon stimulation by extracellular signals appeared in 1955,[57] and progress through the next three decades has been reviewed in several sources.[58–61] In 1979, PKC activation was linked to signal transduction via the phosphatidylinositol cascade as diagrammed in Figure 4.[62–64]

Phosphatidylinositol is the most abundant of the inositol lipids, which as a group make up less than 10% of all phospholipids.[60] Phosphatidylinositol, phos-

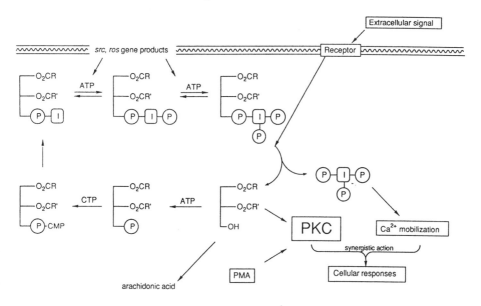

**Figure 4.** The phosphatidylinositol cascade.

phatidylinositol-4-phosphate, and phosphatidylinositol-4,5-bisphosphate are inter-converted through ATP-dependent kinases which are encoded in oncogenes.[65–67] The latter two inositol lipids are found predominantly in tissues enriched in plasma membrane. In response to various external stimuli, such as hormones and neuro-transmitters (Table 3),[68] phosphatidylinositol-4,5-bisphosphate is converted by phospholipase C to inositol-1,4,5-trisphosphate and diacylglycerol (DAG).[69] In-ositol-1,4,5-trisphosphate causes the mobilization of calcium from intracellular stores, thought to be located in the endoplasmic reticulum.[70] Increased $Ca^{2+}$ concentrations are accompanied by translocation of the enzyme to the plasma membrane where PKC is primed for activation.[71,72] The calcium and diacylglycerol synergistically activate PKC, allowing for the phosphorylation of a variety of target proteins (Table 4),[73] resulting in a host of cellular responses (Table 5).[74] The fate of the DAG after its release from PKC is either re-phosphorylation and return through the cycle to phosphatidylinositol, or hydrolysis of the ester at position 2.[75,76] In most of the endogenous DAGs, that position is occupied by arachidonic acid,[77] which is freed by hydrolysis to enter the arachidonic acid cascade, leading to the synthesis of prostaglandins and leukotrienes.[78]

The stimulation of phospholipase C is thought to be mediated by GTP-binding (G-) proteins, which are associated directly with extracellular receptor-based transmembrane events. Phorbol esters have been shown to block inositol-1,4,5-tris-phosphate formation and $Ca^{2+}$ mobilization by affecting the coupling of the G-protein to phospholipase C, and not by affecting the catalytic activity of phos-

***Table 3.*** A Partial List of Hormones, Neurotransmitters, and other Agents
Stimulating Phosphatidyl Inositol Turnover

| | |
|---|---|
| Acetylcholine | f-Methionyl-leucyl-phenylalanine |
| Adrenaline | Mitogens |
| Angiotensin | Norepinephrine |
| Antigen | Pancreozymin |
| Bombesin | Photons |
| Bradykinin | Phytohaemaglutinin |
| Caerulein | Platelet activating factor |
| Collagen | Pseudomonal leukocidin |
| Chemoattractants | Secretatogues |
| Cholecystokinin | Serotonin |
| Dopamine | Spermatozoa |
| Electrical stimulation | Substance P |
| Gastrin | Thrombin |
| Glucose | Thromboxane |
| Histamine | Thyrotropin-releasing hormone |
| $K^+$ Polarization | Vasopressin |

***Table 4.*** A Partial List of In Vitro Target Substrate Proteins for PKC Homogenate

| | |
|---|---|
| Receptor Proteins | Enzymes |
|   Epidermal growth factor |   Glycogen phosphorylase kinase |
|   Insulin receptor |   Glycogen synthase |
|   Somatomedin C receptor |   Phosphofructokinase |
|   Transferrin receptor |   $\beta$-Hydroxy-$\beta$-methylglutaryl-coenzyme A |
|   Interleukin-2 receptor |     reductase |
|   Nicotinic acetylcholine receptor |   Thrysoin hydroxylase |
|   $\beta$-Adrenergic receptor |   NADPH oxidase |
|   Immunoglobulin E receptor |   Cytochrome P450 |
| Membrane and Nuclear Proteins |   Guanylate cyclase |
|   $Ca^{2+}$ transport ATPase |   DNA methylase |
|   $Na^+/K^+$ ATPase |   Myosin light chain kinase |
|   $Na^+$ channel protein |   Initiation factor 2 |
|   $Na^+/H^+$ exchange system | Other Proteins |
|   Glucose transporter |   Fibrinogen |
|   GTP-binding protein |   Retinoid-binding proteins |
|   HLA antigen |   Vitamin D-binding protein |
|   Chromaffin granule-binding proteins |   Ribosomal S6 protein |
|   Synaptic B50 (F1) protein |   GABA modulin |
|   Histones and protamine |   Stress proteins |
|   Phospholamban |   Myelin basic protein |
| Contractile and Cytoskeletal Proteins |   High-mobility group proteins |
|   Myosin light chain |   Middle T antigen |
|   Troponin T and I |   pp60 protein |
|   Vinculin | |
|   Filamin | |
|   Caldesmon | |
|   Cardiac C-protein | |
|   Microtubule-associated proteins | |
|   Gap junction proteins | |

***Table 5.*** A Partial List of Cellular Responses Initiated by Activation of PKC

| | |
|---|---|
| Aldosterone secretion | B-lymphocyte activation |
| Amylase secretion | T-lymphocyte activation |
| Arachidonate release | Lysosomal enzyme release |
| Calcitonin release | Mucin secretion |
| Catecholamine secretion | Muscle contraction and relaxation |
| Dopamine release | Parathyroid hormone release |
| Gastric acid secretion | Pepsinogen secretion |
| Glucose transport | Pituitary hormone release |
| Glycogenolysis | Prolactin release |
| Growth hormone release | Steroidogenesis |
| Hexose transport | Superoxide generation |
| Histamine release | Surfactant secretion |
| Inhibition of gap junction | Thromboxane synthesis |
| Insulin release | Thyrotropin release |
| Leuteinizing hormone release | Transmitter release |
| Lipogenesis | |

pholipase C.[79] This effect is mediated through PKC as supported by the observation that the phosphatidylinositol cascade in cells lacking PKC is unaffected by addition of phorbol esters. In other systems, PKC has been found to uncouple the G-protein $G_i$ from the $\alpha_2$-adrenoreceptor without altering the interaction of $G_i$ with adenylate cyclase.[80] $^{32}$P has been used to identify the G-protein substrate of PKC as an $\alpha$ subunit of $G_z$, similar or identical to $G_z\alpha$, a unit insensitive to the pertussis toxin.[81]

PKC activation is always associated with the appearance of DAG,[82] which does not persist due to the degradation and recycling pathways available to it. It should be noted, however, that, despite the brief activation of PKC, the consequences may persist for significantly longer times depending on the fate of the phosphorylated substrate protein. In addition to the requirement for DAG and $Ca^{2+}$, the activation of PKC requires association with phospholipids.[83] A photoaffinity labeling experiment with an exogenous activator, a phorbol ester, showed that the only detectable interaction of the activator was with the phospholipids and not directly with PKC.[84]

In view of this finding, Bell suggested that the activation of PKC involves an aggregate of $Ca^{2+}$, DAG, and phosphatidylserine (PS) in which the phosphatidylserine is positioned between DAG and PKC. Further data supporting this model were provided by a study of the stoichiometry of activation, which was determined by a mixed micellar assay to require one calcium ion, one DAG, and four molecules of phospholipid.[85] Since the most effective of the phospholipids is phosphatidylserine,[86] the asymmetric distribution of phospholipids in the lipid bilayers of cell membranes may be important. Phosphatidylcholine, phosphatidylinositol, and phosphatidylserine predominate on the inner layer, while phosphatidylethanolamine and sphingomyelin are concentrated on the outside. Nishizuka examined the cooperative roles of membrane phospholipids in the activation of PKC[87] and determined that the addition of phosphatidylethanolamine to phosphatidylserine

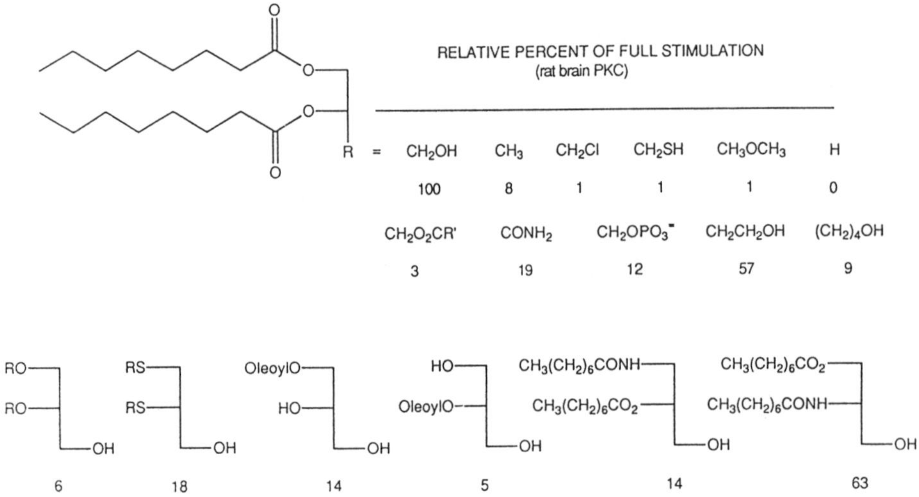

**Figure 5.** Effect on stimulation of PKC homogenate of modification of the glycerol backbone.

resulted in enhanced activity. Combination with either phosphatidylcholine or sphingomyelin resulted in diminished activity, and was competitive with respect to both phosphatidylserine and DAG. The combination of phosphatidylserine with phosphatidylinositol had no effect on activity. Weinstein has proposed a model for PKC activation in which interaction of the activator (through the phospholipids) and PKC at the regulatory domain causes a conformational change in the enzyme that exposes the catalytic site for binding to substrate proteins.[88]

Stereochemical and other structural requirements for the activation of PKC by DAG have been examined by several groups. In 1984, Rando showed that the $S$ isomer of DAG activates PKC while the $R$ isomer does not, thereby establishing that DAG exerts its influence through a stereospecific recognition and not merely a bulk lipid effect.[89] It has also been shown that mono- and triacylglycerols are inactive.[90] Furthermore, the 1,3-DAG isomer is inactive.[91] A variety of studies[92–95] of saturated and unsaturated acyl moieties has shown that there is little consensus about the "best" DAG for PKC activation, which may be a consequence of the heterogeneity of PKC used in the assays and the variation of assay protocols.

In general, though, unsaturation appears to be favorable in the second position but is not required in the first; furthermore, the length of the acyl chain at position 2 is critical, with affinity gradually diminishing when the chains are longer or shorter than the optimal range. For example, Go et al.[94] found that among symmetrical saturated diacylglycerols, dioctanoylglycerol (di-$C_8$) was the most active as measured by displacement of tritiated phorbol dibutyrate ([$^3$H]PDBu), with a $K_a$ of 0.68 micromolar and a $V_{max}$ of 93.5% of that of the standard 1-stearoyl-2-

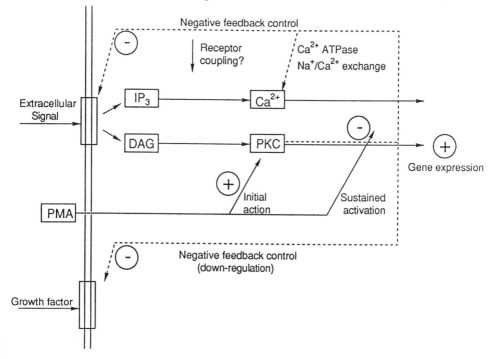

**Figure 6.** Dual action of PKC upon stimulation by endogenous or exogenous activators.

arachidonoylglycerol. The homologous di-$C_{10}$ glycerol had values of 0.75 micromolar and 88.8%, respectively, while di-$C_6$ glycerol was described as far less effective.

Bell also studied the glycerol backbone requirements[90] and found that substitution of the acyl groups by other functional groups resulted in greatly diminished affinity relative to di-C8 glycerol, as did replacement of the free hydroxymethyl group by functionalities of the same, higher, and lower oxidation states (Fig. 5). Finally, he determined that spacing of the free hydroxyl group relative to the vicinal diesters was important, but that it can be varied within a small range.

Several suggestions have been made about the role of PKC in the regulation of signal transduction. One model for the regulation assigns roles both to positive control and to negative feedback control. In response to a short-lived agonist such as DAG, PKC is activated transiently.

Phorbol myristate acetate (PMA), on the other hand, causes sustained activation due in large part to its higher affinity and the lack of degradative mechanisms for PMA. PMA-activated PKC behaves similarly to DAG-activated PKC, but after prolonged activation the former begins to assert negative feedback control, resulting in increased levels of activity of $Ca^{2+}$-ATPase and $Na^+/Ca^{2+}$ exchange proteins in order to decrease the concentration of $Ca^{2+}$ in the cell.[96-98] PKC is also thought to inhibit the frequency of $Ca^{2+}$ concentration fluctuation by inhibiting the receptor-mediated hydrolysis of phosphatidylinositol[99] and by phosphorylating (and thus activating) inositol triphosphate 5'-phosphomonoesterase, which hydrolyzes inositol-1,4,5-trisphosphate to inert molecules.[100] PKC may also play a role in negative feedback control of cell proliferation by phosphorylating the receptor for epidermal growth factor (EGF), resulting in decreased binding of EGF.[101-106]

Another aspect of the duality of PMA activation of PKC is seen in the degradation through calpain-catalyzed proteolysis in the third variable region of PKC on prolonged activation, eventually resulting in removal of the catalytic fragment from the cell through further degradation in a process termed down-regulation of PKC.[13-16] Other positive actions of PKC, unrelieved by subsequent negative feedback control, include its involvement in gene expression and in growth regulation.[107,108] The dual action of PKC is depicted in Figure 6.

## IV.   RELATIONSHIP OF PKC AND CALCIUM ION

The presence of small amounts of PKC activators, such as DAG or PMA, greatly increases the affinity of PKC for $Ca^{2+}$ so that full activation of PKC can occur without detectable cellular mobilization of the ion.[87] As previously noted, however, in response to extracellular signals two PKC-associated second messengers are released: DAG and inositol-1,4,5-trisphosphate.The latter is thought to have its own receptor on the endoplasmic reticulum that is involved with the release of $Ca^{2+}$, and therefore an increased concentration of $Ca^{2+}$ accompanies PKC activation by endogenous factors.[70] This concomitant activity has been shown in some cases to have synergistic effects upon cellular responses.[109,110] Each arm of the signal transduction pathway can be stimulated independently: PKC can be activated by injection of exogenous DAG, and intracellular concentrations of $Ca^{2+}$ can be increased by use of a calcium-selective ionophore, such as A23187.

One example of synergistic activity involves the response of rat mast cells.[111] Treatment with exogenous DAG alone caused the release of small quantities of histamine, while treatment with A23187 alone had little effect on the cell. However, simultaneous treatment with DAG and A23187 caused greatly enhanced histamine release. Several other examples of synergistic responses have also been presented.[110-112]

## V.   PKC AND ION CHANNELING

Among the intriguing activities influenced by PKC, the regulation of chloride ion and potassium ion channels is particularly interesting, especially since the former is implicated in cystic fibrosis and the latter in memory storage. Recent work has examined the roles of protein kinase A (PKA) and PKC in regulation of chloride

**Table 6.** Relative Activities of PKC and PKA in Various Tissues

| Tissue | Protein Kinase C (units/mg of protein) | Protein Kinase A (units/mg of protein) |
|---|---|---|
| Platelets | 6300 | 340 |
| Brain | 3270 | 250 |
| Lymphocytes | 1060 | 320 |
| Small intestinal smooth muscle | 770 | 560 |
| Granulocytes | 530 | 100 |
| Lung | 360 | 290 |
| Kidney | 280 | 150 |
| Liver | 180 | 130 |
| Adipocyte | 170 | 270 |
| Heart | 110 | 230 |
| Skeletal muscle | 80 | 110 |

channels in the airway epithelia of cystic fibrosis (CF) patients.[113,114] While chloride ion channels are normally regulated by PKA, PKC can also stimulate or inhibit the channel, depending on the state of the cell.[114] In normal cells, PKC has been found to activate chloride channels in the presence of ATP and phorbol esters or di-$C_8$ glycerol at $Ca^{2+}$ concentrations less than 10 nanomolar. At concentrations greater than 10 nanomolar, activated channels become inactivated. The chloride channels in cells of CF patients fail to respond to PKA as those in normal cells do, and they are not activated by PKC. These results suggest that the genetic defect in CF results either in the capability of the channel to be activated (if both PKA and PKC act on a common pathway directly involved in chloride ion channel gating) or in the mechanism by which phosphorylation induces channel activation.

In order to place PKC in an appropriate context, one must be aware of its interactions with other enzymes involved in cellular regulation. The above discussion serves as one example of the relative roles of PKA and PKC in regulating cells. PKA is a cyclic nucleotide-dependent protein kinase, also found in various species and organs.[115] The relative activities of PKC and PKA vary according to tissue type as assayed by phosphorylation of histone under similar conditions (Table 6).[116]

PKA and PKC are associated with the two major receptor types for signal transduction and even share target proteins.[117] The type of receptor that activates PKA is involved with cyclic AMP while the one that activates PKC is involved with phosphatidylinositol turnover. The two types are linked in some cells, such as platelets, neutrophils and lymphocytes, through a bidirectional control system in which the cellular responses invoked by the signal to induce phosphatidylinositol turnover are antagonized by the signal to produce cAMP, or vice versa. By contrast, cells such as those in the pituitary, in lymphomas, and in some endocrine tissues appear to have monodirectional control wherein one receptor type may potentiate the other.

Alkon has reported that potassium channels are regulated by cytosolic and membrane-bound PKC in the mollusc *Hermissenda crassicornus*[118,119] and in the rabbit,[119] and that the regulation is related to associative learning. In two examples of Pavlovian conditioning, measurements of $K^+$ currents and translocation of PKC were made for groups in which the trigger (the unconditioned stimulus to which there is an involuntary response) and signal (the conditioned stimulus) were paired, for groups in which they were unpaired, and for control groups which received no stimulation at all.

In one set of experiments, the trigger was a rotation mimicking ocean turbulence to which the snail responded by flexing its muscular foot to anchor itself. At the same time (in the paired group) a light was flashed. Following conditioning, the snail learned to flex its foot in response to a flash of light. This associative memory was related to an increase in membrane-bound PKC (and a corresponding decrease in cytosolic PKC) within neurons, altering their properties so that input signals could more readily trigger responses. The translocation, which was measured relative to the unpaired and control groups, persisted for several days following conditioning. Changes in voltage-dependent $K^+$ currents were also measured during conditioning of the mollusc, and it was observed that conditioning resulted in a persistent reduction of $K^+$ currents as well as increased phosphorylation of a 20 kDa protein which may be part of the channel itself or an important regulator of flux through the channel.

Similar effects on the subcellular distribution of PKC were noted when rabbits were conditioned by the pairing of a tone (conditioned stimulus) with a puff of air on the eye (unconditioned stimulus), causing the rabbit to extend its nictitating membrane.

## VI.  PHORBOL ESTERS AND TUMOR PROMOTION

One of the areas most profoundly affected by the discovery of protein kinase C is carcinogenesis. Cancer is currently a leading cause of death and disability in industrialized nations, ranking second only to cardiovascular disease in the United States and ranking first in Japan. Although generally recognized as a genetic disease, the molecular mechanism by which gene defects are translated into malignant tumors is not well understood. Such knowledge, however, is crucial for the development of new protocols for the diagnosis, treatment, and prevention of cancer.

Efforts to elucidate the molecular mechanism of carcinogenesis span much of this century. Early studies established that the application of certain chemical agents, termed carcinogens, or, more appropriately, initiators, to mouse skin result in tumor formation.[120–122] Later studies began to question whether carcinogenic agents may exert their influence via more than one mechanism.

Friedewald and Rous introduced the terms initiation and promotion to the field of tumorigenesis in 1944[123] in the second of a pair of papers in which they observed

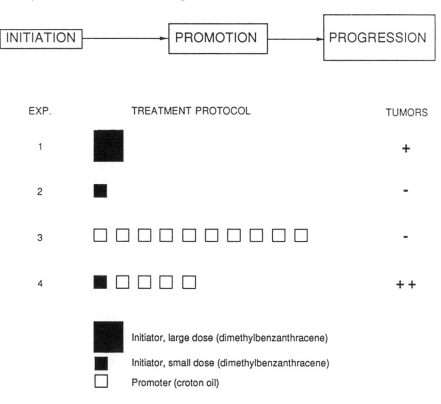

**Figure 7.** Berenblum protocol for the determination of multistage carcinogenesis.

that tarring of rabbit skin caused large benign growths which were "possessed of so little innate vigor as to be wholly dependent for success upon favoring local conditions" and became malignant only when "submitted to encouraging influences." Initiators have thus come to be viewed as compounds capable of inducing neoplastic changes in cells, while promoters were seen to provide the influence which permits further proliferation of the genetic damage.

Promotion has been observed in response to chemical agents or physiological manipulations such as deep skin wounding. It has been noted that our environment and diets are replete with chemicals, both natural and synthetic, that are carcinogenic at the maximum tolerated dose in rodents, but that animals, including humans, have developed natural defenses against low doses of toxic chemicals.[124] This suggests that mitogenesis, which is related to promotion, may increase mutagenesis since a dividing cell is at greater risk of mutation.

The effects of initiation and promotion were more rigorously defined by experiments first developed by Berenblum[125] and subsequently standardized as an assay for two-stage tumor promotion (Fig. 7). In these experiments, the known carcinogen dimethylbenzanthracene was employed as an initiator and croton oil as a promoting agent equivalent to wounding. It was determined that while application of either croton oil (experiment 3), or sub-effective doses (experiment 2), of established carcinogens (experiment 1), to mouse back skin failed to elicit tumor formation, the sequence-specific application of both agents (experiment 4), resulted in an unusually high rate of carcinoma development. From these and related studies,[126] it was determined that initiation is associated with irreversible genetic damage, and that promotion, which is reversible, involves expression of the damaged gene.

Refinement in the understanding of multistage carcinogenesis has come in the identification of first and second stage promoters.[127,128] For instance, structural modification of tumor promoters has provided compounds which possess only low potency as complete tumor promoters for mouse skin.[129] However, if after a single application of an initiator, twice-weekly application of these compounds for two weeks is followed by application of a second stage promoter, tumorigenesis is observed. Therefore, exposure in the proper sequence to compounds which are neither complete carcinogens nor complete promoters may still result in carcinogenesis.[130] Further studies on multistage carcinogenesis revealed that the initiation/promotion process was not peculiar to mouse skin but demonstrable in a variety of organs and species.[131]

Surprisingly, despite their well-established properties as irritants and toxins, members of the *Euphorbiaceae* family occupied a place in relatively modern pharmacopeia,[132–135] having been used as therapeutic agents for migraines, parasitic and bacterial infestations, and skin conditions. They have also seen use as purgatives, abortifacients, and, in fiction[136] as well as fact,[137] as poisons.

Weber and Hecker have provided the first and rather compelling epidemiological evidence on the relevance of tumor promotion to human carcinogenesis.[138] Specifically, this group proposed that the high rate of esophageal cancer among the natives of Curacao is a consequence of their daily consumption of a tea derived from parts of the bush *Croton flavens* L. (Welensali). The basis for this proposal was the finding that this bush tea contains tumor promoting agents similar to those found in croton oil. Thus, the higher incidence of cancer in this population can be attributed to post-initiation, chronic exposure to a tumor promoter, a situation paralleling the aforementioned Berenblum protocol. The identification of tumor promoters is thus clearly as important as the identification of initiators.[124]

Recently, for example, an oil isolated from the seed of *J. Curcas* L., *Saboodam*, intended to be used in Thailand as a substitute for diesel oil and as a component of commercial printing ink, was found to contain croton oil-like components and to possess weak tumor-promoting properties.[139] The identification of this property should be valuable in formulating policy concerning the handling of the seeds and oil so as to minimize human exposure.

In the late 1960s the structure of the most active principle of croton oil, phorbol myristate acetate (PMA, Fig. 8), was established by X-ray crystallography of

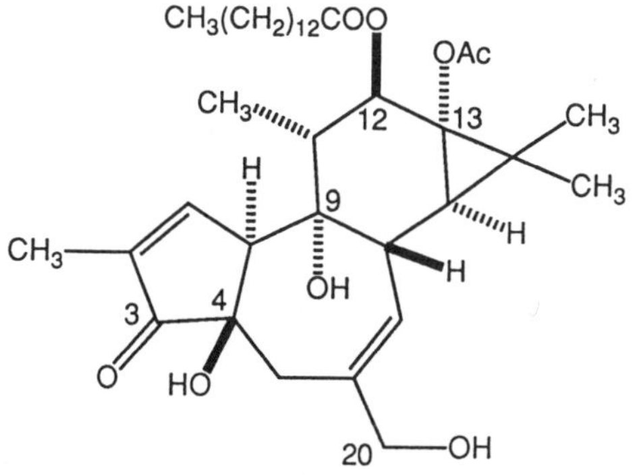

**Figure 8.** The structure of phorbol myristate acetate (PMA).

several ester derivatives.[140-142] This significant advance spawned several proposals about the structural role of plant-derived PMA in mammalian biochemistry. Given the high potency of PMA, it was reasonably suggested that PMA usurps the role of some endogenous factor involved in cell regulation and/or proliferation.[143] PMA was proposed, for example, to be a surrogate for endogenous polyunsaturated fatty acids,[143] while another view drew attention to the structural similarity between PMA and prostaglandins.[144] A rather intriguing correspondence between PMA and corticosteroids was also advanced.[145]

While speculation on the nature of the endogenous factor corresponding to PMA continued, efforts to identify the receptor for PMA were also underway. Early attempts at direct detection of the receptor for the phorbol esters using [³H]PMA were complicated by the observation that binding was nonsaturable and noncompetitive, indicating a high level of non-specific binding. A breakthrough was made in Blumberg's laboratories when the binding of [³H]PDBu, a derivative considerably less lipophilic than PMA but still highly active in vitro and in vivo was studied in chicken embryo fibroblast[146] and mouse skin[147] particulate preparations. In this case, there was rapid and reversible binding in a saturable, competitive manner as determined by double reciprocal and Scatchard plots of the data. Following Blumberg's work, other researchers used [³H]PDBu to find the high affinity receptor in other tissues.[131]

The high-affinity phorbol ester receptor was found to be very similar to the kinase identified in 1977 by Nishizuka,[1,2] having in common the degree of conservation throughout diverse species, tissue distribution, concentration in the brain, and the requirement for $Ca^{2+}$ and phospholipid for activation. The primary difference, the localization of the phorbol ester receptor in the particulate fraction as opposed to the cytosolic localization of Nishizuka's kinase, was explained by the observation that the cytosolic material was an apoenzyme which could be activated by reconstitution with added phospholipid. Several groups[148-151] independently showed that the kinase and receptor activities copurified through several fractionation steps, and Castagna et al.[8] showed that PMA activated PKC in vitro providing strong circumstantial evidence that the high-affinity phorbol ester receptor identified by Blumberg and the kinase identified by Nishizuka are in fact the same enzyme in different states. With the subsequent finding that DAG activates PKC in a fashion competitive with PMA, the endogenous factor for which PMA substitutes was identified as diacylglycerol.[109,152,153]

The exact role of phorbol esters, and therefore of PKC, in tumor promotion has been a subject of research for several decades. The phorbol esters have been shown to elicit in vitro a whole host of changes which may be related to tumorigenesis including morphological changes,[154] increased hexose transport,[155] increased prostaglandin synthesis,[156-158] changes in phospholipid[159] and protein synthesis,[160] increased ornithine decarboxylase activity[161] and polyamine production,[162] altered rates of DNA synthesis,[163-165] induction of secretion[166-167] and superoxide production,[168] alteration of surface receptors,[169-170] induction or blocking of terminal differentiation[171-173] and cell adhesion.[167,174] Several excellent reviews of the implications of these changes for tumor promotion can be found in references 9 and 175–179.

# VII.  AGONISM AND ANTAGONISM OF PKC

The complexity of the signal transduction pathway suggests a myriad of possibilities for PKC agonism and antagonism, as depicted in Figure 9. Several indirect methods of antagonism are listed in the first column. For example, inhibition of DAG formation upstream of PKC (e.g., by inhibition of phospholipase C, which catalyzes the hydrolysis of phosphatidylinositol) or increased activity of a phosphatase downstream of PKC will give a result comparable to direct antagonism. Additionally, the levels of PKC within the cell could be decreased by inhibiting induction and/or expression of the gene(s) responsible for PKC synthesis, or by enhancing the mechanisms of PKC degradation. Finally, an alteration of the lipid bilayer or of its local composition could destabilize the PKC-lipid association and result in inhibition of enzyme activity.

Direct methods of antagonism include interference with translocation of the enzyme from the cytosol to the membrane. Antagonism at the binding sites for ATP, protein substrates, $Ca^{2+}$, and phospholipids are also conceivable. One area that is a current focus of much research involves agonism and antagonism at the activator (i.e., DAG) binding site as this site offers greater potential for selective regulation than would be expected, for example, at the ATP binding site, a relatively conserved region in kinases.

Several of the indirect PKC inhibitors are shown in Figure 10. Retinoids, including retinoic acid, have been shown to inhibit induction of ornithine decar-

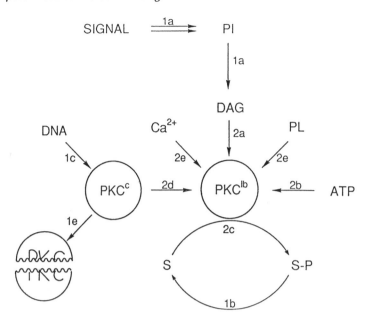

1. INDIRECT
   a. Upstream (e. g., DAG formation)
   b. Downstream (e. g., dephosphatase)
   c. PKC synthesis (gene induction, expression)
   d. Lipid bilayer (conformation effects)
   e. PKC degradation

2. DIRECT
   a. Activator receptor(s)
   b. ATP binding site
   c. Substrate binding site
   d. Translocation
   e. $Ca^{++}$, phospholipid receptors

**Figure 9.** Possible targets for agonism and/or antagonism of PKC.

boxylase activity[180] and tumor promotion[181] by PMA in mouse skin. The mechanism of action appears to be through alteration of the cell-surface membrane although the role of the retinoid receptor is as yet undefined.[182] Cell surface effects have been observed in other systems including changes in the degree of adhesiveness of fibroblast[183,184] and intestinal epithelial[185] cells on treatment with retinoids. Retinoic acid receptors have been identified by several groups,[186–189] but no direct association with modulation of PKC activity has been reported to date for these receptors.

Ascorbic acid has also been shown to alter the membrane interactions essential for PKC activation by stimulating the peroxidation of membrane lipids, resulting in a change in membrane microviscosity.[190] Quercetin, a plant flavonoid found in many fruits and vegetables, is competitive with ATP and on topical application to mouse skin causes decreased [³H]PDBu binding to PKC, although it does not interact with the regulatory domain.[191] Structure-activity relationships among a

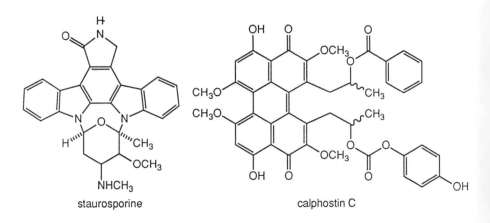

retinoic acid

ascorbic acid

sphingosine

quercetin

staurosporine

calphostin C

**Figure 10.** A partial list of reported PKC inhibitors.

variety of plant flavonoids indicate that the 3', 4', and 7' free hydroxyls are required for PKC inhibition.[192] Staurosporine, an alkaloid produced by *streptomyces,* sp., is a potent but non-specific inhibitor of PKC and PKA[193] that has been the object of considerable attention. Although there is at least one report[194] that it interacts with protein kinase C primarily at the regulatory domain, several studies[195,196] indicate that the interaction is at the catalytic domain, primarily at the ATP binding site. The initial observation that staurosporine is not competitive with ATP[193] has been supported by later workers.[196,197] The exact mechanism of action has not yet been determined. Sphingosine[198] has been shown to inhibit PKC in a manner competitive with DAG and phorbol esters.[199] As is often the case with highly lipophilic molecules, the degree of inhibition depends on its concentration relative to other lipids. It has been suggested recently that the conversion of sphingosine to sphingomyelin, with the concomitant production of DAG, might result in an "on-off" switch for PKC as an inhibitor is exchanged for an activator.[200]

To date, the only class of compounds identified as high affinity inhibitors of PKC at the DAG binding site are the calphostins,[201] compounds isolated from the culture broth of the microbe *Cladosporium cladosporioides*. A member of this class, calphostin C, was found to be a specific inhibitor of PKC and to competitively inhibit [³H]PDBu binding.[202] The structural novelty of these compounds and their unusual effect on PKC clearly merit a detailed investigation of their inhibitory

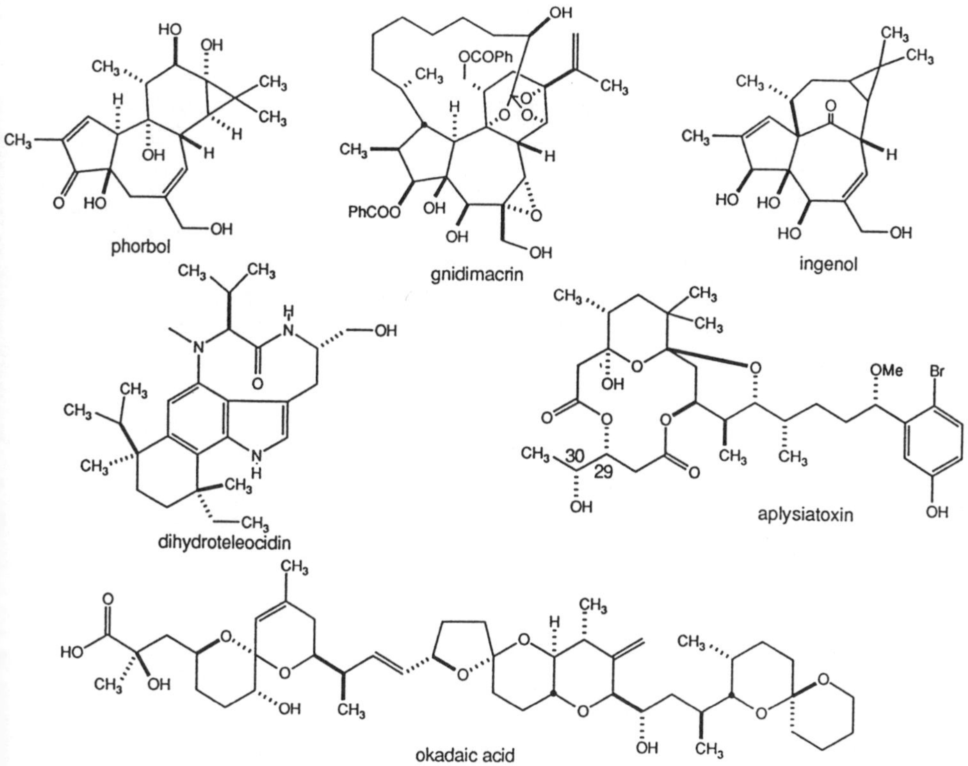

**Figure 11.** A partial list of PMA-type and non-PMA-type tumor promoters.

effects. Related antiretroviral agents hypericin and pseudohypericin are also thought to inhibit PKC through interaction with the regulatory domain, but specific inhibition of [$^3$H]PDBu binding has not been demonstrated.[203]

It is important to note that tumor promotion effects are not peculiar to PKC regulation since downstream effects are also possible. For example, a group of compounds have been identified as non-PMA-type tumor promoters.[204] One of these compounds, okadaic acid (Fig. 11), has been found to be a potent inhibitor of phosphatases.[205,206] The light chain of myosin, for example, is phosphorylated by PKC and by myosin light-chain kinase on different residues. Addition of okadaic acid led to an increase in the maximal level of phosphorylation to 1.7 mol phosphate/mol enzyme (relative to 0.75 mol phosphate/mol enzyme in the absence of okadaic acid).[205] As noted at the outset, abnormal phosphorylation can thus be obtained by increased PKC activation or by modulation of turnover of phosphory-lated PKC substrates.

# VIII.  DEVELOPMENT OF A PHARMACOPHORE MODEL FOR PKC ACTIVATION

It was recognized in the early 1980s that the phorbol esters, long known to be tumor promoters, were potent activators of PKC.[8] Somewhat later, Nishizuka found that PMA competes with DAG in the activation process, establishing PMA as an exogenous analog of DAG.[109,152,153] Other naturally occurring tumor promoters such as gnidimacrin,[129] ingenol,[129] dihydroteleocidin,[207] and aplysiatoxin[208] (Fig. 11) were also identified as PKC activators at the phorbol ester binding site. A comprehensive list of environmental tumor promoters is provided in reference 209. Clearly, elucidation of the structural requirements of PKC activation would provide an important key in understanding the molecular mechanism of tumor promotion and would facilitate the search for other potential promoters and for the development of cancer prevention approaches.

Our laboratory's involvement in this area began in the late 1970s. As noted, the link between the phorbol esters and PKC had not been identified at that time and the endogenous analog of the phorbol esters was not known. It was clear, however, that the structure of this analog and an explanation for the biochemical role of the phorbol esters could be drawn from an identification of the structural features of the phorbol esters which were required for its promoter activity, i.e. the tumor-promoter pharmacophore. This task was expected to be a difficult one since the phorbol esters exert their activity without being chemically changed and therefore carry no chemical imprint of their interaction with PKC. Nevertheless, it was expected that one could draw on established structure activity relationships (SARs) and emerging computer modeling techniques to arrive at an understanding of how the phorbol esters are recognized and function. Such studies were additionally expected to have broad fundamental value since the non-covalent associations

Structural Modifications of the Phorbol Skeleton

**Bold** indicates retention of activity
*Italics* indicates loss of activity

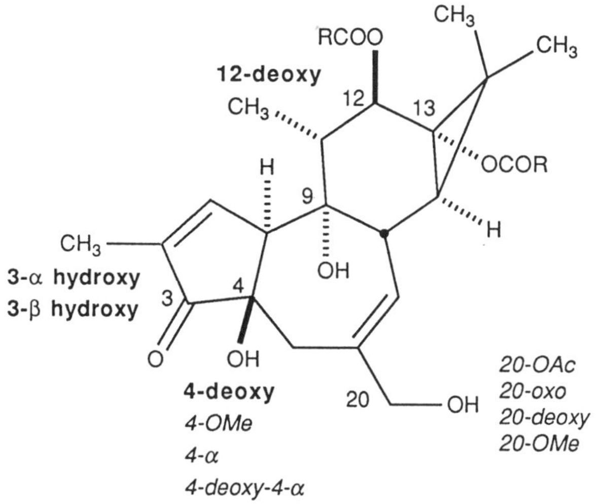

***Figure 12.*** Structure-activity relationships of phorbol derivatives. Various assays have been used to determine activity; retention or loss of activity is meant in only a qualitative sense.

characterizing the molecular recognition between the phorbol esters and PKC are representative of a vast array of biologically significant enzyme-substrate interactions.

Several structure-activity relationships (Fig. 12), drawn from numerous laboratories including those of Hecker, Van Duuren, and Blumberg, were of critical value in our effort to establish the tumor-promoter pharmacophore through computer analysis. It is important to note that the protocol by which a compound is defined as a tumor promoter has changed over the course of the last several decades and therefore caution must be exercised in relating and evaluating data. Early assays used the high correspondence between irritancy and tumor promotion[210] to indicate the promoter potency of compounds since approximately 95% of tumor promoters also have potent activity as irritants.

The mouse-back skin assay has also been used to evaluate qualitatively the tumor-promoting ability of a compound. In this assay, the number and size of papillomas elicited by a compound are taken as a measure of its promoter potency. Neither assay allows quantification, nor indeed confirmation, of interaction with PKC. A more direct assay is the measurement of inhibition of [³H]PDBu binding to PKC, and even this approach can give highly variable results. For example, many experiments are done with the readily available PKC homogenate, and the binding affinity that is obtained is an average of affinities for the isozymes that are present in the mixture. The most rigorous assay would therefore be to test binding to each of the pure PKC isozymes. However, such as assay evaluates only recognition by the isozyme in a cell-free context, thereby obscuring microenvironmental effects within the cell, and does not establish the in vivo promoting ability of a compound. Nevertheless, it provides a standard for reference work, a quick screen of promoter candidates, and the first step in evaluating a compound's molecular role as a tumor promoter. Additional studies would involve evaluating the compound in whole cell and whole organism assays.

## A.  Role of Phorbol Esters

When our work began, it was known that phorbol itself is not a tumor promoter.[211,212] Consequently, the esters at C-12 and C-13 must play a role in the activity exhibited by PMA. It was also reported that 4-α-hydroxy-PMA is inactive as a tumor promoter.[211] Thus, epimerization of a single stereocenter in phorbol esters is sufficient to render the molecule inactive as a promoter. Alkylation of either the oxygen at C-4 or at C-20 was also known to produce inactive compounds[213] (although it is noteworthy that 4-OMe-PMA is a first stage promoter). It has been reported that 4-deoxy-phorbol diesters are active as irritants,[129] but their affinity for PKC has not been evaluated. The roles, if any, that the C-3 and C-9 oxygens might play in recognition were unknown since the biological effects of variations at these sites had not been determined. Finally, it was established that ingenol esters and members of the daphnane family of diterpenes exhibited some activities in common with the phorbol esters.[129]

In work initiated in 1979, through the use of the Merck computer modeling facility, we were able to arrive at a hypothesis for the promoter pharmacophore of PMA, which accommodated the aforementioned structure-activity relationships.[214] This hypothesis was based on three assumptions: (1) compounds with similar activities and potencies are likely to have a subset of recognition features in common; (2) for the diterpene promoters, these structural features are involved in reversible chemical changes and/or non-covalent recognition (H-bonding, van der Waals forces, electrostatic interactions) at the receptor site; and (3) due to the conformational rigidity of the diterpene promoters, the relative locations of these features is similar in both the active (recognized) conformation and the crystal structure conformation, except for the conformationally mobile C-20 hydroxymethyl group, whose position could vary at the expense of a small but calculable energy change in going from the crystal-phase conformation to the bound conformation.

As illustrated in Figure 13, by comparing a variety of crystal structures, allowing for variation only at C-20, a particularly good correlation was found between the

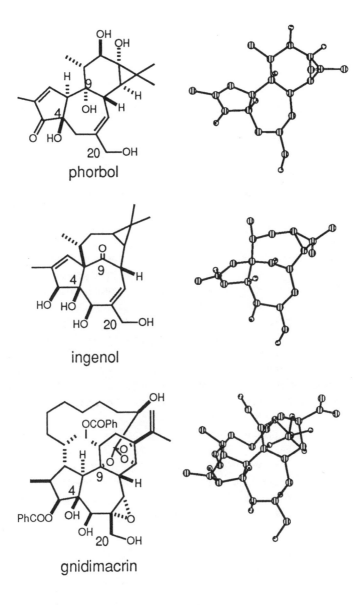

**Figure 13.** ORTEP comparisons of phorbol, ingenol and gnidimacrin. Carbon atoms are represented by striped circles, and oxygen atoms by dotted circles. Hydrogen atoms are omitted for clarity.

27

C-4, C-9, and C-20 oxygens of PMA and the corresponding oxygens in ingenol and in the daphnane promoters. The Mogli computer software used to evaluate such matches is designed to compare the distances between the centers of mass of correlated atoms.

It should be noted that "recognition" involves the shape and electron density of orbitals about such atom centers. Our use of atom center (point) comparisons was based largely on software availability and somewhat by the facility that this approach offered in comparing large numbers of structures. The "best" comparisons obtained by center-of-atom correlations were then inspected for orbital directionality and electron density similarities. The goodness of the fit between two structures based on the atom center comparison is expressed as an average root mean square deviation (RMSD) of the set of atomic coordinates for recognition sites in one structure with the corresponding set in a second structure. If the centers of mass of all correlated recognition sites in two molecules superimpose identically, the average root mean square deviation (RMSD) for this comparison would be 0 Å.

For calibration, the RMSD for two very similar structures was required. Accordingly, phorbol and its 20-bromofuroate ester were selected and found to have an RMSD for the C-4, C-9, and C-20 oxygens of 0.16Å. Structures giving numbers in this range and having similar electron distributions would therefore be representative of good fits. When the C-4, C-9, and C-20 oxygens of phorbol were compared in this way with those of ingenol and of gnidimacrin, compounds that differ in several respects from phorbol esters, average RMSDs of 0.35Å and 0.10Å, respectively, were obtained. Thus, in accord with their similar biological activity, the phorbol, ingenol, and gnidimacrin promoters have recognition elements which are similarly positioned even though their gross structures differ substantially.[215]

Further support for the above hypothesis came from its ability to rationalize not only the similar activities of various tumor promoters but also why some phorbol-derived esters are inactive. For example, when the stereochemical configuration at C-4 in the phorbol esters is inverted, the resulting 4-α-phorbol esters are found to be inactive as tumor promoters and as ligands for PKC. Comparison of the two structures reveals that the relative positions of the C-12, C-13, and C-20 oxygens are virtually unchanged, while the C-3 and C-9 oxygens are changed only slightly (Fig. 14). However, *the relative positions and orbital orientations* of the C-4 oxygens in these two structures differ dramatically, in accord with the hypothesis that the C-4 oxygen plays a role in recognition.[216]

## B.  Similar Roles by Other Structures

The discovery by Fujiki et al.[207,217] that the indole alkaloid dihydroteleocidin (DHT) exhibits activity and potency similar to that found for the phorbol esters provided a further test of the above hypothesis since the two structures are seen to differ significantly in several respects. Nevertheless, when the C-4, C-9, and C-20 oxygens of phorbol esters were compared with N-13, N-1, and O-24 atoms of DHT,

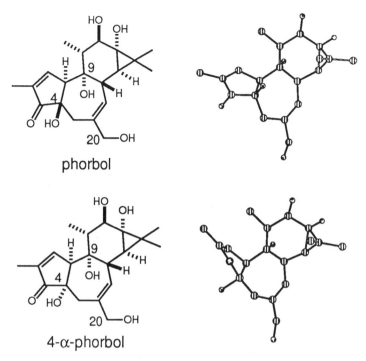

**Figure 14.** ORTEP comparisons of phorbol and the MM2-minimized structure of 4-α-phorbol. Carbon atoms are represented by striped circles, and oxygen atoms by dotted circles. Hydrogen atoms are omitted for clarity.

respectively, an RMSD of 0.33 Å was obtained, suggesting that the heteroatom triad in each molecule can be positioned similarly in the receptor site.[215] Moreover, when these two triads are overlaid in the computer (Fig. 15) the rigidly oriented hydrocarbon portion of DHT falls into a region between the A- and C-rings of the phorbol esters, indicating that the hydrocarbon portion (i.e., the long chain ester) of the phorbol esters assumes a position between the A- and C-rings in the bound conformation.

Of special significance in the further development of this field was the suggestion that the endogenous activator of PKC is a diacylglycerol (DAG).[109,152,153] Since DAG competes with the phorbol esters for the same binding site on PKC, it follows that the two could have recognition features in common. Nishizuka et al. suggested that the C-12–C-13 diester subunit of the phorbol esters is the feature that structurally emulates DAG, a reasonable suggestion in view of their common vicinal diester functionality. However, given that 4-α-phorbol diesters are inactive even though they have the same C-12–C-13 diester subunit found in the active

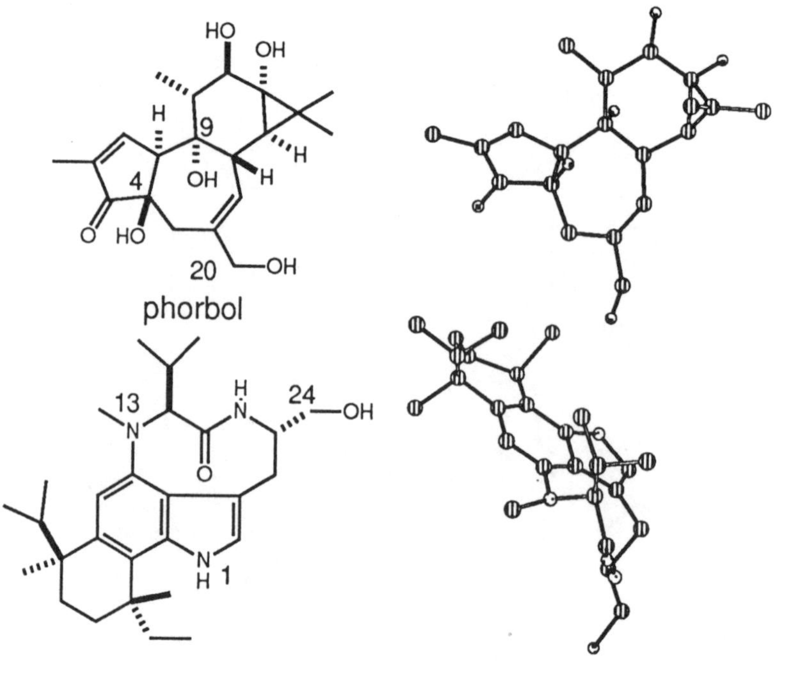

phorbol

dihydroteleocidin

**Figure 15.** ORTEP comparisons of phorbol and dihydroteleocidin. Carbon atoms are represented by striped circles, and oxygen and nitrogen atoms by dotted circles. Hydrogen atoms are omitted for clarity.

phorbol esters and that the active ingenol esters lack C-12–C-13 functionality, it was expected that another phorbol subunit must account for the similarity between the phorbol esters and DAG.

Also supporting this conclusion was Bell's discovery that glycol esters are inactive.[90] The Stanford group provided the initial correlation of the phorbol esters and DAG which accommodates this SAR data.[215] On the basis of the previous analysis, it was expected that the C-4, C-9, and C-20 oxygens of the phorbol esters would correspond spatially to three of the oxygens of DAG, allowing then for the lipophilic groups of each compound to co-locate in space. Evaluation of this comparison proved striking (Fig. 16). An RMSD of 0.03 Å was found between the phorbol esters and a DAG conformer calculated to be only 3.35 kcal/mol above the lowest energy conformation.

Comparisons between DAG and other phorbol subunits were fair or unacceptable based on energy and/or RMSD evaluations. For example, comparison of the C-12, C-13, and C-9 triad of the phorbol esters, an extension of the Nishizuka model, with DAG gave an RMSD of 0.21 Å for a DAG conformation that is 5.58

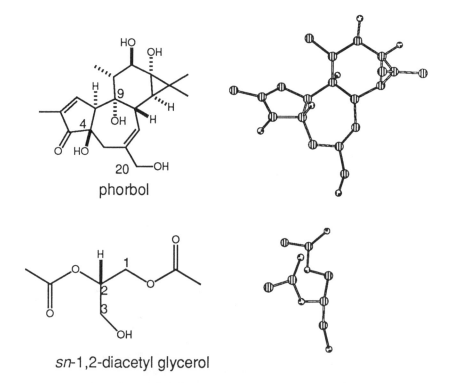

**Figure 16.** ORTEP comparisons of phorbol and a conformationally restricted depiction of *sn*-1,2-diacetyl glycerol. Carbon atoms are represented by striped circles, and oxygen atoms by dotted circles. Hydrogen atoms are omitted for clarity.

kcal/mol above the global energy minimum. Comparisons between DAG and pharmacophore possibilities, including within the triads the phorbol C-3 and/or C-4 oxygens, were similarly off in energy and/or RMSD.

The analysis thus suggests that the recognition features of DAG are best emulated by the C-4, C-9, and C-20 oxygens and the lipophilic group of phorbol esters. It should be noted that the spatial proximity of the C-3 and C-4 oxygens does not allow one to differentiate completely the roles of these two atoms as discussed in an early report.[215] Moreover, the restricted orientation of these phorbol ester features relative to those in the conformationally mobile DAG allows one to rationalize the higher binding affinity of the phorbol esters. Finally, this analysis also explains the binding specificity of enantiomers of DAG established by Rando and Young. Here the importance of our orbital directionality analysis is clearly evident since the centers of mass of oxygen atoms in (*S*)- and (*R*)-DAG are identical while their orbitals are oriented in opposite directions. When phorbol is compared with (S)-DAG, the centers of mass and the orientation of the orbitals on these oxygens, as well as the lipophilic groups in both molecules, show good correlation. The correlation of the two carbonyl oxygens of (S)-DAG with the C-4 and C-9 hydroxyls of phorbol suggests that the phorbol oxygens serve as hydrogen bond acceptors in binding with PKC. The relationship of the hydroxyl group of (S)-DAG with the C-20 hydroxyl group of phorbol, coupled with SAR data, indicates that the role of that oxygen is one of a hydrogen bond donor.

The above studies produced three important results. They provided for the first time a comprehensive explanation for the similar activity of the many structurally dissimilar families of potent tumor promoters or PKC activators. Second, these studies provided the opportunity to identify other tumor promoter candidates on the basis of whether they possess the proposed pharmacophore. Finally, by using the promoter pharmacophore as an architectural template, it became possible to design completely new molecules with potential PKC activator/tumor promoter activity. Up to this point, all known promoters and PKC activators were natural products or their derivatives. However, since the analysis of their structures suggested that they all possess three electron rich regions (heteroatoms) and a lipophilic group all spatially oriented in a well-defined manner, it was expected that one could design *de novo* other molecules that would have these same structural attributes. Such an effort would serve additionally to test the proposed model and to add to the understanding of PKC activation.

In the mid 1980s, several hypothetical molecules were designed which were expected to fit the PKC activator pharmacophore. Of the possibilities, a class of 1,3,5-trisubstituted aromatic compounds (Fig. 17) was selected partly because they fit the model rather well but primarily because they were expected to be readily synthesized. This proved to be an accurate assessment. In collaboration with the Blumberg group, we found that these compounds bound to PKC homogenate (from mouse brain cytosol) with affinities in the range of 12 to 60 μM, i.e., within an order of magnitude of the endogenous PKC activators, as measured by inhibition of [³H]PDBu binding.[215] Comparison of the [³H]PDBu binding in the absence or presence of 6-(N-decylamino)-4-hydroxymethylindole (DHI) indicated that the designed compounds and the phorbol ester standard bound competitively to the same receptor site. In addition, the analogs elicited responses in intact cells similar to those of the phorbol esters. Namely, DHI inhibited epidermal growth factor binding in mouse 3T3 cells with a half-maximal inhibitory concentration of approximately 100 micromolar, and 3-(N-acetylamino)-5-(N-decyl-N-methyl-amino)benzylalcohol (ADMB) at 300 micromolar concentration induced phos-phorylation of a 40 kDa protein in human platelets, effects also observed with phorbol ester treatment. Thus, the first class of non-natural PKC activators became available and their activities were in accord with expectations based on the pharmacophore analysis.

As a further test of this analysis, a search of a standard chemical catalog was made for compounds which would have features in common with the new analogs

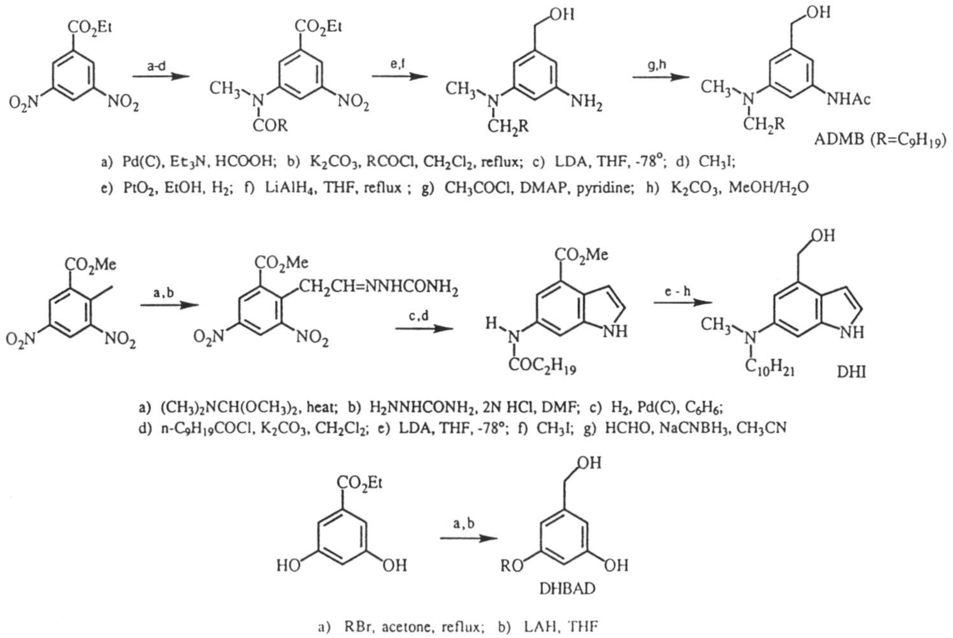

a) Pd(C), Et₃N, HCOOH; b) K₂CO₃, RCOCl, CH₂Cl₂, reflux; c) LDA, THF, -78°; d) CH₃I;
e) PtO₂, EtOH, H₂; f) LiAlH₄, THF, reflux ; g) CH₃COCl, DMAP, pyridine; h) K₂CO₃, MeOH/H₂O

a) (CH₃)₂NCH(OCH₃)₂, heat; b) H₂NNHCONH₂, 2N HCl, DMF; c) H₂, Pd(C), C₆H₆;
d) n-C₉H₁₉COCl, K₂CO₃, CH₂Cl₂; e) LDA, THF, -78°; f) CH₃I; g) HCHO, NaCNBH₃, CH₃CN

a) RBr, acetone, reflux; b) LAH, THF

**Figure 17.** Synthesis of rationally designed 1,3,5-substituted aromatic analogs of phorbol esters displaying PKC binding activity.

and therefore PKC activators such as phorbol esters. Among other compounds, dihydroxybenzoic acid was identified in this fashion and converted to the dihydroxybenzylalcohol diether analogs (DHBAD). In further support of the above model, this trivially derived compound was found to have PKC binding activity at micromolar concentrations.[218]

## C. Potential for a More Comprehensive Approach

Notwithstanding the usefulness of the rapidly developing techniques of computer modeling in the above studies, it became clear to us that the potential exists for the development of a more comprehensive approach to the task of pharmacophore identification. Studies of the type described above are driven by SARs and by the ability to *perceive* structural similarities that fit the SAR data. *Visual recognition* is by definition a somewhat subjective process, potentially resulting in an incomplete or biased analysis. Consequently, we set out to establish a protocol that would attempt to identify *all* structural similarities (potential pharmacophores) between two compounds involved in a comparison.[219] The resulting set of *all* fits could then by systematically reduced to the set of *best* fits through the use of SAR data and computer modeling.

Our starting assumption was that a set of three recognition elements in a molecule (e.g., regions of high electron density, lipophilic groups, etc.) that possess asymmetric directionality (e.g., as defined by orbital orientations) is the minimum number needed to define a pharmacophore candidate which exhibits stereospecificity in recognition. For a molecule with $n$ recognition elements (for example, heteroatoms), there are $n!/(n–3)!3!$ triads that represent possible pharmacophores. Thus, while two molecules might possess four or more features in common, an asymmetric triad is sufficient to provide unique recognition between the enzyme and its ligand. The number of correlations that would be possible for these triads with those of a second molecule interacting at the same site and containing $m$ recognition elements (and therefore $m!/(m–3)!3!$ triads) is given by the equation:

$$\text{number of correlations} = 6 * [n!/(n-3)!3!] * [m!/(m-3)!3!] \qquad (1)$$

The factor of six arises from the number of unique ways to match three objects of one set to three objects of another set. When this analysis is applied to compounds possessing heteroatoms having a rotational degree of freedom, one must also account for the possibility of the atom occupying more than a single relative position in space. Any number of points along the circle swept out by the heteroatom upon rotation about a bond can be selected, but a comprehensive search, for example, with movements of one degree, would come at the cost of time involved in evaluating the correlations. Reasonable values are twelve (every 30 degrees, representing a movement of approximately 0.67 Å for a hydroxyl oxygen atom center and therefore only a small change in RMSD when all else is held constant) and three (representing local energy minima). The number of triads containing the freely rotating heteroatom is given by the expression $(n - 1)!/(n - 3)!2!$; this value is then multiplied by the number of points along the circle selected to represent possible positions for the heteroatom. The number of triads not containing the freely rotating heteroatom, and therefore considered only once each, is $[n!/(n - 3)!3!] - [(n - 1)!/(n - 3)!2!]$. The total number of triads is given by the sum of these two expressions, where three represents the number of possible positions selected for the heteroatom:

$$\#=\{[n!/(n - 3)!3!] - [(n - 1)!/(n - 3)!2!]\} + \{3*[(n - 1)!/(n - 3)!2!]\} \qquad (2)$$

which simplifies to

$$\# = [n!/(n - 3)!3!] + \{2 * [(n - 1)!/(n - 3)!2!]\} \qquad (3)$$

This analysis has been applied to comparisons of phorbol with ingenol and with dihydroteleocidin. Phorbol contains six heteroatoms, or discrete electron-rich regions, and ingenol and dihydroteleocidin each contain five; there is a single heteroatom in each molecule whose relative spatial location could change through relatively unrestricted rotation. Therefore the number of *possible* correlations between phorbol and ingenol (or dihydroteleocidin) is:

$$\# = 6*([n!/(n-3)!3!]+\{2*[(n-1)!/(n-3)!2!]\})*$$

$$\{[m!/(m-3)!3!]+\{2*[(m-1)!/(m-3)!2!]\}\} \qquad (4)$$

or substituting the appropriate values for $n$ and $m$,

$$\# = 6*\{[6!/(6-3)!3!]+\{2*[(6-1)!/(6-3)!2!]\}\}*$$

$$\{[5!/(5-3)!3!]+\{2*[(5-1)!/(5-3)!2!]\}\}=8160 \qquad (5)$$

Had twelve been selected as the number of points along the arc to be considered instead of three, the number of correlations to be examined would have been 110,760. The value of this approach is that it allows one to determine all correlations and then begin to reduce their number systematically through SAR analysis or experimentation.

The problem can be simplified if one factors in knowledge derived from experimental evidence. Specifically, it is well established that modification in any way of the C-20 hydroxyl group of phorbol results in a loss of activity. Therefore, one can assume that the freely rotating hydroxyalkyl group in phorbol is one member of the triad, and that, moreover, it is correlated with the hydroxyalkyl group in the other molecule under consideration. The number of correlations is then reduced to:

$$\#= 2 * [5!/3!2!] * 3 * [4!/2!2!] * 3 = 1080. \qquad (6)$$

Note that when a conformationally mobile group is a member of each triad, the need to separate the triads into those containing the freely rotating heteroatom from those not containing it is obviated, thus simplifying the expression considerably. When these correlations are analyzed for goodness of fit, the triad of phorbol oxygens giving the lowest RMSD to triads of ingenol (Fig. 18) is composed of the C-4, C-9, C-20 oxygens of phorbol. Less good (lower priority) fits are found for the C-3, C-9, C-20 and C-3, C-4, C-20 triads. Similar results are obtained in the correlation of phorbol to dihydroteleocidin. In this case, the C-4, C-9, C-20 and C-3, C-4, C-20 triads are approximately equally good, followed at a short distance by the C-3, C-9, C-20 triad. It is gratifying to see that the systematic analysis assigns highest priority to our original pharmacophore model and identifies variations that have been proposed by others.

For example, the model proposed by Itai et al. is seen to correlate reasonably well.[220] This group looked at two conformations of the teleocidins, the twist form and the sofa form, both of which are observable by NMR.[221] The two conformations were shown to equilibrate rapidly in solution at 37° C,[222] and the lowest energy barrier for interconversion between the two was found to be 18 kcal/mol. For each conformer, a grid was generated around the molecule, and at each point various physical and chemical properties were quantified. PMA was superimposed on the

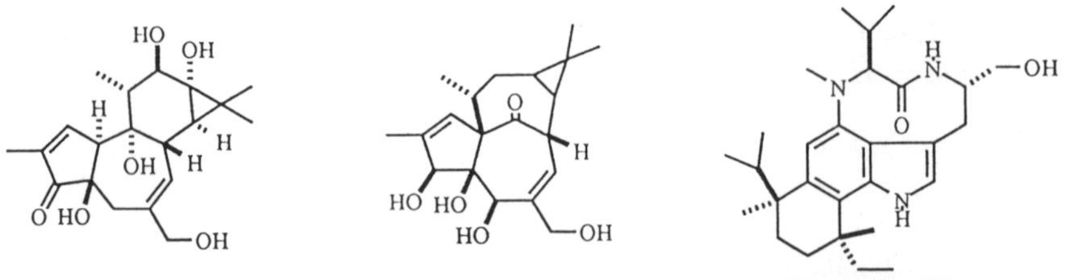

Phorbol                          Ingenol                          Dihydroteleocidin

| RMS DEV (Å) | NUMBER OF CORRELATIONS AS A FUNCTION OF RMS DEVIATION IN COMPARISONS OF PHORBOL WITH INGENOL (AND DIHYDROTELEOCIDIN) | | | | | | |
| | PHORBOL TRIADS | | | | | | |
| | 4,9,20 | 3,4,20 | 3,9,20 | 9,13,20 | 4,13,20 | 9,12,20 | OTHERS |
|---|---|---|---|---|---|---|---|
| .11-.20 | 1 (1) | (1) | | | | | |
| .21-.30 | 3 (2) | (5) | 1 | | | | |
| .31-.40 | 5 (7) | 1 (3) | 3 | (1) | 2 | | |
| .41-.50 | 4 (6) | (4) | 1 | 2 | 1 | | |
| .51-.60 | 4 (8) | 6 (1) | 2 (4) | 2 (1) | 3 | 1 | |
| .61-.70 | 7 (4) | 10 (1) | 13 (9) | 2 (2) | 2 (4) | 2 | 2 (2) |
| .71-.80 | 12 (5) | 8 (9) | 6 (11) | 6 | (4) | | (2) |
| .81-.90 | 5 (3) | 5 (11) | 3 (6) | 4 (7) | 2 (3) | 3 (1) | 4 (7) |
| .91-1.0 | 7 (4) | 7 (2) | 7 (2) | 6 (3) | 3 (2) | 6 (1) | 5 (4) |
| TOTALS | 48 (40) | 37 (37) | 36 (32) | 22 (14) | 13 (13) | 12 (2) | 13 (15) |

*Figure 18.* Results of the correlation of possible pharmacophoric candidates (triads of heteroatoms) of phorbol with ingenol and, parenthetically, with dihydroteleocidin.

grid so as to maximize potential sites of hydrogen bonding between the molecule and the presumed receptor. The sofa form provided the best correlation with phorbol, with hydrogen bond donors at the C-20 and C-4 hydroxyls of phorbol and a hydrogen bond acceptor at the C-3 carbonyl of phorbol (corresponding to the C-24 hydroxyl, amide NH and C-11 carbonyl of teleocidin, respectively).

Independent work by Hecker has also provided a correlation identified by the above protocol. Hecker compared phorbol, mezerein, and ingenol by overlaying the C-7–C-6–C-20 substructures using least-squares techniques, and assessing visually the fit of the diterpene moieties. While the fit based on only the allylic hydroxyl portion of each molecule is reasonably good, substantial improvement is seen when additional structural elements bearing oxygenation (i.e., C-3, C-4, and C-9) are also included in the least-squares analysis. One drawback to this approach is that it does not explain the differences in biological activity between the complete promoters phorbol and ingenol esters, and the second-stage promoter mezerein.[223]

Weinstein et al. correlated the C-3, C-4, C-9 and C-20 hydroxyls of phorbol to the amide carbonyl, amine nitrogen, indole nitrogen and C-24 hydroxyl of teleocidin, respectively, based on a least squares fit of centers of mass.[224] Unfortunately, the best fit was obtained for the *RR* isomer of teleocidin, an unnatural isomer that possesses less potent activity than the natural *SS* isomer. One of the authors later extended the pharmacophore model to include aplysiatoxin, mezerein, and ingenol.[225]

Another group of workers synthesized a series of cyclohexyltriols in an effort to mimic the relative spatial arrangements of the C ring oxygenation of phorbol.[226] A series of compounds with varying relative stereochemistries was modeled and prepared; none showed good spatial correlation with the putative pharmacophore and, consistent with expectation based on our model, none exhibited significant biological activity.

Most recently, a report from Kishi and Rando[227] on debromoaplysiatoxin (DAT) and DAT-related compounds suggested a correlation between the C-3, C-9 and C-20 oxygens of phorbol and the C-1 and C-27 carbonyl oxygens and C-30 hydroxyl of DAT, respectively. This model retains the same phorbol pharmacophoric features previously suggested,[215] therefore C-3 or C-4, C-9 and C-20 oxygens. The basis for this correlation is SAR data on DAT-derived compounds. Of interest was the observation that the C-30 epimer of DAT retained activity as measured by $^{32}P$ incorporation into lysine-rich histone at nearly the same level as natural DAT; inversion of the stereochemistry at C-29 resulted in loss of activity. Finally, 3-deoxy-DAT also retained activity at the same level as DAT. With the exception of the analogs given in reference 215, none of the models has been tested by the *de novo* design and synthesis of PKC agonists or antagonists.

On the basis of the aforementioned pharmacophore analyses, more recent work at Stanford has been directed at the design of high affinity, structurally simple, activators of PKC. Of several approaches, the most promising thus far has arisen from our earlier proposal that the phorbol esters are essentially conformationally restricted analogs of DAG. Therefore, modifications of DAG that would restrict its conformational mobility would in principle lead to analogs with phorbol ester-like potency. Bell et al. were the first to look at structure-activity relationships in the DAG series (*vide supra*).[90] They found that the requirements for recognition were strikingly specific in that all modifications of DAG resulted in less active or inactive analogs. In retrospect, this finding is not surprising given the importance of PKC in cellular function. A less precise recognition process would clearly have a deleterious effect on the proper functioning of PKC regulated cellular functions. Nevertheless, it was our expectation that DAG could be modified in a way that would preserve its pharmacophoric features but restrict their orientation to that found in the more tightly binding phorbol esters.

All modifications of DAG reported to date have focused on alterations in the DAG backbone, either to identify its recognition elements and/or to restrict its conformation. Our approach to developing a conformationally restricted DAG was to modify its lipophilic groups in a way that would restrict the DAG backbone. Through this approach, the exact molecular features of the backbone would be rigorously preserved, thereby satisfying PKC's impressive recognition specificity, but since the backbone would be restricted by the modified lipophilic groups, enhanced affinity would be achieved. Arguing for the success of this approach is sn-1,2-diacetylglycerol                          modified diacylglycerol

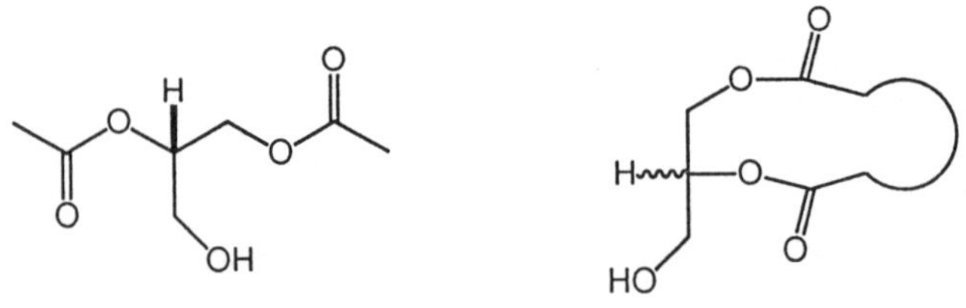

**Figure 19.** sn-1,2-Diacetylglycerol and a generic representation of a conformation-ally restricted DAG.

the general observation that while a lipophilic group is required for PKC activation, the group can be varied considerably.[92–95]

Our approach[228] took a very simple form: by connecting the two lipophilic groups of DAG together a cyclic system would be produced (Fig. 19). It has been well established that cyclic systems are subject to conformational biases depending on functionality and substitution patterns.[229] Therefore, we expected that modifica-tion of the lipophilic groups within the cyclic diester would allow us to adjust the conformation of the exquisitely sensitive backbone without loss of recognition by PKC. Moreover, the adjustments could be made at a coarse or fine level. For instance, incorporation of the backbone into an eight-membered ring would neces-sitate that both ester linkages be of the s-cis conformation. A larger ring, for instance, a 20-membered ring, would allow the esters to relax to their preferred s-trans conformation. Finer adjustments could be made by appropriate placement of alkyl side chains around the ring. Several routes toward the cyclic diesters were explored, the most satisfactory being that shown in Figure 20.

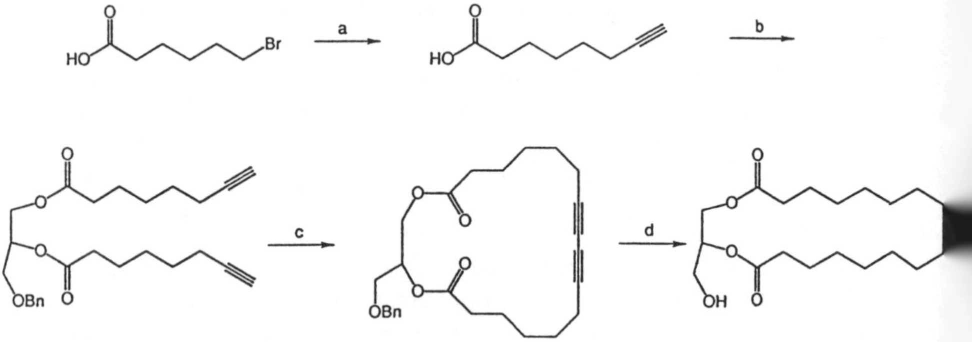

a) sodium acetylide, DMSO (84%); b) i. (COCl)₂, DMF, benzene ; ii. 1-benzyl glycerol, DMAP, pyridine (59%);
c) Cu(OAc)₂·H₂O, 2:1 pyridine:ether  (5mM, 74%); d) H₂, 10%Pd/C (neat, 60% 1,2 isomer, 12% 1,3 isomer, isolated yields)

**Figure 20.** Synthesis of cyclic DAGs, a new class of compounds displaying high affinity for PKC.

A striking advantage of this process is that varied functionality could be introduced in the macrocyclic bis-lactone through the coupling of relatively simple and accessible alkynoic acids.The oxidative diyne coupling proceeded smoothly to give only monomeric products in isolated yields ranging from 45% to 90% for ring sizes from 14 to 26 members. Restriction of the conformational mobility of the glycerol backbone by incorporation into a ring has worked exceptionally well, *leading to simple compounds exhibiting the highest binding affinities yet reported for DAG analogs,* with values for inhibition of binding of [$^3$H]PDBu in the range of 14 nM to 3.5 μM. It should be noted that the possibility that some of the enhancement in binding affinity may be due to differing solubilization or stability under assay conditions relative to natural DAGs has not yet been fully evaluated.

## D.  Design of PKC Inhibitors

With the lessons learned from the previous studies, investigations in our laboratory have also been directed at the design of PKC inhibitors. An important lead in this area came from Pettit's work on marine bryozoa, which resulted in the isolation of a new class of compounds called the bryostatins.[230] The bryostatins have remarkable activities among which is their ability to function as partial phorbol ester antagonists. Specifically, like the phorbol esters, bryostatin induces mitogenesis[231] and activation of human polymorphonuclear leukocytes[232] and blocks the binding of epidermal growth factor to its receptor.[233] Bryostatin has been shown to be a competitive binder with the phorbol esters[231] and in HL-60 cells activates PKC in vitro for the phosphorylation of the same proteins phosphorylated upon phorbol ester treatment, as well as several unique proteins.[234]

However, a subset of the phorbol ester responses are not induced by bryostatin, and, in fact, are blocked. Bryostatin, for instance, shows only weak activity as a tumor promoter and partially blocks tumor promotion by phorbol esters.[235] Bryostatin also fails to induce differentiation in HL-60 cells[236,237] and induces differentiation in Friend erythroleukemia cells,[238] in both cases opposite to the response of the cell lines to phorbol esters. Moreover, bryostatin blocks in an apparently non-competitive manner the normal responses to the phorbol esters.[237,238]

Rationalization for this apparent dichotomy in behavior between phorbol and bryostatin is provided by Blumberg.[239–241] In his laboratory, it has been observed that treatment of PKC with bryostatin causes PKC degradation at an accelerated rate relative to treatment with phorbol esters. In addition, the kinetics of ligand release differ significantly, with a very slow release of bound bryostatin. This result gave rise to the "glue-trap" model for bryostatin action, referring to the anchoring of PKC at the membrane site where it first came into contact with bryostatin. There is also experimental evidence that there may be a second target for bryostatin on PKC that is not efficiently recognized by phorbol esters. Finally, a significant difference between bryostatin and phorbol esters is in their affinities for PKC, with

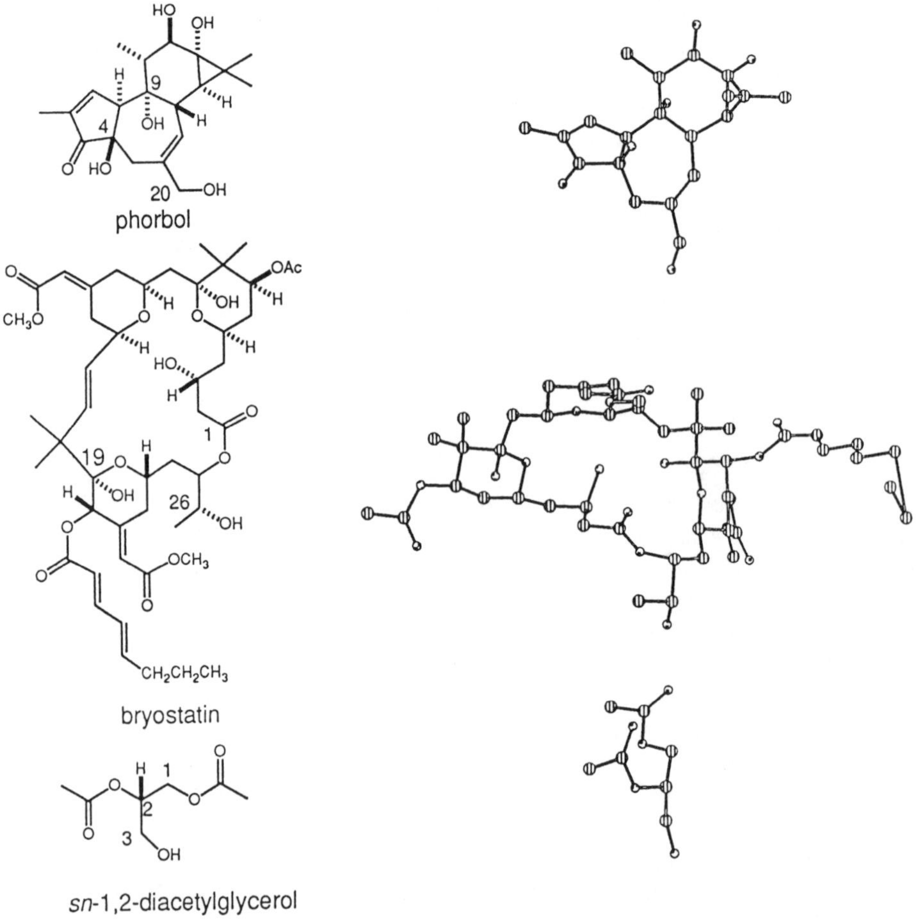

phorbol

bryostatin

sn-1,2-diacetylglycerol

**Figure 21.** ORTEP Comparisons of phorbol, bryostatin and a conformationally restricted depiction of sn-1,2-diacetylglycerol. Carbon atoms are represented by striped circles, and oxygen atoms by dotted circles. Hydrogen atoms are omitted for clarity.

the former possessing binding activity in the picomolar range and the latter in the nanomolar range.

In collaboration with Pettit's group at the University of Arizona and with Blumberg's group at the National Cancer Institute, we have been able to formulate a explanation for bryostatin's *agonist* activity. In short, it is our view that the bryostatins are analogous to conformationally restricted DAGs.[242] Figure 21 shows how the pharmacophoric features of DAG are reproduced in bryostatin. Changes of any of these features in bryostatin afford less active or inactive analogs. The higher affinities of the bryostatins relative to DAGs for PKC are attributed to the fact that the DAG-like features of bryostatin are restricted in an *active* conformation. As yet, a rigorous explanation for the *antagonist* activity of the bryostatins is not available, although for some other enzyme systems the change from agonist to antagonist is attributable to the addition or removal of one or more recognition elements to the basic agonist pharmacophore. It is therefore possible that bryostatin has recognition elements in common with PKC agonists such as DAG but that it also possesses a further recognition element(s) that contributes to its antagonistic activity. A systematic analysis of this hypothesis is now underway.

The ability to rationally design such antagonists is clearly a goal of ongoing studies, particularly since such compounds are expected to shed light on the cellular function of PKC and since bryostatin has already been cleared for trials in humans as a potential cancer chemotherapeutic agent. In addition to the correlation of phorbol esters and bryostatins, the modeling done thus far on the latter provides the basis for the eventual identification of the essential set of functionality required for bryostatin's chemotherapeutic activity and thus for the design of simplified analogues.

# IX. SYNTHETIC STRATEGIES TOWARD PHORBOL AND RELATED COMPOUNDS

Considerable effort by members of our group has been directed toward the synthesis of phorbol and related compounds in order to test our pharmacophore model and further develop this exciting medicinal lead.

In the first synthesis of the complete skeleton of any member of the tigliane, ingenane or daphnane families,[243] an intramolecular Diels-Alder cycloaddition (Fig. 22) was employed to provide the B and C rings of phorbol, leaving in place an ether bridge that would serve the dual purposes of rigidifying the system and offering a basis for stereochemical bias of later manipulations. Cyclopropanation using dibromocarbene proceeded as anticipated from the sterically less encumbered face opposite the ether bridge to establish the D ring. Stepwise construction of the A ring was required to give the thermodynamically disfavored *trans*-fused five-membered ring. Finally, functional group manipulations provided "phorbo-ingenol," a compound possessing the phorbol skeleton and the oxidation pattern of ingenol (excepting the C-5 hydroxyl of ingenol). Not coincidentally, phorbo-ingenol contains only the functional groups required by our pharmacophore model, with the lipophilic group anchored by esterification of the βC-3 hydroxyl group. In preliminary assays, phorbo-ingenol displayed no inhibition of $^3$H-PDBu binding to PKC up to a concentration of 20 μM. However, this lack of activity may be due to trans-esterification of the lipophilic group from the α hydroxyl at C-3 to the C-20 hydroxyl; other possible explanations include the potential lability of the allylic ester, and steric and/or electronic changes caused by reduction of the C-3 ketone

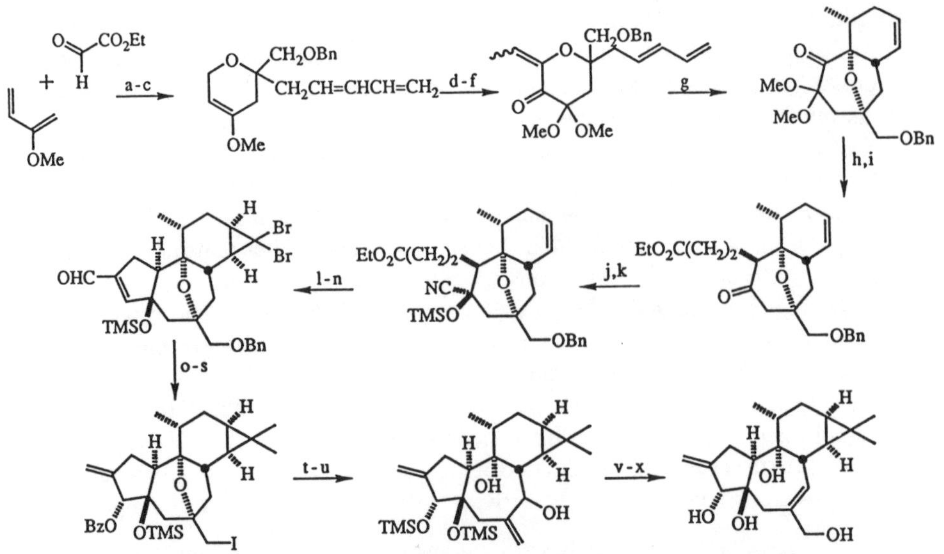

a) 110°C, toluene; b) i) LDA/THF, -78°C; CH₂=CHCH=CHCH₂Br, -78 to 0°C; ii) LiAlH₄, THF, 0°C;
c) BnBr, Bu₄NI, NaH, THF; d) i) m-CPBA, MeOH, 0°C; ii) (COCl)₂, Me₂SO, CH₂Cl₂, -60°C; Et₃N;
e) LiN(TMS)₂, LiBr, THF, -78°C; CH₃CHO, -78°C; f) i) MsCl, Et₃N, CH₂Cl₂; ii) DBU, THF; g) 145°C,
xylene; h) i) Ph₃P=CH₂, toluene, 105°C; ii) (COOH)₂, SiO₂, CH₂Cl₂; i) i) CH₂=C(OTBS)OEt, ZnI₂,
CH₂Cl₂; ii) HF, CH₃CN; j) PhHgCBr₃, benzene, 80°C; k) TMSCN, ZnI₂, CH₂Cl₂; l) DIBAH, toluene,
-78 to 0°C; m) (COCl)₂, Me₂SO, CH₂Cl₂, -60°C, Et₃N; n) Bn₂NH₂CF₃CO₂, benzene; o) DIBAH,
toluene, -78°C; p) Me₂CuCNLi₂, ether, -20°C; MeI; q) i) o-NO₂PhSeCN, Bu₃P, pyridine; ii) H₂O₂,
THF; r) i) Bz₂O, DMAP, Et₃N, CH₂Cl₂; ii) ZnI₂, TMSCN; s) i) Tf₂O, pyridine/CH₂Cl₂; ii) Bu₄NI, HMPA;
t) t-BuLi, THF, -78°C; u) i) TMS-Im, CH₂Cl₂; ii) SeO₂, (CH₃)₃COOH, CH₂Cl₂, 0°C; v) SOCl₂, Et₂O, 0°C;
w) KOAc-TMEDA, AgOAc, CH₃CN; x) TBAF, THF.

***Figure 22.*** First synthesis of the phorbol skeleton.

ketone to the alcohol. The roles of the C-3 and C-4 oxygens of phorbol are a focus
of current efforts in our group.

The first synthesis of phorbol was recently completed in our group.[244,245] The
key step in the preparation of an intermediate, expected to serve as a precursor to
the tiglianes, daphnanes, and ingenanes, was an intramolecular oxidopyryllium-
alkene cycloaddition to give the B and C rings of phorbol with complete selectivity
with respect to the relative stereochemistry at C-8, C-9 and C-11 (Fig. 23).
Annelation of the A ring was accomplished via an internal nitrile oxide cycloaddition. Functional group manipulation then provided the key tricyclic intermediate
in 23 steps and an overall yield of 10%. This compound was further elaborated
(Fig. 24) to phorbol as follows: first, the C ring enone was converted to an
α-acyloxy enone, which upon treatment with the Corey sulfur ylide gave the D ring
directly. Next, conversion of the protected C-20 hydroxyl group to the iodide
allowed base-induced opening of the ether bridge. Reduction of the C-12 ketone
with sodium triacetoxyborohydride provided the desired stereochemistry through
reagent coordination to the C-9 hydroxyl and internal delivery of the hydride
through a six-centered transition state. There remained only elaboration of the A

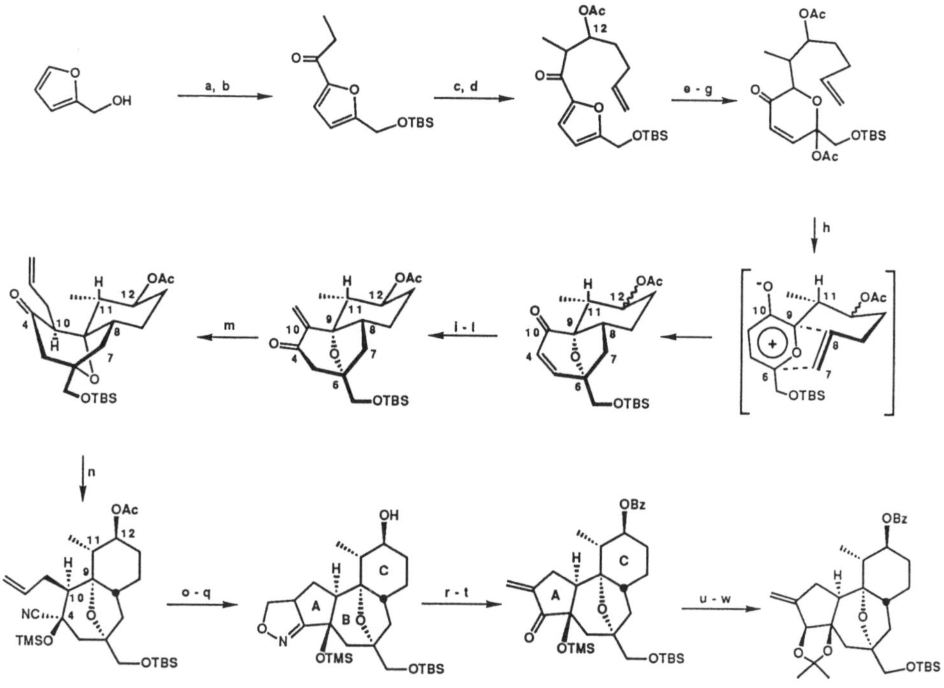

a) TBSCl, DMF, b) n-BuLi; C₂H₅COOLi, THF, c) LIN(TMS)₂, -78°C, 15h, THF; 4-pentenal, d) AcCl, pyr,CH₂Cl₂,
e) NaBH₄, MeOH, f) m-CPBA, THF, g) Ac₂O, DMAP, pyr, h) DBU, CH₂Cl₂, rm. temp., i) sep.; H₂, Pd/C, EtOAc,
j) Ph₃PCH₂, k) SeO₂, t-BuOOH, CH₂Cl₂, l) MnO₂, CH₂Cl₂, m) (CH₂CH)₂CuCNLi₂, THF, n) TMSCN, ZnI₂, o) DIBAH,
PhCH₃, p) NH₂OH, pyr, q) NaOCl, THF, r) H₂, Raney-Ni, acetone/H₂O (4:1), s) Bz₂O, CH₂Cl₂, DMAP, pyr,
) DBU, THF, u) NaBH₄, CeCl₃, MeOH, v) TBAF, Et₂O, w) 2-methoxypropene, PPTS,CH₂Cl₂

***Figure 23.*** Synthesis of a key intermediate in the first synthesis of phorbol.

and B rings' functionality to give phorbol in a total of 52 steps with 93% overall stereoselectivity.

An improvement in the phorbol synthesis[246] has been realized (Fig. 25) in an approach featuring a novel silicon-transfer-induced oxidopyryllium-alkene cycloaddition using a substrate derived from kojic acid through an *O*-alkylation and Claisen rearrangement sequence. The key cycloaddition is effected thermally and proceeds through a chair-like transition state that, remarkably, yielded a single isomer in excellent yield. Interestingly, the same transformation occurs photochemically at room temperature, albeit in only 15% yield. A new approach to A ring formation was also achieved in this second synthesis through a palladium-catalyzed enyne cyclization, representing one of the most complex applications of

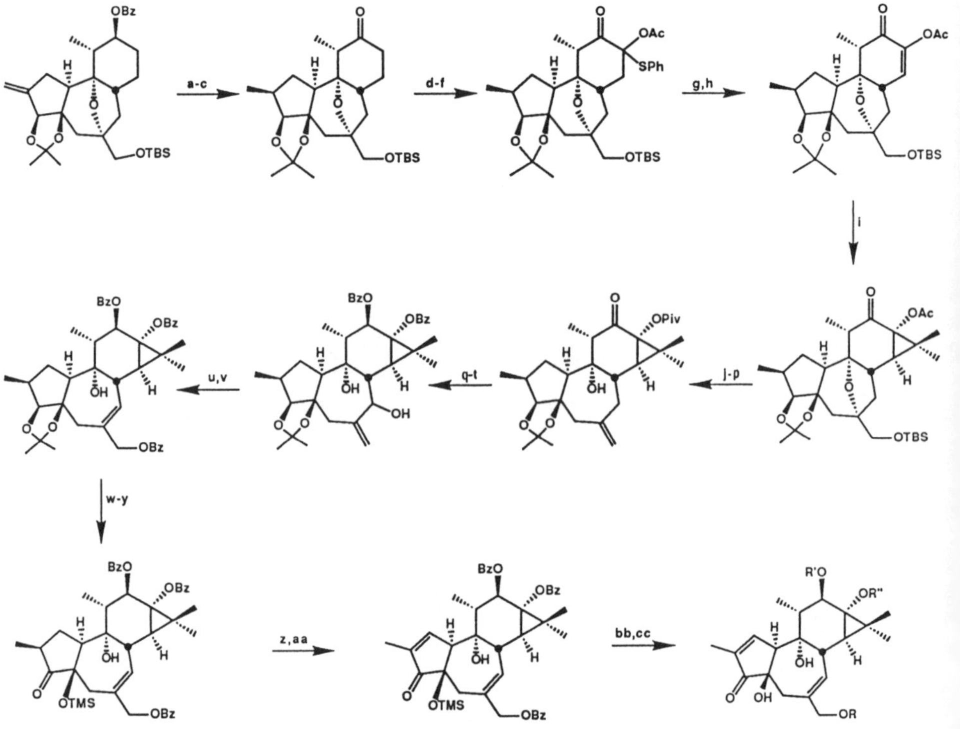

a) $H_2$ (1 atm), RhCl(PPh)$_3$, $C_6H_6$, b) DIBAH, PhCH$_3$, -78°C, c) PCC, CH$_2$Cl$_2$, d) LDA, THF; TMSCl, e) PhSCl, CH$_2$Cl$_2$, f) Pb(OAc)$_4$, C$_6$H$_6$, g) m-CPBA, CH$_2$Cl$_2$, h) 60° C, P(OEt)$_3$, C$_6$H$_6$, i) Ph$_2$SC(CH$_3$)$_2$, -78°C, THF, CH$_2$Cl$_2$, j) DIBAH, PhCH$_3$, k) CO(Im)$_2$, CH$_2$Cl$_2$, l) TBAF, THF, m) Tf$_2$O, Et$_3$N, CH$_2$Cl$_2$, pyr, n) Bu$_4$NI, HMPA, 55°C, o) t-BuLi, Et$_2$O, -78° C, p) PCC, CH$_2$Cl$_2$, q) NaBH(OAc)$_3$, THF, 60°C, r) DIBAH, PhCH$_3$, s) Bz$_2$O, DMAP, pyr, CH$_2$Cl$_2$, t) SeO$_2$, t-BuOOH, CH$_2$Cl$_2$, O°C, u) SOCl$_2$, propylene oxide, Et$_2$O, v) AgOBz, KOBz-TMEDA, CH$_3$CN, w) HClO$_4$, MeOH, Montmorillonite clay (K10), (CH$_2$OH)$_2$, x) SO$_3$·pyr·Et$_3$N, DMSO,y) CF$_3$CON(CH$_3$)TMS,DMAP, CH$_3$CN, z) KN(TMS)$_2$, -78°C; TMSCl, -78°C to rm. temp.; NBS, THF, aa) LiBr-Li$_2$CO$_3$, DMF, 130°C, 3h, bb) HF/CH$_3$CN, cc) KCN/MeOH

***Figure 24.*** Completion of the first phorbol synthesis.

this process in synthesis. Functional group elaboration provided a compound previously described[244] as a key intermediate in the first synthesis of phorbol in a total of 16 steps, thereby establishing a formal synthesis of phorbol.

## X.  CONCLUSION

In summary, protein kinase C has been identified as a key enzyme involved in cellular regulation, having been implicated in signal transduction, cell growth and proliferation, and ion channeling. The amino acid sequences of eight members of

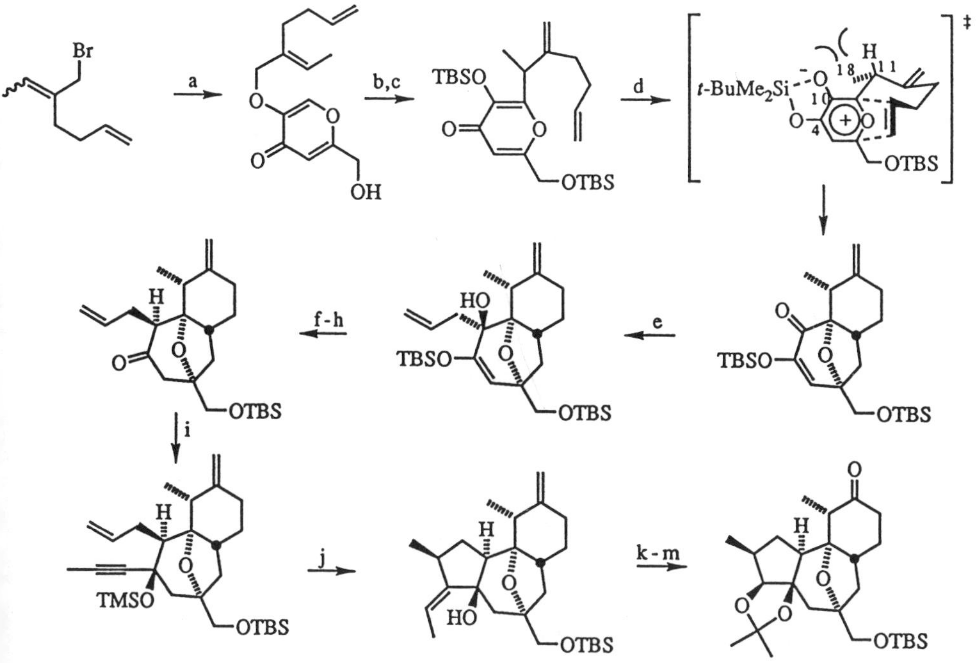

a) kojic acid cesium salt, MeOH; b) EtOH, 78°C, 4 hr; c) TBSCl, imidazole, DMF; d) 200°C, toluene, sealed tube, 48 h; e) CH$_2$=CHCH$_2$MgBr, THF; f) SOCl$_2$, pyridine, Et$_2$O, 0°C; g) TBAF, THF, 0°C; h) Bu$_3$SnH, catalytic AIBN, toluene, 80°C; i) 1-lithio-1-propyne, 4 equiv of LiBr, THF, -78 to 20°C; then TMSCl; j) 0.1 equiv of Pd$_2$(dba)$_3$·CHCl$_3$, 0.2 equiv of tri(o-tolyl)-phosphine, 2 equiv of HOAc, 2 equiv of [(CH$_3$)$_2$SiH]$_2$O, toluene; k) O$_3$, CH$_2$Cl$_2$, MeOH, -78°C; NaBH$_4$, -78 to 20°C; l) 2-methoxypropene, catalytic PPTS, CH$_2$Cl$_2$; m) PCC, NaOAc, CH$_2$Cl$_2$

**Figure 25.** A second generation synthesis of phorbol.

the PKC family have been determined, and show a high degree of homology. Although very little is known about the binding site of PKC activators, computer-guided correlations of the crystal structures of several structurally diverse compounds has produced a working hypothesis for their common mode of action. Each class of compounds is seen to have a set of electron-rich sites similarly situated in space and precisely oriented with respect to a fourth recognition element taking the form of a lipophilic group. This pharmacophoric model has been used to identify potentially new PKC activators/tumor promoters and to rationally design the first class of non-natural PKC activators. It has also been used to identify potentially new PKC activators.

Through these studies, a methodology has been developed that allows for the identification of all possible pharmacophoric candidates in a molecule and of all possible correlatable pharmacophoric candidates between two molecules. Computer comparison programs are then used to rank these candidates according to the quality of their fit. These studies are of fundamental value in elucidating the factors involved in small molecule–large molecule recognition and in rational drug design and are beginning to clarify the structural requirements for tumor promotion. Such information has already figured in the development of approaches to the prevention of cancer and they serve as an important prelude to the development of tumor promoter inhibitors. As more structural information about PKC becomes available, for example, through X-ray crystallographic studies, a better understanding will emerge of the specific interactions of PKC, phospholipids, calcium, and activators, and thus better equip the scientific community to design potential PKC inhibitors.

## ABBREVIATIONS

AA, arachidonic acid
ADMB, 3-(*N*-acetylamino)-5-(*N*-decyl-*N*-methylamino)benzyl alcohol
cDNA, complementary deoxyribonucleic acid
CF, cystic fibrosis
DAG, diacylglycerol
DHBAD, dihydroxybenzyl alcohol diether
DHI, 6-(*N*-decylamino)-4-hydroxymethylindole
DHT, dihydroteleocidin
DMAP, *N,N*-dimethyl-4-aminopyridine
DMHI, 6-(*N*-decyl-*N*-methylamino)-4-hydroxymethylindole
EGF, epidermal growth factor
IP$_3$, inositol-1,4,5-trisphosphate
PDBu, phorbol 12,13-dibutyrate
PKA, protein kinase A
PKC, protein kinase C
PMA, phorbol myristate acetate
PS, phosphatidylserine
RMSD, root mean square deviation
SAR, structure-activity relationship

## REFERENCES

1.  Takai, Y.; Kishimoto, A.; Inoue, M.; Nishizuka, Y. *J. Biol. Chem.* **1977**, *252*, 7603.
2.  Inoue, M.; Kishimoto, A.; Takai, Y.; Nishizuka, Y. *J. Biol. Chem.* **1977**, *252*, 7610.
3.  Lew, K.K.; Chritton, S.; Blumberg, P.M. *Teratogen. Carcinogen. Mutagen.* **1982**, *2*, 19.
4.  Blumberg, P.M.; Delclos, K.B.; Jaken, S. In *Organ and Species Specificity in Chemical Carcinogenesis*; Langenbach, R.; Nesnow, S.; Rice, J.M., Eds.; Plenum: New York, 1983; p. 201.
5.  Kuo, J.F.; Andersson, R.G.G.; Wise, B.C.; Mackerlova, L.; Salmonsson, I.; Brackett, N.L.; Katoh, N.; Shoji, M.; Wrenn, R.W. *Proc. Natl. Acad. Sci. USA* **1980**, *77*, 7039.

6. Hergenhahn, M.; Hecker, E. *Carcinogenesis* **1981**, *2*, 1277.

7. Shoyab, M.; Warren, T.C.; Torado, G.J. *Carcinogenesis* **1981**, *2*, 1273.

8. Castagna, M.; Takai, Y.; Kaibuchi, K.; Sano, K.; Kikkawa, U.; Nishizuka, Y. *J. Biol. Chem.* **1982**, *257*, 7847.

9. For a comprehensive review of tumor promotion, see *Mechanisms of Tumor Promotion*; Slaga, T.J., Ed.; CRC: Boca Raton, FL., 1984; Vols. 1-4.

10. Evans, F.J.; Soper, C.J. *Lloydia* **1978**, *41*, 193.

11. *Naturally Occurring Phorbol Esters*; Evans, F.J., Ed.; CRC: Boca Raton, FL., 1986.

12. Wolf, M.; LeVine, H., III; May, W.S., Jr.; Cuatrecasas, P.; Sahyoun, N. *Nature (London)* **1985**, *317*, 546.

13. Mellgren, R.L. *FEBS Lett.* **1980**, *109*, 129.

14. Murachi, T.; Tanaka, K.; Hatanaka, M.; Murakami, T. *Adv. Enzyme Regul.* **1981**, *19*, 407.

15. Kishimoto, A.; Kajikawa, N.; Tabuchi, H.; Shiota, M.; Nishizuka, Y. *J. Biochem.* **1981**, *90*, 889.

16. Kishimoto A.; Kajikawa, N.; Shiota, M.; Nishizuka, Y. *J. Biol Chem.* **1983**, *258*, 1156.

17. Parker, P.J.; Coussens, L.; Totty, N.; Rhee, L.; Young, S.; Chen, E.; Stabel, S.; Waterfield, M.D.; Ullrich, A. *Science* **1986**, *233*, 853.

18. Coussens, L.; Parker, P.J.; Rhee, L.; Yang-Feng, T.L.; Chen, E.; Waterfield, M.D.; Francke, U.; Ullrich, A. *Science* **1986**, *233*, 859.

19. Knopf, J.L.; Lee, M.-H.; Sultzman, L.A.; Kriz, R.W.; Loomis, C.R.; Hewick, R.M.; Bell, R.M. *Cell* **1986**, *46*, 491.

20. Ono, Y.; Kurokawa, T.; Fujii, T.; Kawahara, K.; Igarashi, K.; Kikkawa, U.; Ogita, K.; Nishizuka, Y. *FEBS Lett.* **1986**, *206*, 347.

21. Ohno, S.; Kawasaki, H.; Imajoh, S.; Suzuki, K.; Inagaki, M.; Yokohura, H.; Sakoh, T.; Hidaka, H. *Nature (London)* **1987**, *325*, 161.

22. Ono, Y.; Kurokawa, T.; Kawahare, K.; Nishimura, O.; Marumoto, R.; Igarashi, K.; Sugino, Y.; Kikkawa, U.; Ogita, K.; Nishizuka, Y. *FEBS Lett.* **1986**, *203*, 111.

23. Makowske, M.; Birnbaum, M.J.; Ballester, R.; Rosen, O.M.A *J. Biol Chem.* **1986**, *261*, 13389.

24. Housey, G.M.; O'Brian, C.A.; Johnson, M.D.; Kirschmeier, P.; Weinstein, I.B. *Proc. Natl. Acad. Sci. USA* **1987**, *84*, 1065.

25. Ono, Y.; Kikkawa, U.; Ogita, K.; Fujii, T.; Kurokawa, T.; Asaoka, Y.; Sekiguchi, K.; Ase, K.; Igarashi, K.; Nishizuka, Y. *Science* **1987**, *236*, 1116.

26. Ono, Y.; Fujii, T.; Ogita, K.; Kikkawa, U.; Igarashi, K.; Nishizuka, Y. *J. Biol. Chem.* **1988**, *263*, 6927.

27. Reviewed by Kikkawa, U.; Kishimoto, A.; Nishizuka, Y. *Annu. Rev. Biochem.* **1989**, *58*, 31.

28. Parker, P.J.; Marais, R.; Bajaj, M.; Mitchell, F.; King, P.; Young, S.; Ullrich, A.; Stabel, S. In *Adv. Exp. Med. Biol.*, Vol 231 (Advances in Post-Translational Modificiation of Proteins and Aging); Zappia, K.; Galletti, P.; Porta, R.; Wold, F., Eds.; Plenum: New York, 1988; p. 417.

29. Blundell, T.L.; Humbel, R.E. Hormone Families: *Nature (London)* **1980**, *287*, 781.

30. Ullrich, A.; Coussens, L.; Hayflick, J.S.; Dull, T.J.; Gray, A.; Tam, A.W.; Lee, J.; Yarden, Y.; Libermann, T.A.; Schlessinger, J.; Downward, J.; Mayes, E.L.V.; Whittle, N.; Waterfield, M.D.; Seeburg, P.H. *Nature (London)* **1984**, *309*, 418.

31. Ullrich, A.; Bell, J.R.; Chen, E.Y.; Herrera, R.; Petruzzelli, L.M.; Dull, T.J.; Gray, A.; Coussens, L.; Liao, Y.-C.; Tsubokawa, M.; Mason, A.; Seeburg, P.H.; Grunfeld, C.; Rosen, O.M.; Ramachandran, J. *Nature (London)* **1985**, *313*, 756.

32. Coussens, L.; Van Beveren, C.; Smith, D.; Chen, E.; Mitchell, R.L.; Isacke, C.M.; Verma, I.M.; Ullrich, A. *Nature* **1986**, *320*, 277.

33. Hanks, S.K.; Quinn, A.M.; Hunter, T. *Science* **1988**, *241*, 42.

34. Hanks, S.K. *Proc. Natl. Acad. Sci. USA* **1987**, *84*, 388.

35. Levin, D.E.; Hammond, C.I.; Ralston, R.O.; Bishop, J.M. *Proc. Natl. Acad. Sci. USA* **1987**, *84*, 6035.

36. Endelman, A.M.; Blumenthal, D.K.; Krebs, E.G. *Annu. Rev. Biochem.* **1987**, *56*, 567.

37. House, C.; Kemp, B.E. *Science* **1987**, *238*, 1726.
38. Ono, Y.; Fujii, T.; Igarashi, K.; Kuno, T.; Tanaka, C.; Kikkawa, U.; Nishizuka, Y. *Proc. Natl. Acad. Sci. USA* **1989**, *86*, 4868.
39. Berg, J.M. *Science* **1986**, *232*, 485.
40. Evans, R.M.; Hollenberg, S.M. *Cell* **1988**, *52*, 1.
41. Nishizuka, Y. *Nature (London)* **1988**, *334*, 661.
42. Shearman, M.S.; Kosaka, Y.; Ase, K.; Kikkawa, U.; Nishizuka, Y. *Biochem. Soc. Trans.* **1988**, *16*, 307.
43. Shearman, M.S.; Ase, K.; Saito, N.; Sekiguchi, K.; Ogita, K.; Kikkawa, U.; Tanaka, C.; Nishizuka, Y. *Biochem. Soc. Trans.* **1988**, *16*, 460.
44. Kikkawa, U.; Kitano, T.; Saito, N.; Kishimoto, A.; Taniyama, K.; Tanaka, C; Nishizuka, Y. In *Calcium and the Cell*; Baker, P., Ed.; Ciba Foundation Symposium; Wiley: Chichester, U.K., 1986; p. 197.
45. Sekiguchi, K.; Tsukuda, M.; Ogita, K.; Kikkawa, U.; Nishizuka, Y. *Biochem. Biophys. Res. Commun.* **1987**, *145*, 797.
46. Ase, K.; Berry, N.; Kikkawa, U.; Kishimoto, A.; Nishizuka, Y. *FEBS Lett.* **1988**, *236*, 396.
47. Kishimoto, A.; Mikawa, K.; Hashimoto, K.; Yasuda, I.; Tanaka, S.; Tominaga, M.; Kuroda, T.; Nishizuka, Y. *J. Biol. Chem.* **1989**, *264*, 4088.
48. Rodriguez-Pena, A.; Rozengurt, E. *Biochem. Biophys. Res. Commun.* **1984**, *120*, 1053.
49. Ballester, R.; Rosen, O. M. *J. Biol. Chem.* **1985**, *260*, 15194.
50. Chida, K.; Kato, N.; Kuroki, T. *J. Biol. Chem.* **1986**, *261*, 13013.
51. Krebs, E.G.; Beavo, J.A. *Annu. Rev. Biochem.* **1979**, *48*, 923.
52. Ashedel, C.L. *Biochim. Biophys. Acta* **1985**, *822*, 219.
53. Jaken, S.; Leach, K.L. *Annu. Rep. Med. Chem.* **1988**, *23*, 243.
54. Gescher, A.; Dale, I.L. *Anti-Cancer Drug Design* **1989**, *4*, 93.
55. Nishizuka, Y. *Cancer (Philadelphia)* **1989**, *63*, 1892.
56. Parker, P.J.; Kour, G.; Marais, R.M.; Mitchell, F.; Pears, C.; Schaap, D.; Stabel, S.; Webster, C. *Mol. Cell. Endocrinol.* **1989**, *65*, 1.
57. Hokin, L.E.; Hokin, M.R. *Biochim. Biophys. Acta* **1955**, *18*, 102.
58. Hawthorne, J.N.; White, D.A. *Vitam. Horm.* **1975**, *33*, 529.
59. Michell, R.H. *Biochim. Biophys. Acta* **1975**, *415*, 81.
60. Michell, R.H. *Trends in Biochem. Sci.* **1979**, *4*, 128.
61. Berridge, M.J.; Irvine, R.F. *Nature (London)* **1984**, *312*, 315.
62. Nishizuka, Y. *Nature (London)* **1984**, *308*, 693.
63. Nishizuka, Y. *Trends in Biochem. Sci.* **1984**, *9*, 163.
64. Nishizuka, Y. *Science* **1984**, *225*, 1365.
65. Preiss, J.; Loomis, C.R.; Bishop, W.R.; Stein, R.; Niedel, J.E.; Bell, R.M. *J. Biol. Chem.* **1986**, *261*, 8597.
66. Macara, I.G.; Marinetti, G.V.; Balduzzi, P.C. *Proc. Natl. Acad. Sci. USA* **1984**, *81*, 2728.
67. Kaplan, D.R.; Whitman, M.; Schaffhausen, B.; Raptis, L.; Garcia, R.L.; Pallas, D.; Roberts, T.M.; Cantley, L. *Proc. Natl. Acad. Sci. USA* **1986**, *83*, 3624.
68. For specific references, see literature cited in Nishizuka, Y. *Science* **1984**, *255*, 1365; and Berridge, M.J.; Irvine, R.F. *Nature (London)* **1984**, *312*, 315.
69. Dawson, R.M.C. *Biochim. Biophys. Acta* **1959**, *33*, 68.
70. Streb, H.; Irvine, R.F.; Berridge, M.J.; Schulz, I. *Nature (London)* **1983**, *306*, 67.
71. Kraft, A.S.; Anderson, W.B. *Nature (London)* **1983**, *301*, 621.
72. Hirota, K.; Hirota, T.; Aguilera, G.; Catt, K.J. *J. Biol. Chem.* **1985**, *260*, 3243.
73. For specific references, see literature cited in Kikkawa, U.; Nishizuka, Y. In *The Enzymes*; Boyer, P.D., Ed.; Academic Press: New York, 1986; Vol 17, p. 167; and Nishizuka, Y. Studies and Perspectives of Protein Kinase C. *Science* **1986**, *233*, 305.

74. For specific references, see literature cited in Nishizuka, Y. *Science* **1984**, *225*, 1365; and Nishizuka, Y. *Science* **1986**, *233*, 305.

75. Fisher, S.K.; Van Rooijen, L.A.A.; Agranoff, B.W. *Trends in Biochem. Sci.* **1984**, *9*, 53.

76. Majerus, P.W.; Wilson, D.B.; Connolly, T.M.; Bross, T.E.; Neufield, E.J. *Trends in Biochem. Sci.* **1985**, *10*, 168.

77. Holub, B.J.; Kukis, A.; Thompson, W. *J. Lipid Res.* **1970**, *11*, 558.

78. Samuelsson, B.; Granstrom, E.; Green, K.; Hamberg, M.; Hammarstrom, S. *Annu. Rev. Biochem.* **1975**, *44*, 669.

79. Brown, J.H.; Orellana, S.A.; Buss, J.E.; Quilliam, L.A. *Adv. Exp. Med. Biol.* **1988**, *236*, 229.

80. Garcia-Sainz, J.A.; Gutierrez-Venegas, G. *FEBS Lett.* **1989**, *257*, 427.

81. Carlson, K.E.; Brass, L.E.; Manning, D.R. *J. Biol. Chem.* **1989**, *264*, 13298.

82. Ieyasu, H.; Takai, Y.; Kaibuchi, K.; Sawamura, M.; Nishizuka, Y. *Biochem. Biophys. Res. Commun.* **1982**, *108*, 1701.

83. Kikkawa, U.; Takai, Y.; Tanaka, Y.; Miyake, R.; Nishizuka, Y. *J. Biol. Chem.* **1983**, *258*, 11442.

84. Delclos, K.B.; Yeh, E.; Blumberg, P.M. *Proc. Natl. Acad. Sci. USA* **1983**, *80*, 3054.

85. Hannun, Y.A.; Loomis, C.R.; Bell, R.M. *J. Biol. Chem.* **1985**, *260*, 10039.

86. Takai, Y.; Kishimoto, A.; Kikkawa, U.; Mori, T.; Nishizuka, Y. *Biochem. Biophys. Res. Commun.* **1979**, *91*, 1218.

87. Kaibuchi, K.; Takai, Y.; Nishizuka, Y. *J. Biol. Chem.* **1981**, *256*, 7146.

88. Weinstein, I.B. *J. Cell. Biochem.* **1987**, *33*, 213.

89. Rando, R.R.; Young, N. *Biochem. Biophys. Res. Commun.* **1984**, *122*, 818.

90. Ganong, B.R.; Loomis, C.R.; Hannun, Y.A.; Bell, R.M. *Proc. Natl. Acad. Sci. USA* **1986**, *83*, 1184.

91. Nomura, H.; Ase, K.; Sekiguchi, K.; Kikkawa, U.; Nishizuka, Y. *Biochem. Biophys. Res. Commun.* **1986**, *140*, 1143.

92. Mori, T.; Takai, Y.; Yu, B.; Takahashi, J.; Nishizuka, Y.; Fujikura, T. *J. Biochem.* **1982**, *91*, 427.

93. Sharkey, N.A.; Leach, K.L.; Blumberg, P.M. *Proc. Natl. Acad. Sci. USA* **1984**, *81*, 607.

94. Go, M.; Sekiguchi, K.; Nomura, H.; Kikkawa, U.; Nishizuka, Y. *Biochem. Biophys. Res. Commun.* **1987**, *144*, 598.

95. Kerr, D.E.; Kissenger, L.F.; Gentry, L.E.; Purchio, A.F.; Shoyab, M. *Biochem. Biophys. Res. Commun.* **1987**, *148*, 776.

96. Limas, C.J. *Biochem. Biophys. Res. Commun.* **1980**, *96*, 1378.

97. Iwasa, Y.; Hosey, M.M. *J. Biol. Chem.* **1984**, *259*, 534.

98. Movsesian, M.A.; Nishikawa, M.; Adelstein, R.S. *J. Biol. Chem.* **1984**, *259*, 8029.

99. Woods, N.M.; Cuthbertson, K.S.R.; Cobbold, P.H. *Biochem. J.* **1987**, *246*, 619.

100. Connolly, T.M.; Lawring, W.J., Jr.; Majerus, P.W. *Cell* **1986**, *46*, 951.

101. Cochet, C.; Gill, G.N.; Meisenhelder, J.; Cooper, J.A.; Hunter, T. *J. Biol. Chem.* **1984**, *259*, 2553.

102. Iwashita, S.; Fox, C.F. *J. Biol. Chem.* **1984**, *259*, 2559.

103. Davis, R.J.; Czech, M.P. *J. Biol. Chem.* **1984**, *259*, 8545.

104. Hunter, T.; Ling, N.; Cooper, J.A. *Nature (London)* **1984**, *311*, 480.

105. Brown, K.D.; Blay, J.; Irvine, R.F.; Helsop, J.P.; Berridge, M.J. *Biochem. Biophys. Res. Commun.* **1984**, *123*, 377.

106. Moon, S.O.; Palfrey, H.C.; King, A.C. *Proc. Natl. Acad. Sci. USA* **1984**, *81*, 2298.

107. Ran, W.; Dean, M.; Levine, R.A.; Henkle, C.; Campisi, J. *Proc. Natl. Acad. Sci. USA* **1986**, *83*, 8216.

108. Grausz, J.D.; Fradelizi, D.; Dautry, F.; Monier, R.; Lehn, P. *Eur. J. Immunol.* **1986**, *16*, 1217.

109. Nishizhka, Y. *J. Natl. Cancer Inst.* **1986**, *76*, 363.

110. Kikkawa, U.; Nishizuka, Y. In: *The Enzymes*; Boyer, P. D., Ed.; Academic: New York, 1986; Vol. 17; p. 167.

111. Nishizuka, Y. *Philos. Trans. R Soc. B* **1983**, *302*, 101.

112. Nishizuka, Y.; Takai, Y.; Kishimoto, A.; Kikkawa, U.; Kaibuchi, K. *Recent Prog. Horm. Res.* **1984**, *40*, 301.

113. Hwang, T.-C.; Lu, L.; Zeitlin, P.L.; Gruenert, D.C.; Huganir, R.; Guggino, W.B. *Science* **1989**, *244*, 1351.

114. Li, M.; McCann, J.D.; Anderson, M.P.; Clancy, J.P.; Liedtke, C.M.; Nairn, A.C.; Greengard, P.; Welsh, M.J. *Science* **1989**, *244*, 1353.

115. For a lead reference, see Rubin, C.S.; Erlichman, J.; Rosen, O.M. *J. Biol. Chem.* **1972**, *247*, 36.

116. Minakuchi, R.; Takai, Y.; Yu, B.; Nishizuka, Y. *J. Biochem.* **1981**, *89*, 1651.

117. Kishimoto, A.; Nishiyama, K.; Nakanishi, H.; Uratsuji, Y.; Nomura, H.; Takeyama, Y.; Nishizuka, Y. *J. Biol. Chem.* **1985**, *260*, 12492.

118. Alkon, D.L.; Naito, S.; Kubota, M.; Chen, C.; Bank, B.; Smallwood, J.; Gallant, P.; Rasmussen, H. *J. Neurochem.* **1988**, *51*, 903.

119. Alkon, D.L. *Scientific American* **1989** (July), 42.

120. Yamagiwa, K.; Ichikawa, K. *Gann* **1914**, *8*, 11.

121. Shear, M.J. Studies in Carcinogenesis. V. *Am. J. Cancer* **1938**, *33*, 499.

122. Sall, R.D.; Shear, M.J. *J. Natl. Cancer Inst.* **1940**, *1*, 45.

123. Friedewald, W.F.; Rous, P. *J. Exp. Med.* **1944**, *80*, 101.

124. Ames, B.N.; Gold, L. S. *Science* **1990**, *249*, 970.

125. Berenblum, I. *Cancer Res.* **1941**, *1*, 807.

126. Boutwell, R.K. *Prog. Exp. Tumor Res.* **1964**, *4*, 207.

127. Slaga, T.J.; Klein-Szanto, A.J.P.; Fischer, S.M.; Weeks, C.E.; Nelson, K.; Major, S. *Proc. Natl. Acad. Sci. USA* **1980**, *77*, 2251.

128. Slaga, T.J.; Fischer, S.M.; Nelson, K.; Gleason, G.L. *Proc. Natl. Acad. Sci. USA* **1980**, *77*, 3659.

129. Hecker, E. In *Mechanisms of Tumor Promotion and Cocarcinogenesis*; Slaga, T.J.; Sivak, A.; Boutwell, R.K., Eds.; Raven: New York, 1978; p. 11.

130. Furstenberger, G.; Berry, D.L.; Sorg, B.; Marks, F. *Proc. Natl. Acad. Sci. USA* **1981**, *78*, 7722.

131. Reviewed by Blumberg, P.M.; Dunn, J.A.; Jaken, S.; Jeng, A.Y.; Leach, K.L.; Sharkey, N.A.; Yeh, E. In *Mechanisms of Tumor Promotion*; Slaga, T.J., Ed.; CRC: Boca Raton, 1984; Vol. 3, p. 143.

132. *The Medicinal and Poisonous Plants of Southern and Eastern Africa*; Watt, J.M.; Breyer-Brandwijk, M.G., Eds.; E. and S. Livingstone: Edinburgh, 1962; p. 395.

133. Seghal, L.; Paliwal, G.S. *Bot. J. Linn. Soc.* **1974**, *68*, 173.

134. Dymock, W.; Warden, C.J.H.; Hooper, D. *Pharmacographica Indica*; Hamdard National Foundation: Karachi, Pakistan, 1890; Vol. 1, p. 247.

135. Ainslie, J.R. *A List of Plants in Native Medicine in Nigeria*; Oxford University: Oxford, 1937; Paper 7, p. 27.

136. In Steinbeck's *East of Eden*, Kate, after insinuating her way into favor with the madam of the brothel where she worked, killed Faye with chronic application of croton oil (stolen from a local doctor) to her green beans in order to inherit her wealth.

137. Castagnou, R.; Boudriment, R.; Gauthier, J. *Compt. Rend.* **1965**, *260*, 4109.

138. Weber, J.; Hecker, E. *Experientia* **1978**, *34*, 679.

139. Hirota, M.; Suttajit, M.; Suguri, H.; Endo, Y,; Shudo, K.; Wongchai, V.; Hecker, E.; Fujiki, H. *Cancer Res.* **1988**, *48*, 5800.

140. Hoppe, W.; Brandl, F.; Strell, I.; Roehrl, M.; Gassman, J.; Hecker, E.; Bartsch, H.; Kreibich, G.; Szczepanski, C. *Angew. Chem.* **1967**, *79*, 824.

141. Pettersen, R.C.; Ferguson, G.; Crombie, L.; Games, M.L.; Pointer, D.J. *Chem. Commun.* **1967**, *14*, 716.

142. Hecker, E.; Schmidt, R. In *Fortschritte de Chemie Organischer Naturstoffe*; Herz, W.; Grisebach, H.; Kirby, G.W., Eds.; Springer-Verlag: New York, 1974; Vol. 31, p. 378.

143. Rohrschneider, L.R.; Boutwell, R.K. *Nature New Biology* **1973**, *243*, 212.

144. Smythies, J.R.; Benington, F.; Morin, R.D. *Psychoneuroendocrinology* **1975**, *1*, 123.

145. Wilson, R.; Huffman, J.C. *Experientia* **1976**, *32*, 1489.
146. Driedger, P.E.; Blumberg, P.M. *Proc. Natl. Acad. Sci. USA* **1980**, *77*, 567.
147. Delclos, K.B.; Nagle, D.S.; Blumberg, P.M. *Cell* **1980**, *19*, 1025.
148. Niedel, J.E.; Kuhn, L.J.; Vandenbark, G.R. *Proc. Natl. Acad. Sci. USA* **1983**, *80*, 36.
149. Sando, J.J.; Young, M.C. *Proc. Natl. Acad. Sci. USA* **1983**, *80*, 2642.
150. Leach, K.L.; James, M.L.; Blumberg, P.M. *Proc. Natl. Acad. Sci. USA* **1983**, *80*, 4208.
151. Ashendel, C.L.; Staller, J.M.; Boutwell, R.K. *Cancer Res.* **1983**, *43*, 4333.
152. Nishizuka, Y. *Science* **1986**, *233*, 305.
153. Nishizuka, Y. *Trends in Biochem. Sci.* **1986**, *9*, 163.
154. Dreidger, P.E.; Blumberg, P.M. *Cancer Res.* **1977**, *37*, 3257.
155. Lee, L.-S.; Weinstein, I.B. *J. Cell Physiol.* **1979**, *99*, 451.
156. Levine, L.; Hassid, A. *Biochem. Biophys. Res. Commun.* **1977**, *79*, 477.
157. Ohuchi, K.; Levine, L. *J. Biol. Chem.* **1978**, *253*, 4783.
158. Yamasaki, H.; Mufson, R.A.; Weinstein, I.B. *Biochem. Biophys. Res. Commun.* **1979**, *89*, 1018.
159. Wertz, P.W.; Mueller, G.C. *Cancer Res.* **1978**, *38*, 2900.
160. Cabral, F.; Gottesman, M.M.; Yuspa, S.H. *Cancer Res.* **1981**, *41*, 2025.
161. Yuspa, S.H.; Litchi, U.; Ben, T.; Patterson, E.; Hennings, H.; Slaga, T.J.; Colburn, N.; Kelsey, W. *Nature (London)* **1976**, *262*, 402.
162. Huberman, E.; Weeks, C.; Herrmann, A.; Callaham, M.F.; Slaga, T. J. *Proc. Natl. Acad. Sci. USA* **1981**, *78*, 1062.
163. Touraine, J.L.; Hadden, J.W.; Touraine, F.; Hadden, E.M.; Estensen, R.; Good, R.A. *J. Exp. Med.* **1977**, *145*, 460.
164. Huberman, E.; Callaham, M.F. *Proc. Natl. Acad. Sci. USA* **1979**, *76*, 1293.
165. Rovera, G.; O'Brien, T.G.; Diamond, L. *Science* **1979**, *204*, 868.
166. White, J.G.; Estensen, R.D. *Am. J. Pathol.* **1974**, *75*, 45.
167. White, J.G.; Rao, G.H.R.; Estensen, R.D. *Am. J. Pathol.* **1974**, *75*, 301.
168. DeChatelet, L.R. *J. Reticuloendothel. Soc.* **1978**, *24*, 73.
169. Lee, L.-S.; Weinstein, I.B. *Science* **1978**, *202*, 313.
170. Shoyab, M.; De Larco, J.E.; Todara, G.J. *Nature (London)* **1979**, *279*, 387.
171. Lotem, J.; Sachs, L. *Proc. Natl. Acad. Sci. USA* **1979**, *76*, 5158.
172. Koeffler, H.P.; Bar-Elli, M.; Territo, M. *J. Clin. Invest.* **1980**, *66*, 1101.
173. Pegoraro, L.; Abrahm, J.; Cooper, R.A.; Levis, A.; Lange, B.; Meo, P.; Rovera, G. *Blood* **1980**, *55*, 859.
174. Zucker, M.B.; Troll, W.; Belman, S. *J. Cell. Biol.* **1974**, *60*, 325.
175. Diamond, L.; O'Brien, T.G.; Rovera, G. *Life Sciences* **1978**, *23*, 1979.
176. Diamond, L.; O'Brien, T.G.; Baird, W.M. *Adv. Cancer Res.* **1980**, *32*, 1.
177. Diamond, L. *Pharmac. Ther.* **1984**, *26*, 89.
178. Parkinson, E.K. *Br. J. Cancer* **1985**, *52*, 479.
179. Nakadate, T. *Japan J. Pharmacol.* **1989**, *49*, 1.
180. Takagi, K.; Suganuma, M.; Kagechika, H.; Shudo, K.; Ninomiya, M.; Muto, Y.; Fujiki, H. *J. Cancer Res. Clin. Oncol.* **1988**, *114*, 221.
181. Verma, A.K.; Shapas, B.G.; Rice, H.M.; Boutwell, R.K. *Cancer Res.* **1979**, *39*, 419.
182. Lichti, U.; Patterson, E.; Hennings, H.; Yuspa, S.H. *Cancer Res.* **1981**, *41*, 49.
183. Adamo, S.; De Luca, L.M.; Akalovsky, I.; Bhat, P.V. *J. Natl. Cancer Inst.* **1979**, *62*, 1473.
184. Jetten, A.M.; Jetten, M.E.R.; Shapiro, S.S.; Poon, J.P. *Exp. Cell Res.* **1979**, *119*, 289.
185. Shapiro, S.S.; Poon, J.P. *Exp. Cell. Res.* **1979**, *119*, 349.
186. Petkovich, M.; Brand, N.J.; Krust, A.; Chambon, P.A. *Nature (London)* **1987**, *330*, 444.
187. Giguere, V.; Ong, E.S.; Segui, P.; Evans, R.M. *Nature (London)* **1987**, *330*, 624.
188. Benbrook, D.; Lernhardt, E.; Pfahl, M. *Nature (London)* **1988**, *333*, 669.
189. Hashimoto, Y.; Petkovich, M.; Gaub, M.P.; Kagechika, H.; Shudo, K.; Chambon, P. *Mol. Endocrinol.* **1989**, *3*, 1046.

190. Delclos, K.B.; Blumberg, P.M. *Cancer Res.* **1982**, *42*, 1227.
191. Horiuchi, T.; Fujiki, H.; Hakii, H.; Suganuma, M.; Yamashita, K.; Sugimura, T. *Jpn. J. Cancer. Res.* **1986**, *77*, 526.
192. Ferriola, P.C.; Cody, V.; Middleton, E., Jr. *Biochem. Pharmacol.* **1989**, *38*, 1617.
193. Tamaoki, T.; Nomoto, H.; Takahashi, I.; Kato, Y.; Morimoto, M.; Tomita, F. *Biochem. Biophys. Res. Commun.* **1986**, *135*, 397.
194. Junco, M.; Diaz-Guerra, M.J.M.; Bosca, L. *FEBS Lett.* **1990**, *263*, 169.
195. Nakadate, T.; Jeng, A.Y.; Blumberg, P.M. *Biochem. Pharmacol.* **1988**, *37*, 1541.
196. Ruegg, U.T.; Burgess, G.M. *TIPS* **1989**, *10*, 218.
197. Fujita-Yamaguchi, Y.; Kathuria, S. *Biochem. Biophys. Res. Commun.* **1988**, *157*, 955.
198. Merrill, A.H., Jr.; Stevens, V.L. *Biochim. Biophys. Acta* **1989**, *1010*, 131.
199. Hannun, Y.A.; Loomis, C.R.; Merrill, A.H., Jr.; Bell, R.M. *J. Biol. Chem.* **1986**, *261*, 12604.
200. Hampton, R.Y.; Morand, O.H. *Science* **1989**, *246*, 1050.
201. Kobayashi, E.; Ando, K.; Nakano, H.; Tamaoki, T. *J. Antibiotics* **1989**, *42*, 153.
202. Kobayshi, E.; Nakano, H.; Morimoto, M.; Tamaoki, T. *Biochem. Biophys. Res. Commun.* **1989**, *159*, 548.
203. Takahashi, I.; Nakanishi, S.; Kobayashi, E.; Nakano, H.; Suzuki, K.; Tamaoki, T. *Biochem. Biophys. Res. Commun.* **1989**, *165*, 1207.
204. Fujiki, H.; Sugunuma, M.; Suguri, H.; Yoshizawa, S.; Hirota, M.; Takagi, K.; Sugimura, T. In: *Prog. Clin. Biol. Res.* Vol. 298 (Skin Carcinogenesis); Slaga, T.J.; Klein-Szanto, A.J.P.; Boutwell, R.K.; Stevenson, D.E.; Spitzer, H.L.; D'Motto, B., Eds.; Alan R. Liss: New York, 1989; p. 281
205. Erdodi, F.; Rokolya, A.; Di Salvo, J.; Barany, M.; Barany, K. *Biochem. Biophys. Res. Commun.* **1988**, *153*, 156.
206. Haavid, J.; Schelling, D.L.; Campbell, D.G.; Andersson, K.K.; Flatmark, T.; Cohen, P. *FEBS Lett.* **1989**, *251*, 36.
207. Fujiki, H.; Mori, M.; Nakayasu. M.; Terada, M.; Sugimura, T.; Moore, R.E. *Proc. Natl. Acad. Sci. USA* **1981**, *78*, 3872.
208. Fujiki, H.; Suganuma, M.; Nakayasu, M.; Hoshino, H.; Moore, R.E.; Sugimura, T. *Gann* **1982**, *73*, 495.
209. Fujiki, H.; Sugunuma, M.; Sugimura, T. *Envir. Carcino. Revs. (J. Envir. Sci. Hlth.)* **1989**, *C7*, 1.
210. Hergenhahn, M.; Furstenberger, G.; Opferkuch, H.J.; Adolf, W.; Mack, H.; Hecker, E. *J. Cancer Res. Clin. Oncol.* **1982**, *104*, 31.
211. Van Duuren, B.L.; Tseng, S.-S.; Segal, A.; Smith, A.C.; Melchionne, S.; Seidman, I. *Cancer Res.* **1979**, *39*, 2644.
212. Yamasaki, H.; Weinstein, I.B.; Van Duuren, B.L. *Carcinogenesis* **1981**, *2*, 537.
213. Hecker, E. *Arzneim.-Forsch.* **1985**, *35*, 1890.
214. Wender, P.A. (Unpublished results.) The assistance of G. Smith and the Merck organization in this effort is greatly appreciated.
215. Wender, P.A.; Koehler, K.F.; Sharkey, N.A.; Dell'Aquila, M.L.; Blumberg, P.M. *Proc. Natl. Acad. Sci. USA* **1986**. *83*, 4214.
216. Wender, P.A.; Cribbs, C.M. (Unpublished results.)
217. Fujiki, H.; Tanaka, Y.; Niyake, R.; Kikkawa, U.; Nishizuka, Y.; Sugimura, T. *Biochem. Biophys. Res. Commun.* **1984**, *120*, 339.
218. Wender, P.A.; Cribbs, C.M. (Unpublished results.)
219. Wender, P.A.; Cribbs, C.M. (Unpublished results.)
220. Itai, A.; Kato, Y.; Tomioka, N.; Iitaka, Y.; Endo, Y.; Hasegawa, M.; Shudo, K.; Fujiki, H.; Sakai, S.-I. *Proc. Natl. Acad. Sci. USA* **1988**, *85*, 3688.
221. Endo, Y; Shudo, K.; Itai, A.; Hasegawa, M.; Sakai, S. *Tetrahedron* **1986**, *42*, 5905.
222. Cardellina, J.H., II; Marner, F.-S.; Moore, R.E. *Science* **1979**, *204*, 193.
223. Lotter, H.; Hecker, E. *Fresenius S. Anal. Chem.* **1985**, *321*, 640.

224. Weinstein, I.B.; Gattoni-Celli, S.; Kirschmeier, P.; Lambert, M.; Hsiao, W.; Backer, J.; Jeffrey, A. In: *Biochemical Basis of Chemical Carcinogenesis*; Greim, H.; Jung, P.; Kramer, M.; Marquardt, H.; Oesch, F., Eds.; Raven: New York, 1984; p. 200.

225. Jeffrey, A.M.; Liskamp, R.M.J. *Proc. Natl. Acad. Sci. USA* **1986**, *83*, 241.

226. Laughton, C.A.; Dale, I.L.; Gescher, A. *J. Med Chem.* **1989**, *32*, 428.

227. Nakamura, H.; Kishi, Y.; Pajares, M.A.; Rando, R.R. *Proc. Natl. Acad. Sci. USA* **1989**, *86*, 9672.

228. Wender, P.A.; Cribbs, C.M. (Unpublished results).

229. Still, W.C.; Galynker, I. *Tetrahedron* **1981**, *37*, 3981.

230. Pettit, G.R.; Leet, J.R.; Herald, C.L.; Kamano, Y.; Boettner, F.E.; Baczynskyj, L.I. *J. Org. Chem.* **1987**, *52*, 2854 and references therein.

231. Smith, J.B.; Smith, L.; Pettit, G.R. *Biochem. Biophys. Res. Commun.* **1985**, *132*, 939.

232. Berkow, R.L.; Kraft, A.S. *Biochem. Biophys. Res. Commun.* **1985**, *131*, 1109.

233. Sako, T.; Yuspa, S.H.; Herald, C.L.; Pettit, G.R.; Blumberg, P.M. *Cancer Res.* **1987**, *47*, 5445.

234. Warren, B. S.; Herald, C.L.; Pettit, G.R.; Blumberg, P.M. *J. Cell Biochem.* **1987**, *11A*, 47.

235. Hennings, H.; Blumberg, P.M.; Pettit, G.R.; Herald, C.L.; Shores, R.; Yuspa, S.H. *Carcinogenesis* **1987**, *8*, 1343.

236. Stone, R.M.; Sariban, E.; Pettit, G.; Kufe, D.W. *Blood* **1986**, *68*, 193a.

237. Kraft, A.S.; Smith, J.B.; Berkow, R.L. *Proc. Natl. Acad. Sci. USA* **1986**, *83*, 1334.

238. Dell'Aquila, M.L.; Nguyen, H.T.; Herald, C.L.; Pettit, G.R.; Blumberg, P.M. *Cancer Res.* **1987**, *47*, 6006.

239. Blumberg, P.M. *Cancer Res.* **1988**, *48*, 1.

240. de Vries, D.J.; Herald, C.L.; Pettit, G.R.; Blumberg, P.M. *Biochem. Pharmacol.* **1988**, *37*, 4069.

241. Blumberg, P.M.; Pettit, G.R.; Warren, B.S.; Szallasi, A.; Schuman, L.D.; Sharkey, N.A.; Nakakuma, H.; Dell'Aquila, M.L.; de Vries, D.J. In: *Prog. Clin. Biol. Res.* Vol. 298 (Skin Carcinogenesis); Slaga, T.J.; Klein-Szanto,A.J.P.; Boutwell, R.K.; Stevenson, D.E.; Spitzer, H.L.; D'Motto, B., Eds.; Alan R. Liss: New York, 1989; p. 201.

242. Wender, P.A.; Cribbs, C.M.; Koehler, K.F.; Sharkey, N.A.; Herald, C.L.; Kamano, Y.; Pettit, G.R.; Blumberg, P.M. *Proc. Natl. Acad. Sci. USA* **1988**, *85*, 7197.

243. Wender, P.A.; Keenan, R.M.; Lee, H.Y. *J. Am. Chem. Soc.* **1987**, *109*, 4390.

244. Wender, P.A.; Lee, H.Y.; Wilhelm, R.S.; Williams, P.D. *J. Am. Chem. Soc.* **1989**, *111*, 8954.

245. Wender, P.A.; Kogen, H.; Lee, H.Y.; Munger, J.D., Jr.; Wilhelm, R.S.; Williams, P.D. *J. Am. Chem. Soc.* **1989**, *111*, 8957.

246. Wender, P.A.; McDonald, F.E. *J. Am. Chem. Soc.* **1990**, *112*, 4956.

# CHEMISTRY AND BIOLOGY OF THE IMMUNOSUPPRESSANT (−)-FK-506

Ichiro Shinkai and Nolan H. Sigal

Advances in Medicinal Chemistry,
Volume 1, pages 55–108.
Copyright © 1992 by JAI Press Inc.
All rights of reproduction in any form reserved.
ISBN: 1-55938-170-1

# I. INTRODUCTION

The modern era of transplantation therapy began nearly ten years ago with the first clinical trials of cyclosporin A (CsA). Its efficacy in organ and bone marrow transplantation, as well as in autoimmune diseases, has established the value of nonspecific immunosuppression in medical practice.[1,2] Cyclosporin A is a very potent drug that markedly synergizes with the steroid prednisone, resulting in a dramatic improvement in the quality of life for transplant recipients. In contrast to the azathioprine/prednisone era, there is now less morbidity from high-dose steroids; moreover, there is a reduction in hospital admissions for bone marrow failure, acute rejection episodes, and infectious complications. During the past ten years we have gained not only an understanding of CsA's mechanism of action, but also considerable knowledge about the biochemical events that lead to lymphocyte activation and differentiation. Although researchers have not yet pieced together a complete picture, the insights to date have led us to the rational design of novel immunosuppressive drugs, with the aim of extending the therapeutic usefulness of such agents.

In the search for safer and more effective immunosuppressants, a more recent discovery was that of FK-506.[3,4] Although CsA and FK-506 are structurally unrelated, they do share several mechanistic similarities at the cellular and molecular levels, which we will address in detail in Section III. Besides CsA and FK-506 there are many compounds, including corticosteroids and antimetabolites (e.g., azathioprine), which can inhibit a lymphocyte proliferative response in vitro and thereby serve as potent immunosuppressives in vivo. However, it is the selectivity of CsA and FK-506 for certain types of activation events that clearly sets them apart.

# II. CHEMISTRY

## A. Discovery of FK-506

FK-506 was discovered in 1985 by scientists at the Fujisawa Pharmaceutical Company in a fermentation broth of a soil organism, *Streptomyces tsukubaensis* No. 9993.[3,4] The organism is characterized by a gray mycelium color, rectiflexible spore chains with smooth spore surfaces, non-chromogenicity, and a limited carbohydrate utilization pattern. FK-506 exhibits antifungal activity against *Aspergillus fumigatus* and *Fusarium oxysporum*, but has little inhibitory effects on bacteria or yeasts. The first published studies demonstrated that FK-506 was a potent inhibitor of murine and human T-lymphocyte proliferation, with an $ED_{50}$ of approximately 0.1 nM, close to 100-times more active than CsA.[4–7] The selectivity of FK-506 was also quickly shown to be quite similar to that of CsA with respect to its ability to inhibit lymphokine production by activated T cells, while having

***Table 1.*** Physicochemical Properties of FK-506[9]

| | |
|---|---|
| Appearance | Colorless prisms |
| mp (°C) | 127–129° |
| Molecular formula | $C_{44}H_{69}NO_{12} \cdot H_2O$ |
| Mass spectrum (m/z) | 804 (M + 1, SI–MS) |
| Optical rotation | $[\alpha]_D$ −84.4° (c = 1.02, $CHCl_3$) |
| Elemental analysis (%) | |
|     Calculated for $C_{44}H_{69}NO_{12} \cdot H_2O$ | C 64.29, H 8.71, N 1.70 |
|     Found | C 64.20, H 8.86, N 1.72 |
| UV spectrum (EtOH) | End absorption |
| IR spectrum ($CHCl_3$) $cm^{-1}$ | 3530, 1750, 1730, 1710, 1650, 1100 |

no effect on bone marrow colony formation or lymphokine-dependent proliferation.[4–8] Fermentation and isolation of FK-506 has been reported.[9] The physicochemical properties of FK-506 are summarized in Table 1.

It is soluble in methanol, ethanol, acetone, ethyl acetate, chloroform, and diethyl ether; sparingly soluble in hexane and petroleum ether; and insoluble in water. Color reactions are as follows: positive in ceric sulfate, sulfuric acid, Ehrlich, Dragendorff and iodine vapor tests; negative in ferric chloride, ninhydrin and Molisch tests.

FK-506 was found to have a novel 23-membered macrocyclic lactone ("macrolide") ring structure containing an unusual hemiketal-masked $\alpha,\beta$-diketoamide linkage ($C_8$–$C_{10}$) as shown in Figure 1. A $^{13}C$ NMR study revealed that FK-506 exists in chloroform as an equilibrium mixture (3:1) of two rotamers (Z and E), due to restricted rotation around the amide C–N bond. The structure of FK-506 was elucidated by NMR experiments on it and on its chemical degradation products[10,11] (see later section for details). Based on the chemical shifts of the methyl groups attached to $C_{19}$ and $C_{27}$, the stereochemistry of the two trisubstituted double bonds were assigned as the E geometry. The total structure, including the relative stereochemistry of the ring and its substitution groups, was determined by single-crystal X-ray crystallography.[12] The absolute configuration was inferred from the configuration at $C_2$ of the L-pipecolic acid unit (Fig. 2).

All the six-membered rings adopt a chair conformation. All the groups attached to the six-membered rings are equatorial except for $C_1$ on the piperidine ring and the $C_{10}$ hydroxyl on the tetrahydropyran ring. The six atoms $N_7$, $C_2$, $C_6$, $C_8$, $C_9$ and oxygen at $C_8$ are coplanar to within 0.06 Å. The two planes of the ketone groups are approximately perpendicular to each other with an O=$C_8$–$C_9$=O torsion angle of 89.4°. All hydrophilic groups bonded to the macrolide ring point outward from the ring. There are intermolecular hydrogen bonds between the hydroxyl at $C_{24}$ and the oxygen at $C_{22}$ and between the hydroxyl at $C_{32}$ and the oxygen at $C_{13}$, but there are no intramolecular hydrogen bonds. The water molecule forms three hydrogen bonds: OH (water)—O=$C_9$, oxygen (water)—OH at $C_{10}$; and OH (water)—O=$C_8$. FK-506 has an allyl group attached to $C_{21}$ of the macrolide ring. Rapamycin,

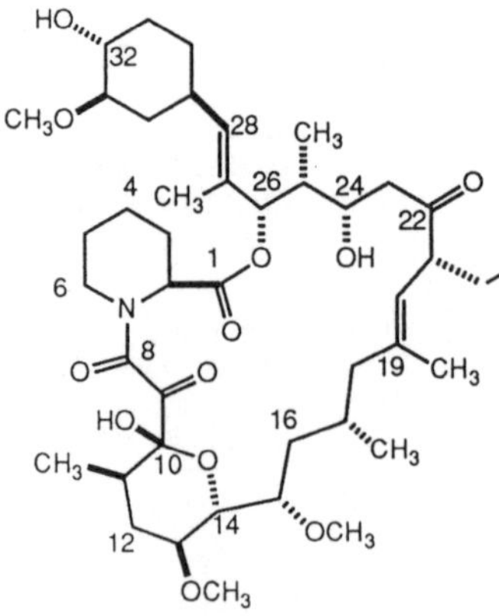

FK-506

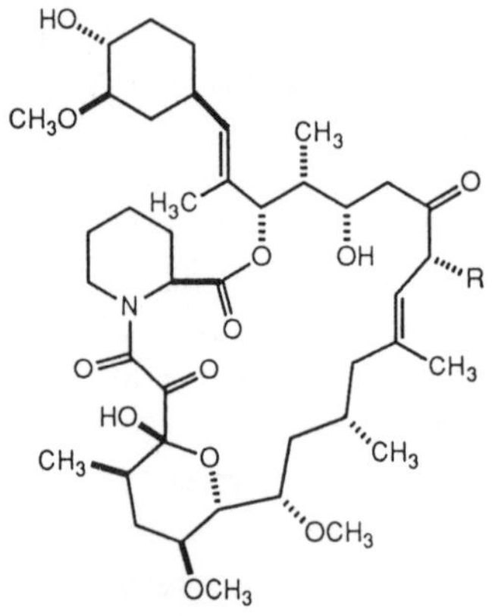

FR-900520 R=-CH$_2$CH$_3$
FR-900523 R=-CH$_3$

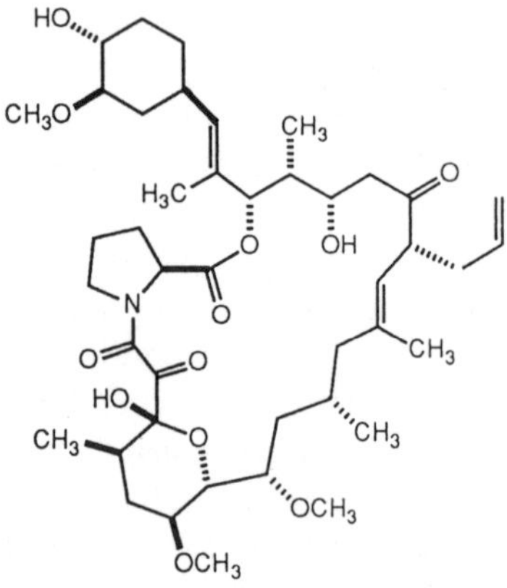

FR-900525

**Figure 1.** FK-506 and derivatives.

oligomycin, milbemycin, avermectin, cytovaricin and rhizoxin, all neutral lactone-ring-based antibiotics, lack such an allyl group. The existence of this group appears to be specific for FK-506 and FR-900525; even the FK-506 analogues FR-900520 and FR-900523 (see Fig. 1) also lack an allyl group.[13,14] Although most of the molecule appears to result from a mixed acetate/propionate polyketide biosynthesis, several features, such as the substituted cyclohexyl group and the pipecolic acid unit, are unusual in polyketide natural products.

Two minor components also isolated from a culture broth of *S. tsukubaensis* No. 9993 also suppressed immune response in vitro and are structurally related to

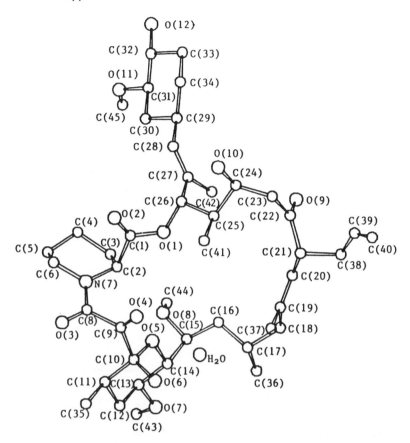

**Figure 2.** X-ray structure of (−)-FK-506.

FK-506. The components were identified as FR-900525 and FR-900520. The structure of FR-900525 differs from that of FK-506 by replacement of the piperidine-2-carboxylic acid moiety with proline.[15] The structure of FR-900525 was confirmed by the isolation of proline after acid hydrolysis (6N HCl, 110 °C, 16 hr) compared to pipecolic acid in the case of acid hydrolysis of FK-506. The other component, FR-900520, was identified as one of the components isolated from culture broth of *Streptomyces hygroscopicus subs. yakushimaensis* No. 72386. FR-900520 (major) and another FK-506 analogue FR-900523 (minor) were isolated from the broth by several reversed-phase chromatographic separations. On the basis of spectroscopic analyses, the structures of FR-900520 and FR-900523 were assigned to new classes of 23-membered macrocyclic lactones related to FK-506. FR-900520 and FR-900523 possess ethyl and methyl groups (R), respectively, instead of the allyl group at $C_{21}$ in FK-506 (Fig. 1).

## B.  Synthesis of the $C_1$–$C_{15}$ Segment

FK-506 contains a peculiar tricarbonyl system masked as a hemiketal, α,β-diketo amide group. This tricarbonyl system is also found in the antifungal antibiotic rapamycin. Many synthetic organic chemists were attracted to this system and have therefore studied model routes to 1,2,3-tricarbonyl derivatives (Fig. 3). These approaches can be divided into two categories. In one, a β-keto amide 1,3-dicarbonyl derivative, which is derived from N-substituted pipecolate, is synthesized first. The center carbonyl group is then incorporated either by direct oxidation of the central methylene group or by a two-step protocol, such as formation of enamine, enone or phosphorane, followed by an oxidation sequence. The other approach is direct insertion of two carbonyl groups by nucleophilic addition to oxalate derivatives. In the former approach trisubstituted olefins undergo reaction with singlet oxygen at rates competitive with the enamine, enone or phosphorane, a possible limitation of this method in the real system. The latter approach has a serious problem: the metalation of highly functionalized systems. Therefore, many of the routes to the tricarbonyl system reported here are only applicable for model studies.

The first synthetic route to the chiral α,β-diketo amide segment of FK-506 was reported by Williams and Benbow[16] in late 1988. Treatment of the (R)-aldehyde 1 with the enantiomerically pure Grignard reagent 2 in anhydrous ether at −78 °C gave the desired alcohol 3, and its corresponding 3R isomer, in the ratio of 8.4 to 1 (Scheme 1). O-Methylation and debenzylation of 3 gave the primary alcohol 4, free of the $C_4$ diastereomer. Jones' oxidation, followed by conversion to the acid chloride 5 was carried out without epimerization or ketal hydrolysis. Coupling of 5 with the magnesium enolate of the pipecolic amide derivative 6 afforded the C-acylation product 7 in 65% yield. After the removal of the benzyl ester group by catalytic hydrogenation, and subsequent decarboxylation, the remaining carboxylic acid was esterified with diazomethane to give the 1,3-dicarbonyl derivative 8. The central $C_9$ carbonyl was introduced by treatment of 8 with selenium dioxide

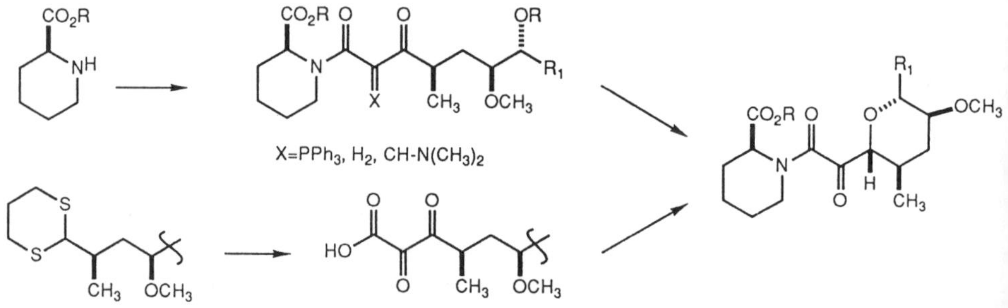

**Figure 3.**  Approaches to the tricarbonyl system.

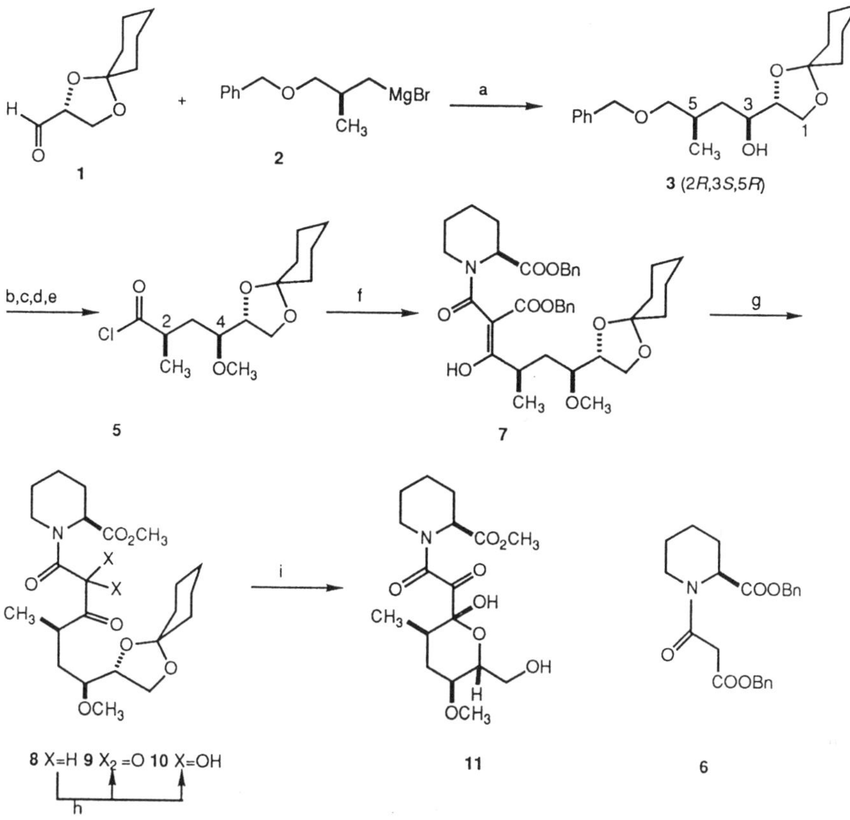

**Scheme 1.** (a) Et$_2$O, −78 °C, 75 %; (b) dimsyl sodium, DMSO, MeI (5 equiv), 84%; (c) Na (3 equiv), THF, NH$_3$(l), 90%; (d) THF, Celite, 0 °C, Jones' reagent, 78%; (e) (COCl)$_2$, benzene, 5–22°C, 98%; (f) pipecolic amide (**6**), isopropylmagnesium chloride, −10–0°C; **5**, THF, 65%; (g) 10% Pd–C, H$_2$, EtOAc; CH$_2$N$_2$, Et$_2$O, 100%; (h) SeO$_2$ (1.2 equiv), dioxane, 76%; (i) hydrolysis.

to give a 2:1 mixture of the desired α,β-diketoamide **9** and its corresponding hydrate **10**. Final hemiketal formation to give **11** was achieved upon hydrolysis of the cyclohexylidene ketal of **9**. An NMR study of **11** showed that all substituents on the tetrahydropyranyl ring are equatorial, with the exception of the C$_{10}$ hydroxyl.

Kocienski and his co-workers reported a synthesis of the C$_1$-C$_{15}$ segment from tri-O-acetyl-D-glucal and methyl (*S*)-pipecolate.[17] The aldehyde **12** was synthesized in six steps (27% overall yield) from the diol **13**, which itself was prepared in three steps (81% overall yield) from commercial tri-O-acetyl-D-glucal (Scheme 2). A key step in the fabrication of the aldehyde **12** was the highly diastereoselective alkylation of the SAMP-hydrazone **14**, which gave an inseparable 97:3 mixture of

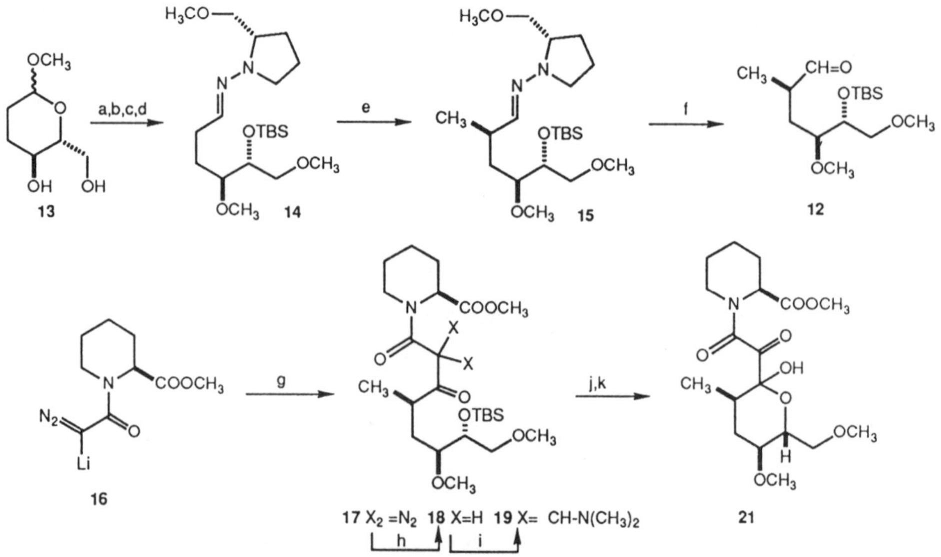

**Scheme 2.** (a) MeI, NaH, THF, 92%; (b) AcOH, $H_2O$, 97%; (c) N-Amino-2-methoxymethylpyrrolidine$CH_2Cl_2$, 86%; (d) TBSCl, imidazole, DMF, 75%; (e) LiAlH$_4$, MeI, THF, 64%, (f) $O_3$, $CH_2Cl_2$, −78°C , 73% (g) THF, −78°C; **12**, 57%; (h) Rh$_2$(OAc)$_4$, DME, 20 °C, 100%; (i) (MeO)$_2$CHNMe$_2$ neat, 80 °C, 93%, (j) $O_3$, $CH_2Cl_2$, −30°C, 70%; (k) HF, MeCN, 95%.

diastereomers (**15**, major product) in 64% yield. Ozonolysis of **15** gave **12** without epimerization of the $C_{11}$ stereogenic center. An aldol condensation of the unstable lithiated N-diazoacetyl (S)-pipecolate **16** with the aldehyde **15** gave the condensation product **17** in 57% yield. Treatment of **17** with a catalytic amount of Rh(II) in dimethoxyethane eliminated nitrogen cleanly to give the desired β-keto amide **18** in quantitative yield. The formation of the enamine derivative **19** was achieved by the reaction of the β-keto amide with two equivalents of dimethylformamide dimethylacetal in 93% yield. Subsequent ozonolysis of **19** gave the tricarbonyl segment **20** in 70% yield. Removal of the TBS silyl protecting group with HF afforded the hemiketal **21**. An NMR study of **21** indicated that it exists in solution as an equilibrium mixture of two rotamers due to restricted rotation around the amide bond as observed in FK-506 and rapamycin.

Danishefsky and Egbertson achieved their $C_1$–$C_{16}$ segment by coupling of a complex lithiodithiane to N-methoxalylpipecolate.[18] A strategy wherein the (S)-pipecolyl residue with its amide bond to $C_8$ would already be in place during this coupling was applied in their $C_9$–$C_{10}$ bond formation. The dithiane **22**, obtained by total synthesis,[19] was treated with n-BuLi followed by N-methoxalyl (S)-pipecolate

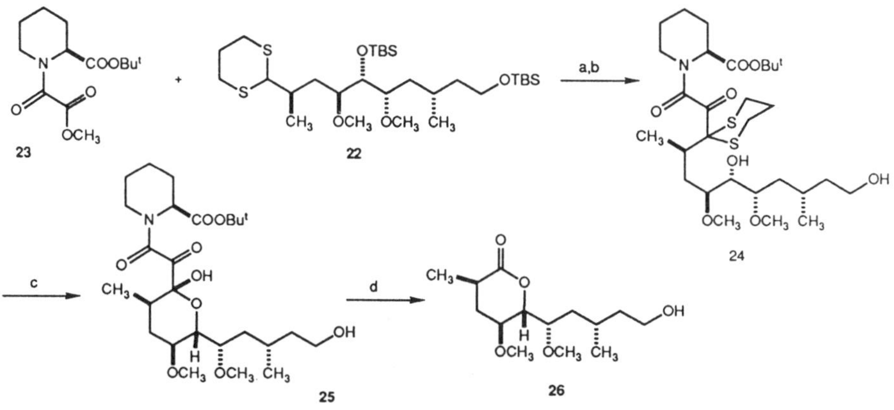

**Scheme 3.** (a) *n*-BuLi, THF, −20 °C; **23**, −78°C; (b) HF, MeCN, 48%; (c) NBS, aq. acetone, 45%; (d) Pb(OAc)₄, MeOH.

**23** (Scheme 3). Treatment of the crude product with HF gave the coupled product **24** in 48% yield after isolation. The dithiane was hydrolyzed to give the lactol **25** in 45% yield. The structural proof was provided by cleavage of the lactol with lead tetraacetate in methanol to produce the ester **23** and the lactone **26**.

An alternative synthesis of the $C_1$–$C_{15}$ segment has also been achieved from D-glucose by Ireland and his group.[20] Methyl α-D-glucoside was converted to the lactone **27** in 56% overall yield (Scheme 4). No condensation product was observed

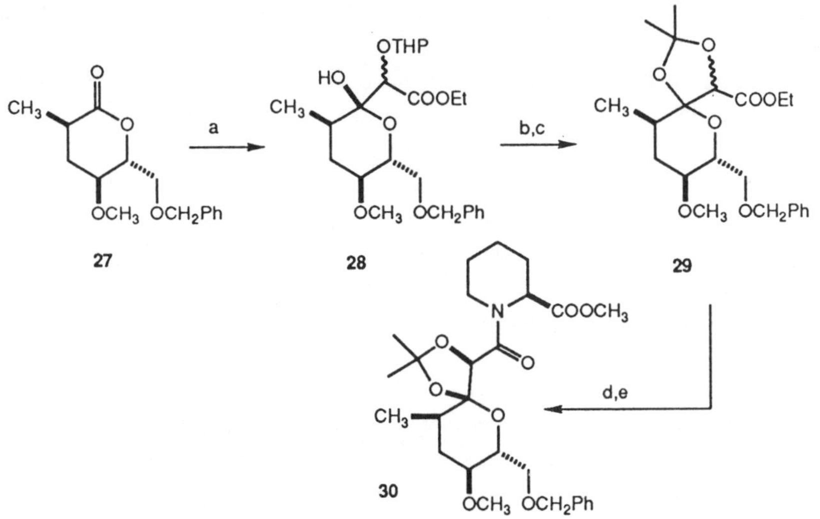

**Scheme 4.** (a) LDA, THF, THPOCH₂CO₂C₂H₅, −78°; EtOH; (b) Pyridine-TsOH, H₂O, 55°C; (c) P₂O₅, acetone; (d) LiOH, H₂O, MeOH; (e) SO₂Cl, pyr, NEt₃, methyl pipecolate HCl.

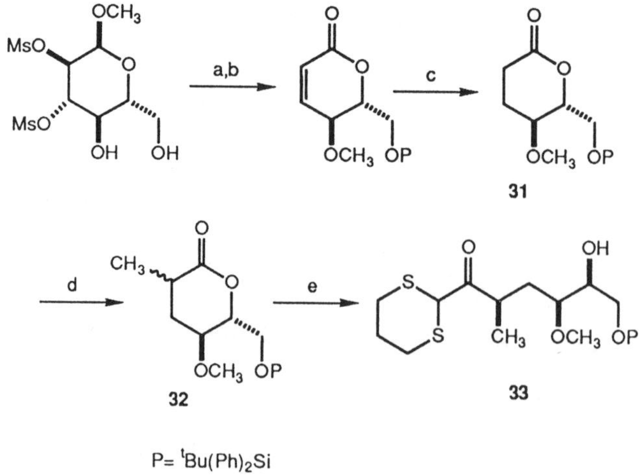

P= $^t$Bu(Ph)$_2$Si

**Scheme 5.** (a) $t$-Bu(Ph)$_2$SiCl; KI, Zn-Cu, DMF; NaH, MeI; (b) MCPBA, BF$_3$·OEt$_2$; (c) H$_2$/Pd-C; (d) LDA, MeI, THF; (e) Li-1,3-dithiane.

when the ketal ester was added to the lactone **27**. However, when tetrahydropyranyl (THP) ethyl glycolate was used the desired condensation product **28** was obtained. While adduct **28** was labile, acetonide formation proceeded smoothly to give the ketal ester **29** in 69% yield. Hydrolysis of the ester with LiOH, followed by the coupling with methyl pipecolate, gave the C$_1$–C$_{15}$ segment **30** of FK-506 in 50% yield.

Similarly, Sirahama and co-workers reported the conversion of methyl-α-D-glucoside to the C$_1$–C$_{15}$ segment.[21] Methylation of lactone **31** gave a 1:1 mixture of α- and β-methylated lactones (**32**, Scheme 5). The α-isomer was efficiently isome rized to the desired β-isomer. Opening of the lactone β-**32** with dithiane anion gave the alcohol **33**. The subsequent coupling reaction with an (S)-pipecolinic acid derivative is under investigation.

Williams and Wasserman reported[22] a modified coupling of the acid chloride **5** with the ylide amide **36** derived from (S)-pipecolinic acid (Scheme 6). This approach involves the coupling of **5** with an acylphosphoranylidene in the presence of bis(trimethylsilyl)acetamide (BSA) to form a keto ylide carboxylate that may be oxidized with ozone or singlet oxygen to generate the tricarbonyl system. Treatment of $t$-butyl (S)-pipecolate with α-bromoacetic acid gave the bromoacetyl derivative **35**, which was converted to the phosphonium salt with triphenylphosphine. Deprotonation of the salt with dilute NaOH gave the ylide **36**, which was coupled with the acid chloride **5** in the presence of BSA to give the keto ylide derivative **37** in 91% yield. Oxidative cleavage of **37** with ozone or singlet oxygen

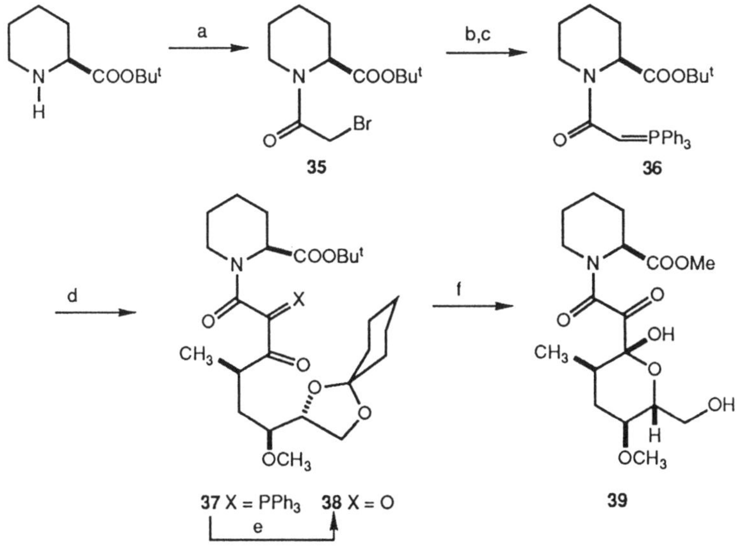

**Scheme 6.** (a) BrCH$_2$COOH, DCC, 99%;(b) Ph$_3$P, 97%; (c) NaOH (aq), 99%; (d) BSA, **5**, 91%; (e) $^1$O$_2$ (69%) or O$_3$ (88%); (f) HCl (dil.), MeOH.

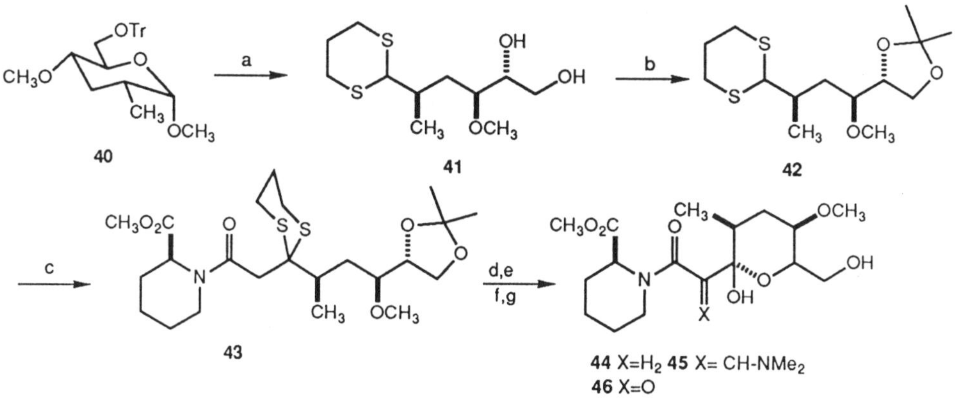

**Scheme 7.** (a)1,3-propandithiol (1.1 equiv), BF$_3$OEt$_2$, CH$_2$Cl$_2$, 0°C; (b) (Me)$_2$C(OMe)$_2$, TsOH (0.2 equiv); (c) $n$-BuLi, THF, –15°C; methyl $N$-bromoacetylpipecolate, –78°C; (d) BF$_3$·OEt$_2$, HgO (2 equiv), THF, H$_2$O, **43**, 0°C; (e) (MeO)$_2$CHNMe$_2$, 80°C; (f) TsOH (0.2 equiv), 10% aq. MeOH, 55°C; (g) O$_3$, CH$_2$Cl$_2$, –78°C.

afforded the $\alpha,\beta$-diketo amide **38** in 88% yield. Deketalization of **38** gave the desired hemiketal **39**.

Recently, Rao reported a similar approach to the $C_1$–$C_{15}$ segment by using an alkylation of dithiane followed by oxidation of the 1,3-dicarbonyl derivatives.[23] The key intermediate **40** was synthesized from methyl $\alpha$-(D)-glucopyranoside (Scheme 7). Treatment of **40** with 1,3-propanedithiol in the presence of tosic acid opened the pyranoside ring to provide the dithioketal **41**. Lithiation of **42** with $n$-BuLi followed by treatment with methyl $N$-($\alpha$-bromoacetyl)pipecolate gave the $C_1$–$C_{15}$ segment **43**. After deprotection of the dithiane, 1,3-dicarbonyl derivative **44** was reacted with two equivalents of $N,N$-dimethylformamide dimethylacetal to give the enamine **45**. Ketalization followed by ozonolysis of the enamine provided the model $\alpha,\beta$-diketo amide segment **46**.

## C.  Synthesis of $C_{10}$–$C_{24}$ Segment

The first diastereospecific, non-racemic synthesis of the $C_{10}$–$C_{18}$ segment of FK-506 was reported by the Merck team.[24] The synthesis of the $C_{10}$–$C_{18}$ segment **47** was envisioned to proceed by a diastereoselective lactonization of the diester **48** (Scheme 8). The $C_{11}$ and $C_{17}$ methyl substituents of **48** would be introduced by simultaneous, stereospecific methylation of the non-racemic bis-lactone **49**. Bis-lactone **49** would be available from the bis-epoxide **50** by benzylation, simul-taneous acetate enolate alkylation, and lactonization. However, an approach wherein the stereocenters are created sequentially was chosen for initial investiga-tion. An alternative approach of this type also provides direct access to a selectively protected species, that is, without the need for additional terminus differentiation methodology, as would be required for diester **48**.

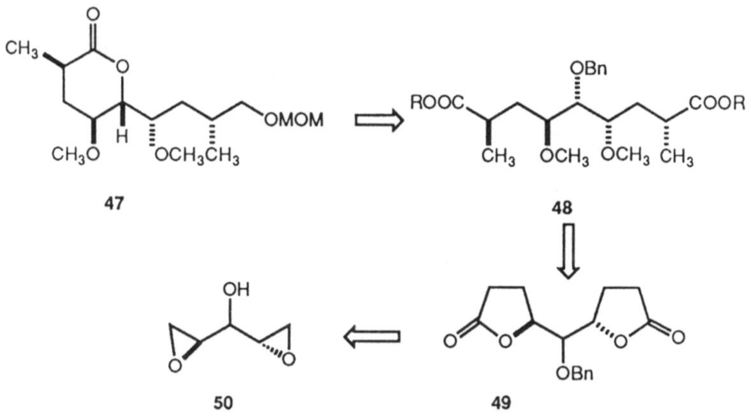

*Scheme 8.*

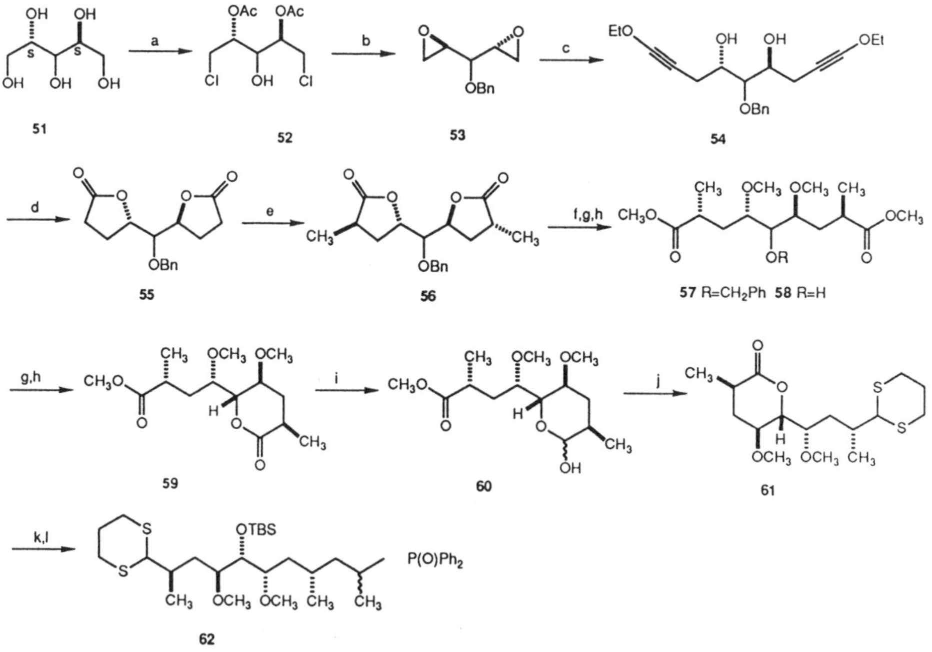

**Scheme 9.** (a) CH$_3$COO(CH$_3$)$_2$COCl, MeCN; (b) NaOCH$_3$; PhCH$_2$Br, NaH; (c) LiC≡C-OEt, BF$_3$·OEt$_2$, (d) HCl, MeOH; (e) LDA, MeI; (f) NaOH (2 equiv); MeI, NaH; (g) H$_2$, Pd(OH)$_2$; (h) Pyridine-TsOH; (i) L-Selectride; (j) 1,3-propanedithiol, BF$_3$·OEt$_2$; (k) LiAlH$_4$; Ph$_3$P, I$_2$; TBSOTf; (l) LiCH$_3$CHP(O)Ph$_2$.

A two-directional chain synthesis strategy was reported by Schreiber and his co-workers.[25] This approach offers certain advantages over conventional chain synthesis strategies provided the problem of terminus differentiation of the two-directionally homologated chains can be surmounted. Treatment of L-arabinitol **51** with the Moffatt reagent provided the dichloro diacetate **52**, which was converted to the bisepoxide benzyl ether **53** in 45% overall yield (Scheme 9). The bisepoxide was simultaneously homologated in two directions by reaction with lithium-2-ethoxyacetylide in the presence of BF$_3$ etherate to give the bisalkyne diol **54**, which was directly transformed into the bislactone **55** by treatment with methanolic HCl in 62% overall yield from **53**. Alkylation of the bisenolate with iodomethane gave a mixture of products from which the desired dimethyl bislactone **56** was isolated in 54% yield. The desired dimethyl ester **57** was obtained by the saponification of the lactones followed by the exhaustive methylation. Negligible epimerization occurred during the methylation process. A group-selective lactonization of **58** gave a 6:1 mixture of lactones with **59** as the major isomer in 65% yield. Selective reduction of the lactone with L-Selectride gave a 1:1 mixture of lactoanomers (**60**)

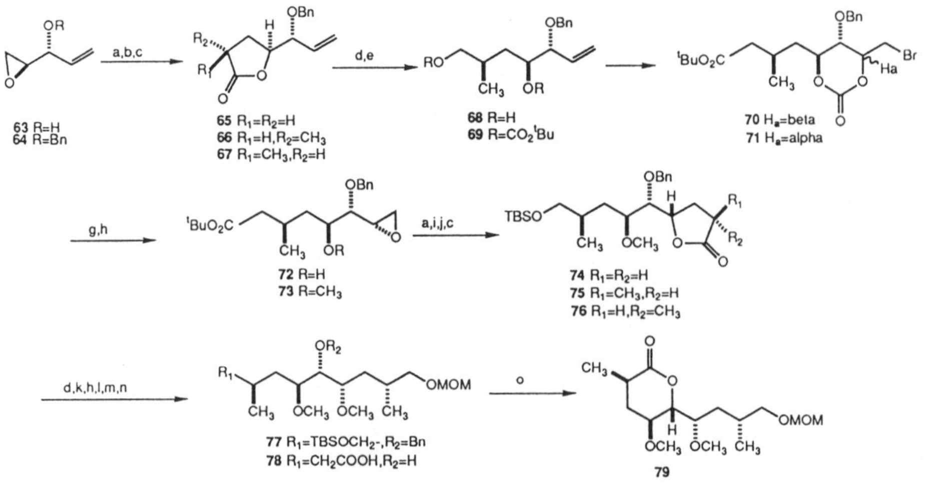

**Scheme 10.** (a) LiCH₂CN, THF, –78–25°C; (b) 12 N HCl, MeOH, reflux; (c) LDA, THF, –78–50°C; Mel, –78°C; (d) LAH (4 equiv), THF; (e) 2-(t-butoxycarbonyloxyimino)-2-phenylacetonitrile (2.05 equiv), NaH, THF 60°C, (f) Br₂ (1.5 equiv), K₂CO₃ (1.5 equiv) CH₂Cl₂, –80°C; (g) NaOMe (2.5 equiv), MeOH; (h) Mel (18 equiv), NaH, THF; (i) 2N HCl dioxane, reflux; (j) TBSCl (1.5 equiv), midazole, DMF; (k) MOMBr (1.5 equiv), DIEA (1.5 equiv), CH₂Cl₂, 0°C; (l) TBAF (4 equiv), THF; (m) PDC (15 equiv), DMF; (n) H₂, Pd(OH)₂, EtOH; (o) Ac₂O, dichloroethane, reflux.

in quantitative yield. Installation of the dithiane-protecting group produced (90% yield) the crystalline lactone dithiane **61**, which was converted to the target phosphine oxides **62**, a mixture of isomers as depicted.

The Merck approach[24] used the monoepoxy alcohol **63**, which is available in high diastereomeric (99%) and enantiomeric (>97%) purity. Treatment of the benzyl ether **64** with lithioacetonitrile and subsequent lactonization gave the butyrolactone **65** in 76% yield (Scheme 10). Methylation of **65** resulted in a mixture of lactones **66** and **67** with **66** as the predominant species (**66:67** = 87:13, as determined by capillary GC analysis). Reduction of **66** with LiAlH₄ gave the desired diol **68**, which was converted to the bis-t-butyl-carbonate **69**. Treatment of **69** with bromine in the presence of potassium carbonate at low temperature produced the bromocarbonates **70** and **71** (74% yield from **66**; **70:71** = 11:1 as determined by ¹H NMR). Selective saponification of the cyclic carbonate group with sodium methoxide/methanol followed by treatment with methyl iodide produced the methyl ether **73** in 71% overall yield from **70**. Conversion of **73** to the butyrolactone **74** was accomplished, as before, by epoxide opening, followed by protection of the primary hydroxyl group. *Trans*-selective methylation of **74** gave the mixture of lactones **75** and **76** in 93% yield (**75:76** = 94:6). Standard

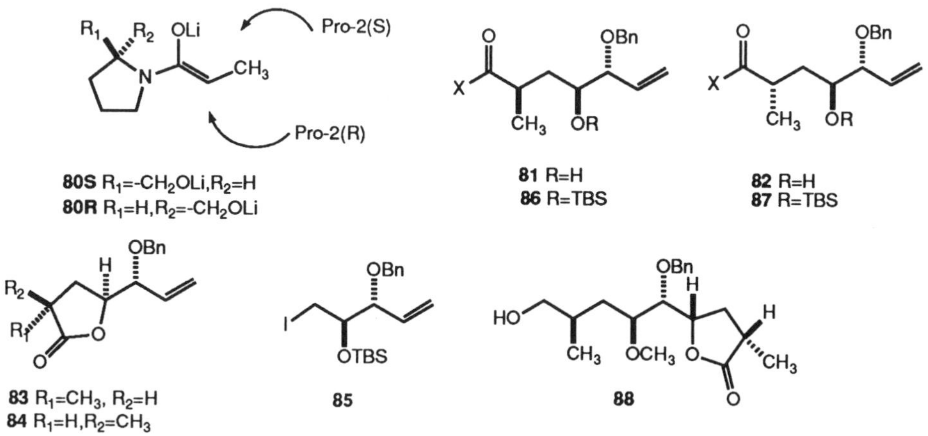

**Scheme 11.**

transformations of **75** furnished the desired lactone **79** via intermediates **77** and **78** in 76% overall yield from **77**.

An interesting alternative approach to the key intermediates **66** and **75** was achieved by alkylation of chiral prolinol propionamide enolates with epoxides. The alkylation of terminal epoxides with a chiral propionamide enolate has been demonstrated, however, these reactions typically suffer from low diastereoselectivity. The alkylation of chiral prolinol amide enolates with achiral and chiral halides has also been investigated; the stereochemistry of such alkylations occur from the enolate face opposite the alkoxymethyl group. We have investigated[26] the use of a chiral prolinol propionamide enolate (**80**) for these transformations and have found a surprising reversal of the expected facial selectivity in the reaction with terminal epoxides (Scheme 11). Treatment of epoxide **64** with five equivalents of enolate **80S**, which possesses a very strong facial bias towards alkylation from the pro-2(*R*) face, followed by hydrolysis of the resulting amides **81** and **82** gave the lactones **83** and **84** in 85% yield. Surprisingly, the undesired *cis* lactone **84** was obtained as the major product. This result implies attack on the epoxide by enolate **80S** from the more hindered pro-2(*S*) face. Similarly, alkylation of the antipodal enolate **80R** with **64** under the same conditions gave reversed facial attack from the pro-2(*R*) face, affording almost exclusively (98:2) the desired *trans*-lactone **83** in 90% yield. In contrast, alkylation of enolate **80S** with iodohydrin **85** gave the amides **86** and **87**, which were cyclized to give a mixture of lactones **83** and **84** with the desired trans-isomer **83** as the major product. Epoxide **73** was also reacted with enolate **80S**, and the product was hydrolyzed to give predominantly (96:4) the *cis*-lactone **88**. This compound arises from attack of the enolate from the pro-2(*S*) face. The reversed selectivity observed with enolates **80S** and **80R** with epoxides

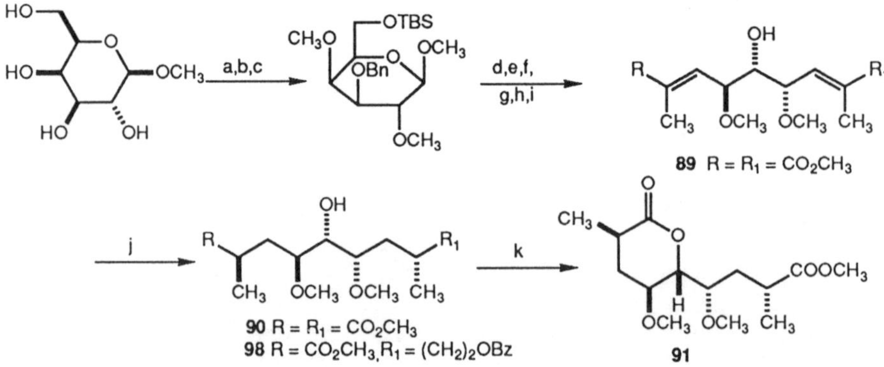

**Scheme 12.** (a) TBSCl, Et$_3$N, cat. DMAP, CH$_2$Cl$_2$, 72%, (b) ($n$-Bu$_3$Sn)$_2$O, toluene, reflux; BnBr, TBAI, 80°C, 98%; (c) MeI, NaH, THF, 93%; (d) 1:1 HOAc-H$_2$O, THF, 95%; (e) PPh$_3$I$_2$, pyridine , PhH, reflux, 77%; (f) Zn, 95% EtOH, reflux, 95%; (g) O$_3$, CH$_2$Cl$_2$, −78°C ;PPh$_3$; (h) Ph$_3$P=C(Me)CO$_2$Me, CH$_2$Cl$_2$ 0-rt, 64% (g and h); (i) TMSI (trace of HI), CH$_2$Cl$_2$, 90%; (j) H$_2$, [Rh(NBD)DIPHOS-4]BF$_4$, 1000 psi, CH$_2$Cl$_2$, 89%; (k) Pyridine-TsOH, CH$_2$Cl$_2$, 63%.

could be due to intermolecular chelation of the basic epoxide oxygen to the lithio-alkoxymethyl group.

Danishefsky and Villalobos reported a stereoselective synthesis of the C$_{10}$–C$_{19}$ segment of FK-506 by rhodium (I)-catalyzed hydrogenation of dienes.[19] This approach uses the stereochemistry of D-galactose. The 1,3-$syn$ relationship between the methyl group at C$_{11}$ and the C$_{13}$ methoxy function is duplicated for the same functions attached to C$_{17}$ and C$_{15}$, respectively. Both relationships could be established by concurrent hydrogenation of a diolefin with guidance of the rhodium(I) catalyst by a homoallylic C$_{14}$ hydroxyl group. Reduction of diolefin **89** using [Rh(NBD)DIPHOS-4]BF$_4$ under 1000 psi hydrogen gave tetrahydro compound **90** as the principal product (Scheme 12). The ratio of **90** to the two other tetrahydro compounds was 20:2.5:1. The fourth hypothetical permutant was not detected. The lactonization of **90** afforded primarily the all equatorial product **91**. However, the selectivity of this diastereotopic end group differentiation was disappointing (ca. 5:1).

For the synthesis of dithiane-sulfone (**92**), we explored an alternate route involving the differentiated termini inherent in compound **93** (Scheme 13). Acetonide ester **93**, obtained from the differentially blocked D-galactose derivative was homologated by a conventional sequence to give the alcohol **94**. After debenzylation, treatment with acetic acid followed by sodium periodate gave the β-hydroxy aldehyde **95**. Smooth Wittig condensation with the ylide **96** and monobenzoylation gave the bis-olefin **97** in overall 55% yield. Homogeneous

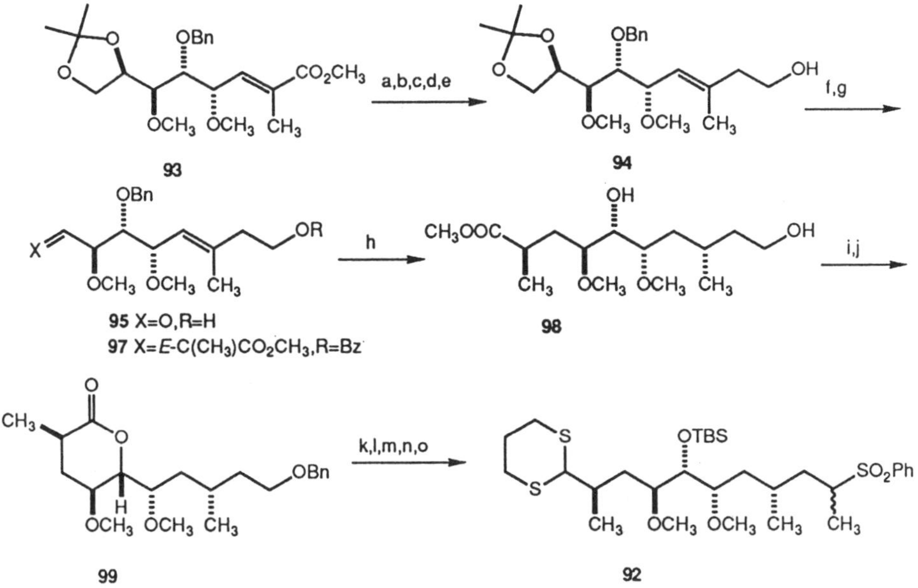

**Scheme 13.** (a) LiEt₃BH, THF, −78–20°C, 100%; (b) MsCl, LiCl, *s*-collidine, DMF, 0°C, 67%;(c) NaCN, DMF, 69%; (d) (i) (*i*-Bu)₂AlH, CH₂Cl₂, −78°C; (ii) NaBH₄, EtOH, <rt, 69% overall; (e) Na, NH₃(l), −78°C, 90%; (f) (i) 3:1 HOAc-H₂O, THF, reflux; (ii) NaIO₄, 1:1 THF-H₂O; (g) Ph₃P=C(CH₃)CO₂Me, CH₂Cl₂ 0–rt; BzCl, pyr, THF, 55% overall; (h) H₂, [Rh(NBD) DIPHOS-4]BF₄, 1000 psi, CH₂Cl₂, 94%; (i) TsOH, 4-Å molecular sieves, CH₂Cl₂, 93%; (j) Li(*s*-Bu)₃BH, THF, −78°C, 94%; (k) HS(CH₂)₃SH, BF₃·OEt₂, CH₂Cl₂, −78–0°C, 85%; (l) TBSOTf,2,6-lutidine, CH₂Cl₂, 97%; (m) K₂CO₃ MeOH<rt, 84%; (n) (i) MsCl, TEA, CH₂Cl₂, 0°C, 84%; (ii) NaI, acetone, reflux, 97%; (o) PhSO₂⁻Na⁺, DMF, 86%; *n*-BuLi, THF, −78°C, MeI, 88%.

catalytic hydrogenation gave the tetrahydro product **98** in 94% yield; **98** lactonized to give **99** in 93% yield. The lactone **99** was converted to the desired dithiane-sulfone **92** by standard operations. The use of this secondary sulfone in a Julia-type coupling was also explored. In a model reaction, lithiation of **92** followed by the addition of isobutylaldehyde gave a mixture of hydroxysulfones which, upon acetylation and reductive elimination, gave an olefin mixture. The *E/Z* ratio by this protocol ranged from 2:1 to 2.5:1.

A highly stereocontrolled synthesis of the C₁₀–C₂₃ segment of FK-506 by using an opening of a symmetrical *trans*-disubstituted epoxide with a trisubstituted vinylalane ate complex was accomplished by Smith and Hale.[27] The reaction of enantiomerically-pure iodide **100** with KCN followed by the *i*-Bu₂AlH reduction gave the aldehyde **101**, which was converted to the alkyne **102** through the corresponding dibromo olefin followed by treatment with *n*-butyllithium (Scheme 14). Carboalumination was achieved by treating alkyne **102** with two equivalents

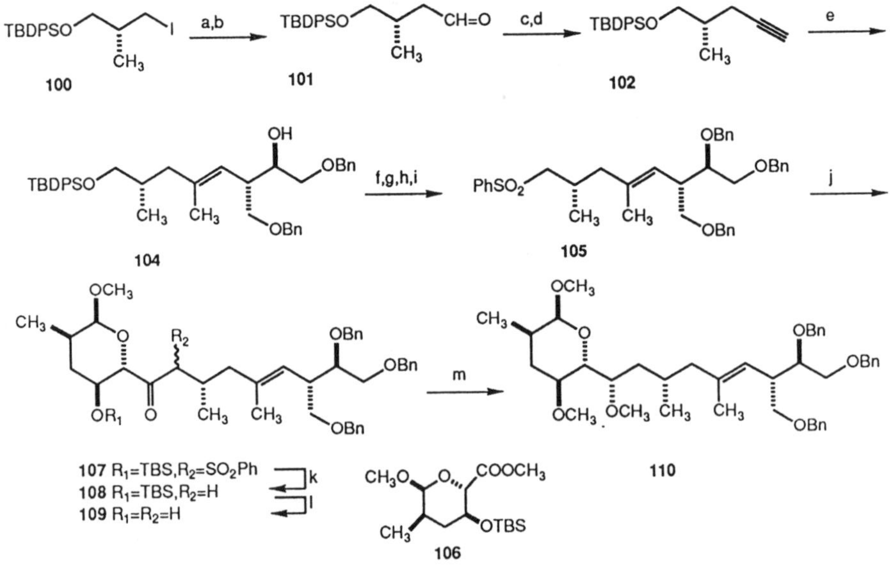

**Scheme 14.** (a) KCN, DMF, 85°C, 87%; (b) DIBAL-H, CH$_2$Cl$_2$,–78°C, 94%, (c) PPh$_3$, CBr$_4$, CH$_2$Cl$_2$, (d) n-BuLi, THF, –78–rt, 96%; (e) Me$_3$Al, Cp$_2$ZrCl$_2$, 1,2-dichloroethane; n-BuLi, hexane, 0°C, add epoxide **213** (see Scheme 31), benzene, 66%; (f) NaH, BnBr, TBAI, THF, 71%; (g) TBAF, THF, 89%; (h) (PhS)$_2$, Bu$_3$P, DMF, 86%; (i) oxone, THF, MeOH, H$_2$O, 91%; (j) n-BuLi, THF, –78°C, add **106**, –78–0°C, 74%, (k) Bu$_3$SnH, AIBN, toluene, reflux, 84%, (l) HF-pyridine, THF 92%; (m) Me$_4$NBH(OAc)$_3$, AcOH, MeCN, –40°C, 84%; NaH, MeI, DMF, 76%.

of trimethylaluminum in the presence of a catalytic amount of zirconocene dichloride. The vinylalane was converted to an ate complex by treatment with n-butyllithium. The coupling of the ate complex with the epoxide gave a 1:1 mixture of ring-opened products. Therefore, the symmetrical *trans*-disubstituted epoxide (see compound **213**, page 85) was investigated. As expected, reaction of the ate complex derived from **102** with enantiomerically-pure epoxide (see compound **213**, page 85) gave coupled product **104** in 66% yield. Phenylsulfone **105** obtained from olefin **104** was treated with n-butyllithium then coupled with ester **106**, which was obtained from methyl α-D-glucopyranoside, to give **107** in 74% yield. Removal of the sulfone moiety was accomplished with tri-n-butyltin hydride in 84% yield. After TBS deprotection of the C$_{13}$ hydroxyl group, the final stereogenic center of the C$_{13}$–C$_{23}$ segment **110** was introduced by a hydroxyl-directed reduction of the C$_{15}$ carbonyl group with tetramethylammonium triacetoxyborohydride. A 20:1 mixture of the *anti* and *syn* 1,3-diols was obtained in 84% yield. Methylation of the major *anti* diol completed the synthesis of **110**.

An alternative stereoselective synthesis of the C$_{10}$–C$_{24}$ segment of FK-506 by

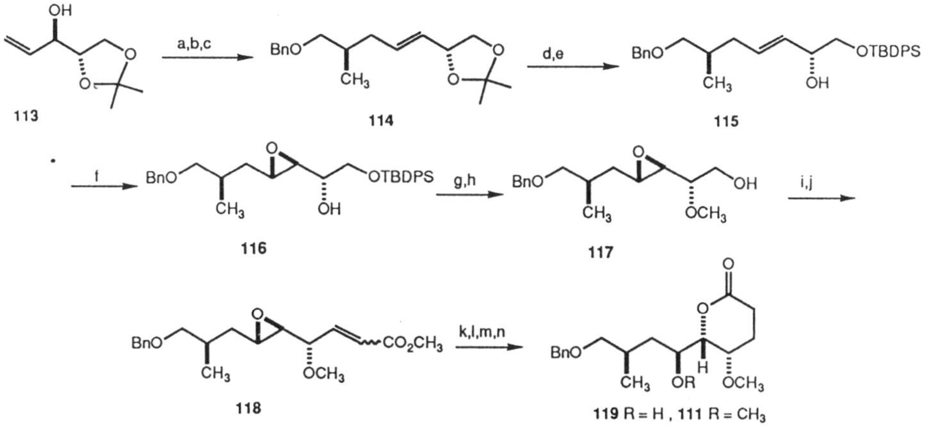

**Scheme 15.** (a) (EtCO)₂O, Et₃N, cat. DMAP, CH₂Cl₂; (b) LDA (2 equiv), TMSCl (10 equiv), THF, −78°C; Et₃N, LiAlH₄; (c) NaH, BnBr, cat. Et₄NI, THF<reflux; (d) 3:1:1 HOAc-H₂O-THF, 50°C; (e) TBDPSCl, imidazole, DMF; (f) Ti(*i*-PrO)₄, (+)-DIPT, *t*-BuOOH, CH₂Cl₂, −20°C; (g) NaH, MeI, THF, 0°C; (h) *n*-Bu₄NF, THF; (i) Dess-Martin; (j) Ph₃P=CHOOMe, THF; (k) H₂, Pd/C, EtOAc; (l) NaOH, EtOH, reflux; (m) cat. CSA, CH₂Cl₂; (n) CH₂N₂, BF₃·OEt₂, CH₂Cl₂, 0°C.

using a stereoselective alkylation of lactone **111** with allylic iodide **112** was reported by Wang[28] of Merck Frosst Laboratories. The benzyl ether **114** was obtained from the readily available allylic alcohol **113** by Ireland-Claisen rearrangement followed by reduction and benzylation (Scheme 15). Hydrolysis of the acetonide group followed by selective monosilylation of the resulting diol gave the olefin **115**. Sharpless epoxidation of **115** using (+)- DIPT/Ti(*i*-OPr)₄ afforded the corresponding epoxide **116** with the desired stereochemistry. Methylation of **116** followed by removal of the *t*-BuPh₂Si group led to **117**. Oxidation of **117** gave an unstable aldehyde that was condensed with the Wittig reagent to afford the α,β-unsaturated ester **118** as a 2:1 *E/Z* mixture. After catalytic hydrogenation, a regioselective intramolecular epoxide opening with the carboxylate anion generated from sodium hydroxide in ethanol gave **119**. Methylation of **119** furnished the required lactone **111**.

The Evans asymmetric aldol methodology was employed to install the C₂₁ stereocenter in alkylating agent **112**. The boron enolate of **120** was condensed with 3-(*p*-methoxybenzyloxy)propionaldehyde to provide **121**, which was oxidatively cyclized to the benzylic acetal **122** with DDQ (Scheme 16). Removal of the chiral auxiliary in **122** was followed by reduction with LiAlH₄ to give the alcohol **123**. Subsequent Swern oxidation of **123** and treatment of the crude aldehyde with the Wittig reagent afforded the α,β-unsaturated ester **124** with complete *E* selectivity. Ester **124** was converted to required iodide **112** by a conventional sequence.

The alkylation of lactone **111** with iodide **112** and sodium hexamethyldisilazide

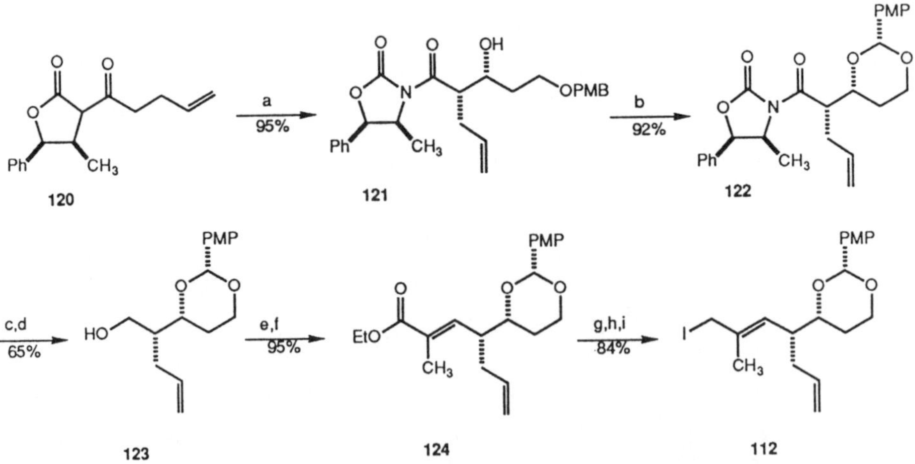

**Scheme 16.** (a) (c-C$_5$H$_9$)$_2$OTf, DIEA; PMBOCH$_2$CH$_2$CH=O, CH$_2$Cl$_2$, −78°C; (b) DDQ, CH$_2$Cl$_2$; (c) LiOH, H$_2$O$_2$, THF-H$_2$O; (d) LiAlH$_4$, THF, 0°C; (e) (COCl)$_2$, DMSO, Et$_3$N, CH$_2$Cl$_2$, −78°C; (f) Ph$_3$P=C(Me) COOEt, THF; (g) LiAlH$_4$, Et$_2$O, 20°C; (h) Ph$_3$P, CBr$_4$, CH$_2$Cl$_2$; (i) NaI, acetone.

gave the desired coupled product **125** and its epimer in a ratio of 8:1 (Scheme 17). The major product was then reduced with LiAlH$_4$ to afford the diol **120**, which, upon monotosylation of the primary hydroxyl group, silylation of the secondary hydroxyl group, and finally reduction with Superhydride (LiEt$_3$BH), completed the properly protected C$_{10}$–C$_{24}$ segment **127** of FK-506.

Recently, a copper-catalyzed migratory insertion reaction was used by

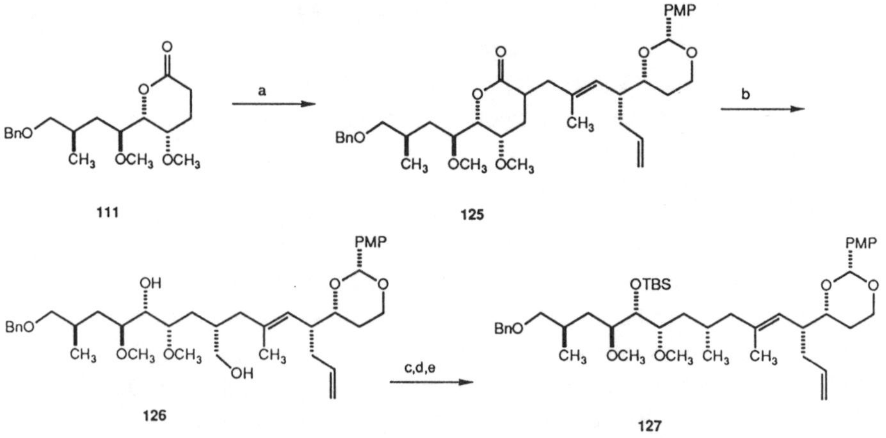

**Scheme 17.** (a) LiN (TMS)$_2$, THF, −78°C, **112**; (b) LiAlH$_4$, THF, −20°C; (c) TsCl, Et$_3$N, cat. DMAP, CH$_2$Cl$_2$; (d) TBSOTf,2,6-lutidine, CH$_2$Cl$_2$; (e) LiEt$_3$BH,THF.

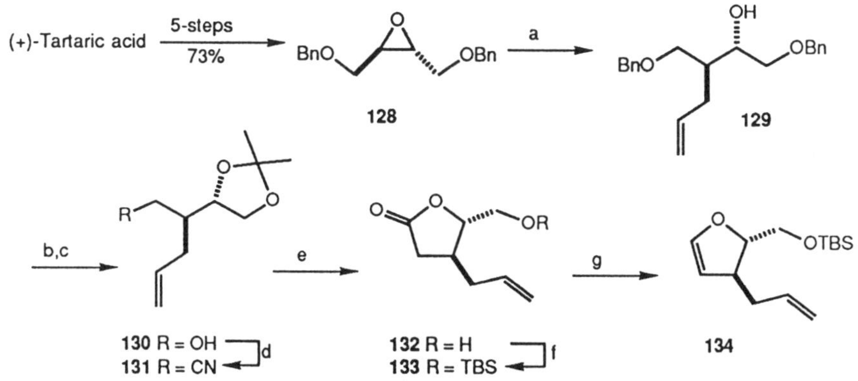

**Scheme 18.** (a) allylMgBr, THF, −20°C, 93%; (b) Na/THF-NH₃; (c) acetone, TsOH, 63%; (d) TsCl, pyr; NaCN, DMSO, 75%; (e) 10% HCl, MeOH, 74%; (f) TBSCl, imidazole, CH₂Cl₂, 98%; (g) DIBAL-H, −78°C; MsCl, Et₃N, THF, 30–50°C, 86%.

Kocienski[29] to construct the tri-substituted olefin of the $C_{16}$-$C_{23}$ segment of FK-506. The synthesis began with (+)-tartaric acid, which was converted to the epoxide **128** in 73% overall yield (Scheme 18). Nucleophilic opening of the epoxide with allylmagnesium bromide gave the alcohol **129**, from which the benzyl groups were reductively removed. The triol was converted to the acetonide **130** in 63% yield. The alcohol **130** was transformed into the nitrile **131**, which was lactonized under hydrolysis conditions to give lactone **133** after protection of the primary alcohol with *t*-butyldimethylsilyl chloride. The lactone **133** was reduced with *i*-Bu₂AlH to a mixture of diastereomeric lactols, which were then dehydrated via the methanesulfonate ester in THF to give the desired enantiomerically-pure dihydrofuran **134** in 86% overall yield from lactone **133**.

The key step in this synthesis was a migratory insertion of the higher-order cuprate intermediate **139**, prepared by reaction of the lithiated dihydrofuran **135** with the homocuprate **138** (Scheme 19). The migratory insertion of **139** took place with clean inversion of stereochemistry to give the putative higher-order oxycuprate **140**. Intermediate **140** underwent alkylation with retention of configuration at low temperature to give the trisubstituted olefin **141** in 57% overall yield from **134**. The principal problem, the thermal instability of the cuprate **138**, was eventually resolved by the addition of dimethyl sulfide to the reaction mixture to stabilize the cuprate **138**.

The protecting group was removed from **141** and the resultant diol was selectively converted to the terminal methansulfonate ester (Scheme 20). Treatment with potassium carbonate gave epoxide **142**, which was converted regioselectively to the secondary alcohol **143** with LiAlH₄. Oxidation with the Dess-Martin periodinane reagent gave the desired ketone **144** without racemization.

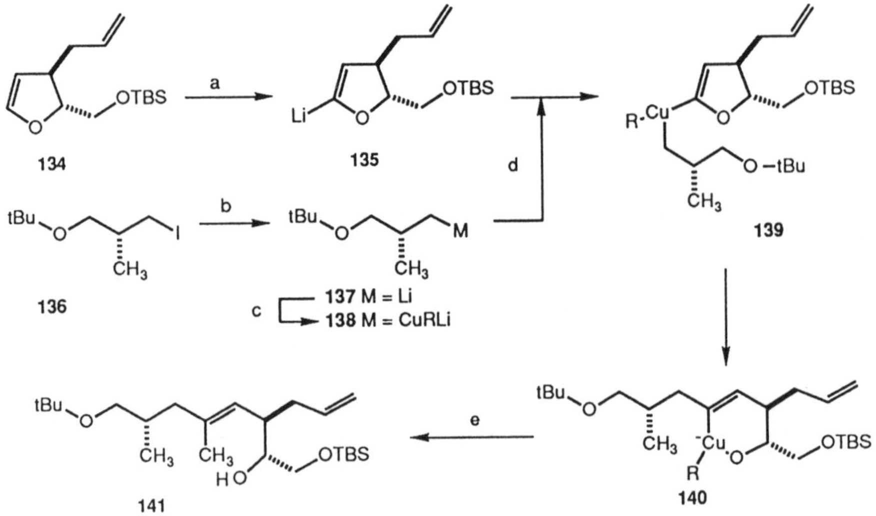

**Scheme 19.** (a) t-BuLi (1.2 equiv), Et₂O-THF (20:1), –78–0°C; (b) t-BuLi (7.7 equiv), Et₂O, –78–0°C; (c) CuBr-Me₂S, Et₂O, Me₂S, –60°C; (d) warm solution of **138** to 35°C and add **135**; (e) MeI, HMPA, –70°C.

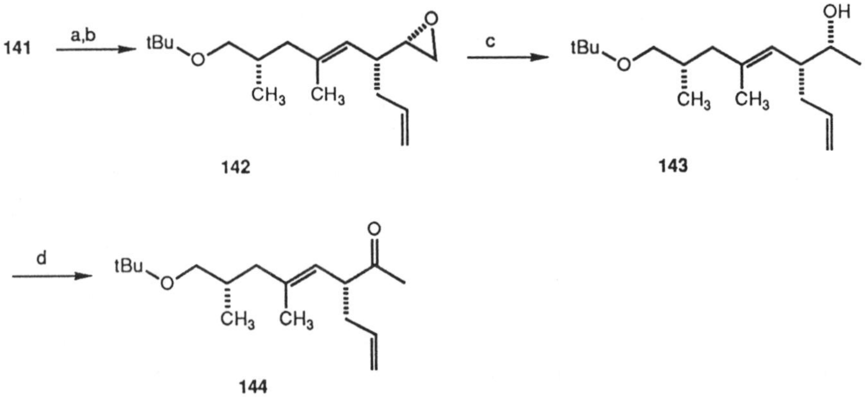

**Scheme 20.** (a) TBAF, THF, 97%; (b) MsCl, Et₃N, THF, –70–20°C; K₂CO₃, MeOH, 91%; (c) LiAlH₄, THF, 96%; (d) Dess-Martin, 92%.

## D.  Synthesis of $C_{24}$–$C_{34}$ Segment

A diastereospecific, non-racemic synthesis of the $C_{24}$–$C_{34}$ segment of FK-506 was first reported by the Merck group.[30] We were particularly intrigued by the $C_{20}$–$C_{34}$ carbon fragment (145) of FK-506 and its similarity to the analogous moiety of the macrolide antifungal antibiotic, rapamycin. A key structural feature, which is common to both molecules, is the $C_{28}$–$C_{34}$ cyclohexane carboxyaldehyde–derived fragment. We envisioned aldehyde 152 to be obtainable from naturally occurring quinic acid 146 by the stereospecific and regioselective reductive cleavage of the hydroxyl groups at $C_1$ and $C_5$ (Scheme 21). The lactone 147 was converted to the bisthiocarbonyl lactone 148 in 74% yield. Treatment of 148 with tributyltin hydride produced the thiocarbonate 149 in 80% yield. Direct treatment of the crude reaction mixture with additional tributyltin hydride in the presence of azobisisobutylonitrile (AIBN) gave the non-racemic lactone 150 efficiently. After protection of the $C_{32}$ hydroxyl group as the triisopropylsilyl (TIPS) ether, the crude reaction mixture was subjected to methylchloroaluminum $N$-methoxy-$N$-methylamide to give the $N$-methoxy-$N$-methylamide 151 in 85% overall yield from 150. Methylation and reduction gave aldehyde 152 in 85% yield. Aldehyde 152 was then condensed with 2-lithio-2-triethylsilylpropanal $t$-butylimine to give α,β-

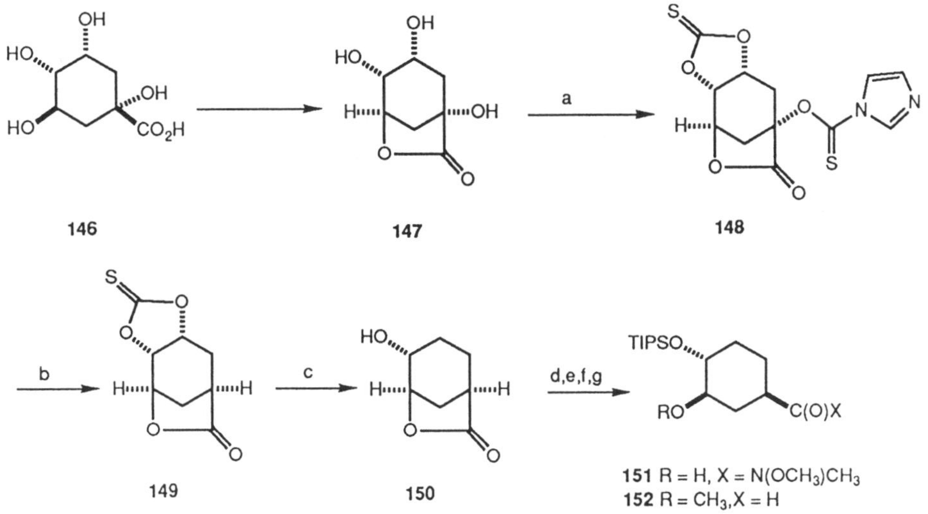

**Scheme 21.** (a) thiocarbonyldiimidazole (3.0 equiv), dichloroethane, reflux; (b) Bu$_3$SnH (1.5 equiv), xylenes, 140°C; (c) Bu$_3$SnH (2.0 equiv), AIBN, xylene, 140°C; (d) triisopropylsilyl triflate (1.3 equiv), 2,6-lutidine (2.2 equiv), CH$_2$Cl, 0°C; (e) Me(MeO)NAl(Me) (2.0 equiv), toluene; (f) methyl triflate,2,6-di-$t$-butyl-4-methyl-pyridine, CH$_2$Cl$_2$; (g) ($i$-Bu)$_2$AlH (1.5 equiv), THF, −75°C.

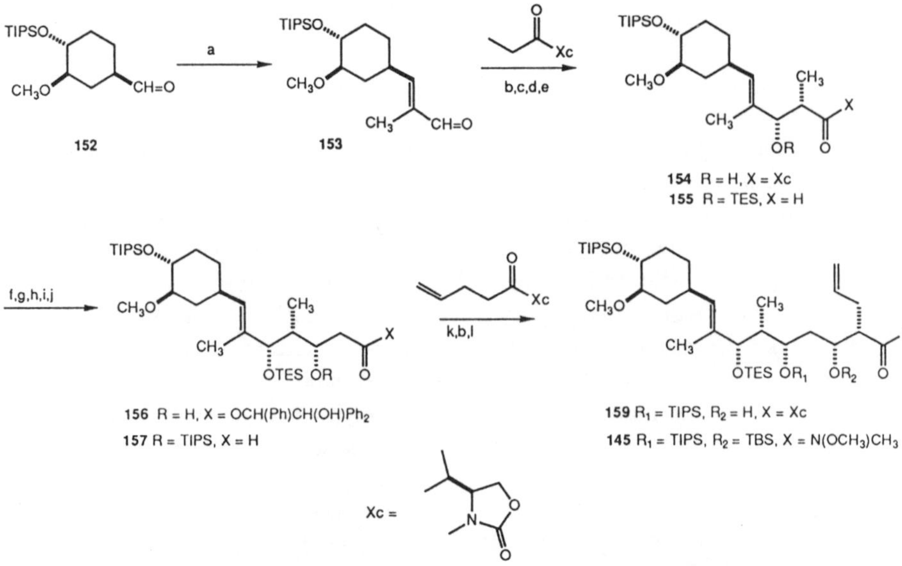

**Scheme 22.** (a) 2-lithio-2-triethylsilylpropanal *t*-butylimine (2 equiv), THF, −78 to −10°C; (b) propionic amide, *n*-Bu$_2$BOTf, DIEA, −78–0°C, CH$_2$Cl$_2$; (c) Me(MeO)NAl(Me)Cl (2.0 equiv), −70°C –rt, CH$_2$Cl$_2$; (d) triethylsilyl triflate (1.5 eq.),2,6-lutadine (2.5 equiv), CH$_2$Cl$_2$, 0°C; (e) LiOH/H$_2$O; (f) S(−)2-acetoxy-1,2,2-triphenylethanol (1.5 equiv), LDA (1.5 equiv), THF, −78°C; (g) NaOMe, MeOH, 0°C; (h) triisopropylsilyl triflate (1.3 equiv), 2,6-lutidine (2.2 equiv) (i) (*i*-Bu)$_2$AlH (2.5 eq.), THF, 0°C; (j) SO$_3$-pyr, Et$_3$N (4.0 eq.), DMSO-CH$_2$Cl$_2$; (k) **158**, *n*-BuBOTf, DIEA, CH$_2$Cl$_2$, −78–0°C; (l) TBDMSOTf (2.0 equiv),2,6-lutidine (3.0 equiv), CH$_2$Cl$_2$, 0°C.

unsaturated aldehyde **153** in 78% yield (Scheme 22). After hydrolysis, the *E,Z* mixture of olefins was chromatographically purified. Alternatively, anhydrous acidic treatment of *E,Z* mixtures of α-methyl-α,β-unsaturated imines followed by hydrolysis gave the corresponding aldehydes with > 100:1 *E:Z* ratios.[31] This new modified process eliminates the need for chromatographic purification of the crude olefin. Treatment of aldehyde **153** with the boron enolate of oxazolidone imide afforded alcohol **154**. Alcohol **154** was converted by standard synthetic transformations to the fully protected aldehyde **155** in 88% yield from **153**. Addition of the lithium enolate of (*S*)-2-acetoxy-1,1,2-triphenylethanol to **155** gave the alcohol **156** in 60% yield. After transesterification of **156** with methanol, protection of the free hydroxyl as its TIPS-ether, reduction of the carboxymethyl group, and oxidation with SO$_3$-pyridine, the resulting aldehyde **157** was treated with imide **158** to give the hydroxy-imide **159** in 95% yield. Weinreb aminolysis and protection of the free hydroxyl as its TBS-ether then gave the highly functionalized and fully protected C$_{20}$–C$_{34}$ segment, **145**, of FK-506.

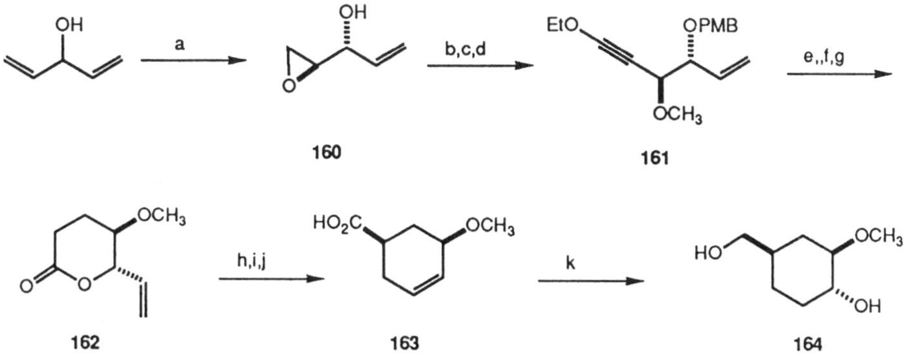

**Scheme 23.**   (a) Sharpless catalytic epoxidation, 97% ee, 58%; (b) PMBBr, NaH, 94%; (c) EtOC≡CLi, BF₃·OEt₃; (d) MeI, NaH, 87%; (e) EtOH, HgCl₂; (f) DDQ, 78% (g) TsOH, 85%; (h) TBSOTf, Et₃N; (i) Toluene, reflux; (j) HCl, 71%; (k) BH₃-THF, H₂O₂, 79%.

A synthesis of the cyclohexyl segment by a group-selective epoxidation has appeared.[32] This approach, based on the group-and face-selective epoxidation of divinylcarbinol, was achieved by the catalytic procedure of Sharpless. The epoxy alcohol **160** was protected as the *p*-methoxybenzyl (PMB) ether in 94% yield (Scheme 23). The methyl ether **161** was prepared in 87% yield by the regioselective epoxide opening with the lithium anion of ethoxyacetylene and BF₃ etherate followed by the methylation. Ethanolysis of **161** with catalytic HgCl₂ followed by deprotection of the *p*-methoxybenzyl ether with DDQ afforded the hydroxy ester in 78% yield. The ester was cyclized to lactone **162** in 85% yield. The Ireland-Claisen rearrangement of the silyl ketene acetal proceeded smoothly to give the carboxylic acid **163** in 71% overall yield. The hydroboration and oxidation of **163** provided diol **164** in 79% yield.

A stereocontrolled synthesis of the C₂₂–C₃₄ segment of FK-506 has been reported by Danishefsky, Schreiber and their co-workers.[33] The key feature of the synthesis is the method of preparation of the *anti* aldol product **166**. Catalytic asymmetric hydrogenation of the β-keto ester **165** gave the corresponding β-hydroxy ester (Scheme 24). Alkylation of the dianion of the β-hydroxy ester with allyl bromide provided **166** in 85% yield when the reaction was performed in 10:10:1 THF/Et₂O/HMPA and when the LiNiPr₂ was prepared from lithium/styrene. Reduction of **166** with LiAlH₄ followed by oxidative cyclization with DDQ provided the benzylidene acetal **167**. Iodo etherification with iodine/sodium bicarbonate, opening of the acetal with diisobutylaluminum hydride, and subsequent oxidation gave aldehyde **168**, in which the side-chain allyl was masked as an iodo ether. The chelation-controlled addition of triphenylcrotylstannane (as a 1:1 *cis/trans* mixture) was achieved with stannic chloride at −78 °C to provide the alcohol **169** in 60% yield. The silyl ether of **169**

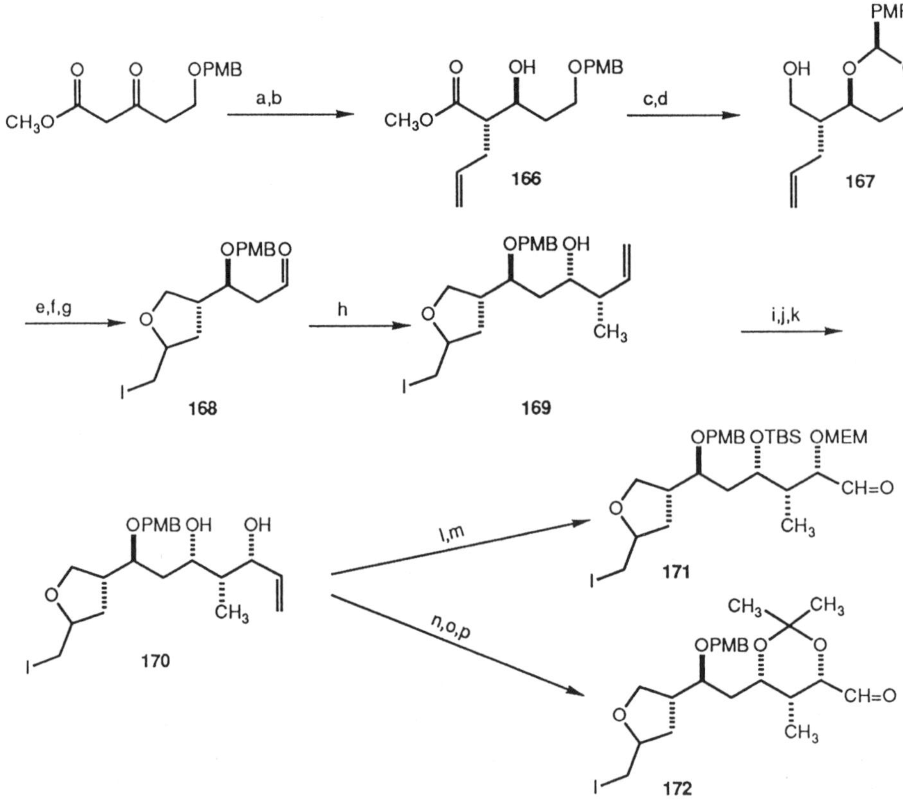

**Scheme 24.** (a) Ru$_2$Cl$_4$[s-BINAP]$_2$ (1 mole%), Et$_3$N, H$_2$, 90%; (b) LDA (2.5 equiv), allyl bromide, HMPA, 85%; (c) LiAlH$_4$, 86% (d) DDQ, 70%; (e) NaHCO$_3$, I$_2$, 76%; (f) DIBAL-H, 75%; (g) Dess-Martin, 94%; (h) MeCH=CHCH$_2$SnPh$_3$, SnCl$_4$, 60%; (i) TBSOTf, 95%; (j) O$_3$, Ph$_3$P; (k) CH$_2$=CHMgCl, CH$_2$Cl$_2$, 78% (3:1); (l) MEMCl, DIEA, 86%; (m) O$_3$, DMS, 72%; (n) Bu$_4$NF; (o) Me$_2$C(OMe)$_2$ CSA; (p) O$_3$, DMS, MeOH; K$_2$CO$_3$, 74%.

was converted to the allylic alcohol. Vinyl Grignard addition gave a 3:1 mixture of readily separable stereoisomers **170** based on a modest level of "Cram-type" selectivity. The methoxyethoxymethyl (MEM) ether of **170** was oxidized to the aldehyde **171**. Alternatively, the allyl alcohol **170** could be converted to the acetal **172** with >20:1 selectivity by the method of "ancillary stereocontrol." A proton NMR study of **172** revealed the syn-stereochemical relationship.

A model coupling sequence has also been demonstrated. Lithiation of **173** and addition to model aldehyde **174** gave a diastereomeric mixture of β-hydroxy-sulfones in good yield (Scheme 25). Oxidation with the Dess-Martin reagent provided the α-sulfonyl ketone, which was desulfonylated with sodium amalgam to provide

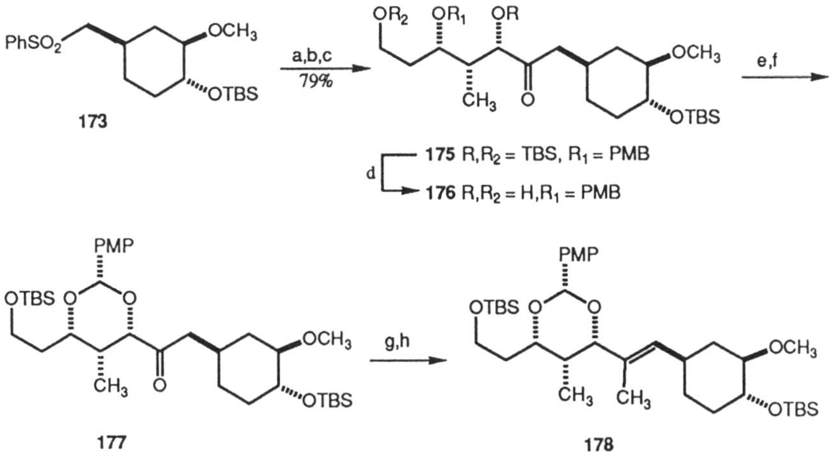

**Scheme 25.** (a) Sulfone, *n*-BuLi, aldehyde **174**; (b) Dess-Martin periodinane; (c) Na(Hg); (d) *n*-Bu₄NF; (e) TBSCl; (f) DDQ; (g) MeMgBr, 92% (20:1); (h) Et₃NSO₂NCOOEt, 96%.

the ketone **175** in 79% overall yield. Protecting group interchange produced the cyclic acetal **177**. The addition of methylmagnesium bromide gave the tertiary alcohol, which underwent elimination with the Burgess zwitterionic reagent to provide the *E*-trisubstituted olefin **178** and the undesired methylidene isomer not shown in a 4:1 ratio.

Corey and his co-workers have also investigated the enantioselective synthesis of the C$_{18}$–C$_{35}$ segment of FK-506 by a sequence of three enantioselective C–C bond-forming reactions.[34] The cyclohexane unit (**179**) was produced by a TiCl₄-catalyzed Diels-Alder reaction of butadiene with the acrylate ester of (*S*)-(+)-pan-tolactone (Scheme 26). Hydrolysis, epoxidation and lactonization gave **180** in 61% overall yield from the acrylate ester. Bicyclic lactone **180** was silylated to hydroxy ester **182**, and methylated. The methyl ester **182** was reduced by *i*-Bu₂AlH to the desired aldehyde **183**. Trisubstituted olefin **184** was obtained in 87% yield by reaction of **183** with lithio derivative of 2-triethylsilylpropanal-cyclohexylimine followed by a modified work-up.

Chain extension was enantioselectively controlled by the aldol reaction of **184** with 1,2-diphenyl-1,2-diaminoethane reagent (**185**, Scheme 27). The *syn* aldol product **186** (92% d.e.) was readily isolated by flash chromatography in 85% yield. The bis-sulfonamide by-product was recovered in high yield. The hydroxy group in **186** was protected as the MOM ether in 94% yield, and the protected aldol product afforded aldehyde **187** in 95% yield after reduction with *i*-Bu₂AlH. Further elaboration of **187** was carried out using a new method for 2-acetoxyallylation that is equivalent to enantioselective carbonyl addition of a functionalized acetone unit

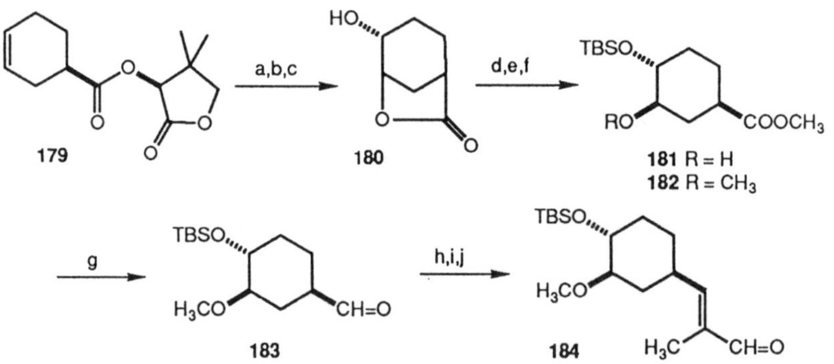

**Scheme 26.** (a) LiOH, H₂O, THF; (b) MCPBA (1.1 equiv), CCl₄, 0°C; (c) Et₃N, 55°C, 61% overall; (d) TBSCl, DMAP, CH₂Cl₂; (e) NaHCO₃ (0.5 equiv), MeOH, 98%; (f) methyl triflate (3.3 equiv), 2,6-di-t-butyl-4-methylpyridine (3.3 equiv), CH₂Cl₂, 94%; (g) DIBAL-H (1.2 equiv), hexane, –78°C; (h) CH₃CH(TMS) CH=N(Li)-C₆H₁₁, THF, –20°C; (i) TFA, 0°C; (j) H₂O, 0°C, 87%.

and that is also based on the 1,2-diphenyl-1,2-diaminoethane controller group. The specific allylation reagent employed was the cyclic borane **188**, which was prepared by reaction of the corresponding S,S-bromoborane with 2-acetoxyallyl-tri-n-butylstannane. Homoallylic alcohol **189** was isolated in 92% yield from the reaction of **188** with aldehyde **187**. Silylation of the hydroxyl group and bromina-

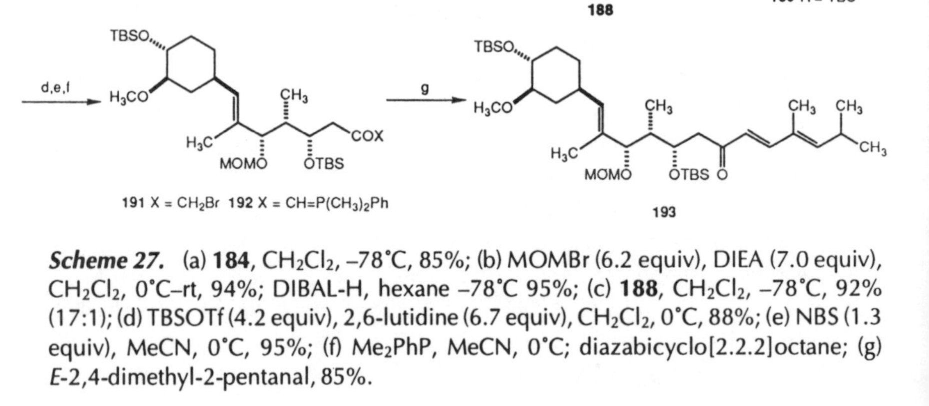

**Scheme 27.** (a) **184**, CH₂Cl₂, –78°C, 85%; (b) MOMBr (6.2 equiv), DIEA (7.0 equiv), CH₂Cl₂, 0°C–rt, 94%; DIBAL-H, hexane –78°C 95%; (c) **188**, CH₂Cl₂, –78°C, 92% (17:1); (d) TBSOTf (4.2 equiv), 2,6-lutidine (6.7 equiv), CH₂Cl₂, 0°C, 88%; (e) NBS (1.3 equiv), MeCN, 0°C, 95%; (f) Me₂PhP, MeCN, 0°C; diazabicyclo[2.2.2]octane; (g) E-2,4-dimethyl-2-pentanal, 85%.

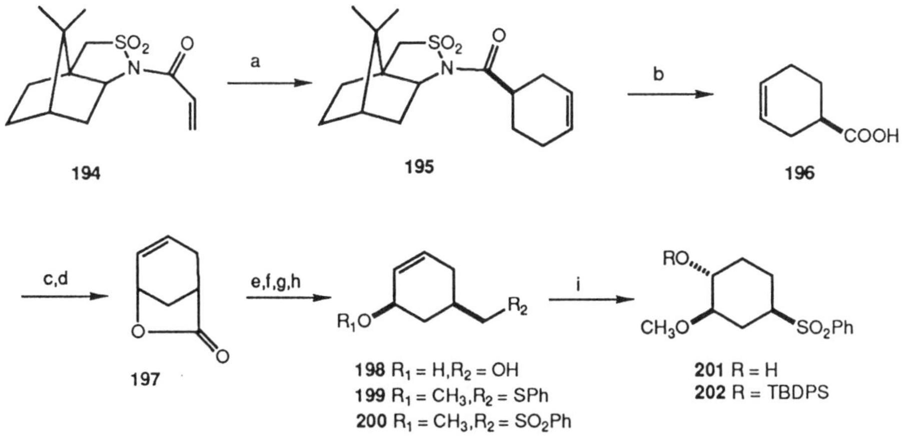

**194** **195** **196**

**197** **198** R₁ = H,R₂ = OH **201** R = H
**199** R₁ = CH₃,R₂ = SPh **202** R = TBDPS
**200** R₁ = CH₃,R₂ = SO₂Ph

**Scheme 28.** (a) 1,3-butadiene, EtAlCl₂, CH₂Cl₂, −78°C, 86%; (b) LiOH, H₂O, THF, 87%; (c) I₂/KI, NaHCO₃, H₂O, 0°C, 93%; (d) DBU, THF reflux, 96%; (e) LiAlH₄, Et₂O, 0°C, 93%; (f) Bu₃P, (PhS)₂, DMF, rt, 70%; (g) CH₂N₂, Et₂O, BF₃·OEt₂, 0°C, 94%; (h) oxone, THF, MeOH, H₂O, 0°C, 87%; (i) BH₃·THF, NaOH, t-BuOOH, −78-rt, 71%; t-BuPh₂SiCl, imidazole, DMF, 70%.

tion with NBS gave bromoketone **191** in 84% overall yield. The bromoketone was reacted with dimethylphenylphosphine and then treated with base to generate ylide **192**, which was reacted with E-2,4-dimethyl-2-pentanal to give dienone **193** in 85% yield.

A similar approach to the cyclohexyl moiety, by using an asymmetric Diels-Alder reaction of a an enantiomerically pure sultam, has been developed by Smith and co-workers.[35] The acid **196** was obtained in good yield by a Lewis acid-catalyzed reaction of 1,3-butadiene with **194** followed by hydrolysis of the amide **195** (Scheme 28). Iodolactonization and DBU-induced elimination of iodide gave the desired unsaturated lactone **197**, which, in turn, was reduced to diol **198** with LiAlH₄. The primary alcohol was selectively converted to the corresponding thiophenyl ether **199**. The remaining hydroxyl was methylated and the thioether was oxidized with oxone to give the phenylsulfone **200**. Hydroboration gave the desired alcohol **201** as the major product (the ratio of the three products was 17.5:1.7:1.0) in 71% yield after flash chromatography. Silylation completed the synthesis of cyclohexyl phenyl sulfone **202**, which is ready for the olefination.

Sharpless' asymmetric epoxidation of crotyl alcohol [**203**, and subsequent silyation produced epoxide **204** in 76% yield (Scheme 29)]. Treatment of **204** with excess 2-lithio-1,3-dithiane led exclusively to the C₃-alkylated product **205** in 76% yield. Desilylation, formation of p-methoxybenzylidene acetal and regioselective reduction with i-Bu₂AlH gave the protected dithiane **207** (5:1 ratio). Oxidation of

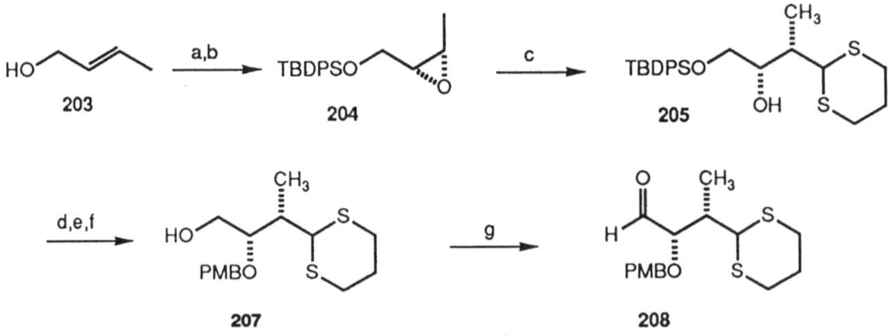

**Scheme 29.** (a) Sharpless epoxydation, D(–)-DET, CH$_2$Cl$_2$; (b) *t*-BuPh$_2$SiCl, Et$_3$N, DMAP, DMF, 76%; (c) 2-lithio-1,3-dithiane, DMU, THF, 0°C, 76%; (d) TBAF, THF; (e) p-MeOC$_6$H$_4$CH$_2$ (OMe)$_2$, DMF, TsOH, 55°C, 95% overall; (f) DIBAL-H, CH$_2$Cl$_2$, –78°C, 90%; (g) SO$_3$-pyr, DMSO, Et$_3$N,CH$_2$Cl$_2$, 0°C.

the alcohol to aldehyde **208** was successfully completed by using SO$_3$-pyridine-DMSO complex.

Coupling of aldehyde **208** with the anion derived from sulfone **202** gave a mixture of hydroxy sulfones **209** in 89% yield (Scheme 30). Oxidation of **209** produced β-keto sulfone **210**, which was reduced to ketone **211** with aluminum analgam. Deprotonation and the addition of *N*-phenyltrifluoromethane sulfonimide in the presence of DMPU afforded exclusively (> 99:1) the requisite *Z* enol triflate, which was treated with lithium dimethylcuprate to give two major products (7:1). The desired *E*-isomer **212** could be isolated in 73% yield.

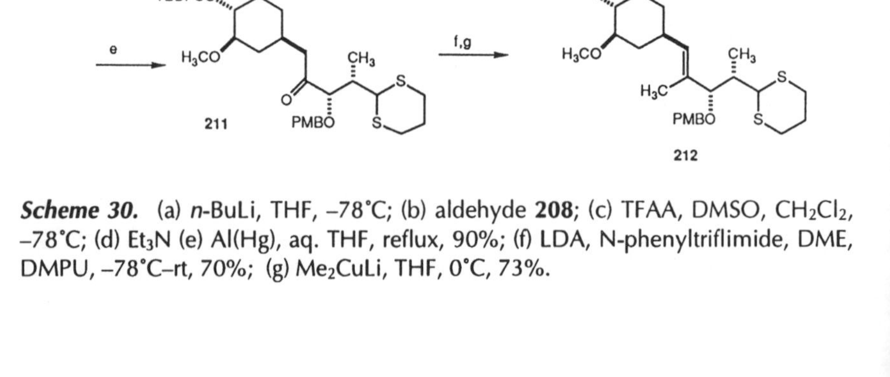

**Scheme 30.** (a) *n*-BuLi, THF, –78°C; (b) aldehyde **208**; (c) TFAA, DMSO, CH$_2$Cl$_2$, –78°C; (d) Et$_3$N (e) Al(Hg), aq. THF, reflux, 90%; (f) LDA, N-phenyltriflimide, DME, DMPU, –78°C–rt, 70%; (g) Me$_2$CuLi, THF, 0°C, 73%.

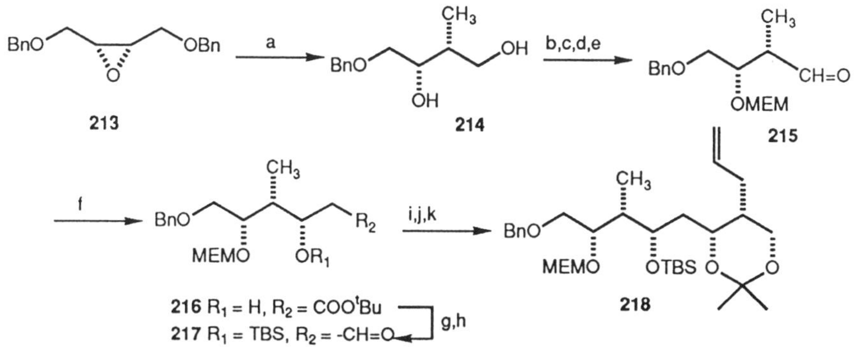

**Scheme 31.** (a) Me$_2$CuLi (4 equiv), Et$_2$O, −40°C; NaIO$_4$ (2 equiv), THF, H$_2$O; (b) pivCl (1.1 equiv), pyr; (c) MEMCl, DIEA, CH$_2$Cl$_2$, rt; (d) MeLi, ether, 0°C; (e) (COCl)$_2$, DMSO, Et$_3$N, CH$_2$Cl$_2$, −78°C; (f) BF$_3$·OEt$_2$, CH$_2$Cl$_2$, −78°C; (g) TBDMSCl, imidazole, DMAP (cat.), DMF; (h) DIBAL-H, CH$_2$Cl$_2$, −78°C; (i) **158**, n-Bu$_2$BOTf, DIEA, CH$_2$Cl$_2$, −78–0°C; (j) LiBH$_4$ (2.2 equiv), THF, 0°C–rt; (k) 2,2-dimethoxypropane, TsOH (cat.).

Sharpless' asymmetric epoxidation was used by Rao and his group[36] in an alternative stereocontrolled synthesis of the C$_{20}$–C$_{27}$ segment. Treatment of the epoxide **213** with lithium dimethylcuprate gave the 1,3-diol **214** (Scheme 31). Protection of both hydroxyls, deprotection of the primary alcohol and Swern oxidation provided aldehyde **215** in 60% overall yield from **214**. Reaction of the t-butyldimethylsilylenol ether derived from t-butylacetate with **215** in the presence of BF$_3$ etherate gave syn-adduct **216** as the major product (ratio *syn:anti*= 95:5). Hydroxy protection, followed by reduction with i-Bu$_2$AlH gave aldehyde **217**. Evans' aldol reaction, followed by standard transformation furnished the C$_{20}$–C$_{27}$ segment **218** of FK-506.

## E.   Completion of the Synthesis of FK-506

Our strategy[37] to form the macrocycle hinged on the initial construction of the C$_{19}$–C$_{20}$ trisubstituted olefin. The C$_{10}$–C$_{18}$ fragment **219** obtained by LiAlH$_4$ reduction of lactone **75** was refunctionalized to the dithiane phosphine oxide **62** by standard transformations (Scheme 32). Treatment of the N-methoxy amide **145** with i-Bu$_2$AlH gave the desired aldehyde **220**. The two segments were coupled by deprotonation of phosphine oxide **62** with n-butyllithium in the presence of TMEDA followed by addition of aldehyde **220**. The resulting mixture of isomers was separated to provide the two desired adducts (38% and 39% of higher and lower R$_f$ components, respectively). The two adducts were treated with potassium hexamethyldisilazide to give (E)-olefin **221** and its (Z) counterpart in 82% and 84%

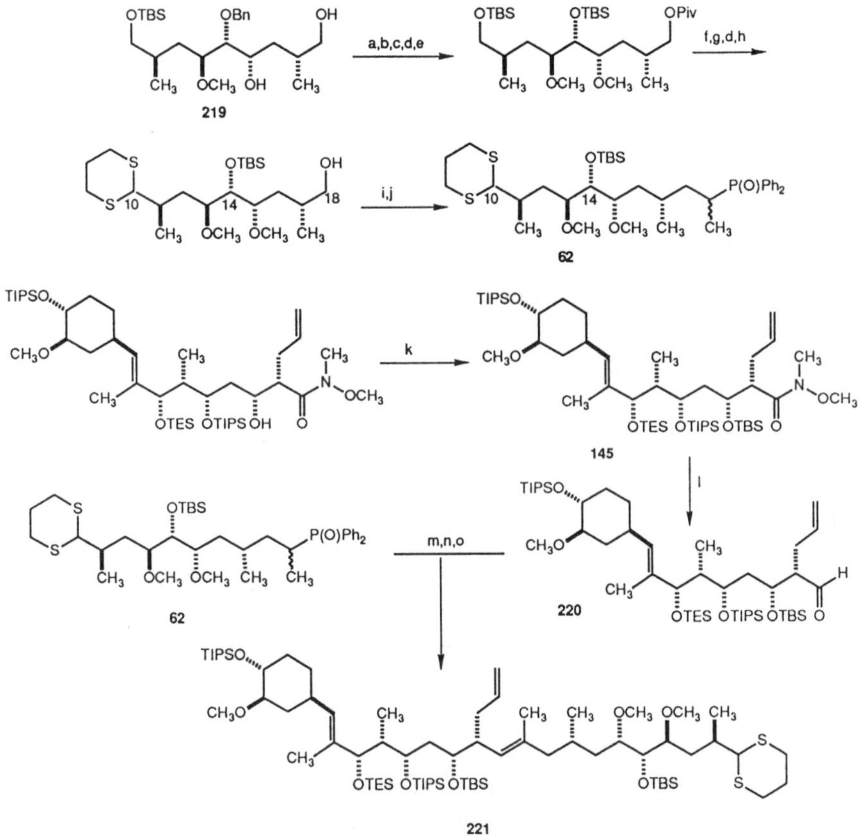

**Scheme 32.** (a) pivCl, pyr, 0°C; (b) NaH, MeI, THF, 0°C; (c) H₂, Pd(OH)₂/C, EtOAc; (d) TBSOTf, 2,6-lutidine, CH₂Cl₂; (e) TFA, H₂O, THF, 20°C; (f) (COCl)₂, DMSO, CH₂Cl₂, –78°C, Et₃N; (g) CH₂(CH₂SH)₂, BF₃·OEt₂, CH₂Cl₂, 0°C; (h) LiAlH₄, THF, 0°C; (i) PhSO₂Cl, pyr, 0°C; (j) Ph₂P(O) Et, n-BuLi, THF, –78°C – 0°C; (k) TBSOTf, 2,6-lutidine, CH₂Cl₂, 0°C; (l) DIBAL-H, THF, –30°C; (m) n-BuLi, THF, –78°C; **220**; (n) column chromatographic separation; (o) KHMDS, THF, 0°C.

yield, respectively. The olefin geometry was assigned by the $^{13}$C chemical shift of the C$_{19}$–CH$_3$ at 16.6 ppm for the E olefin (23.6 ppm for Z olefin).

Our strategy for installing the key tricarbonyl system involved deprotonation of the dithiane followed by acylation with diethyl oxalate. A deprotonation model study of **222** revealed that metallation occurs at C$_{14}$ and is followed by elimination of the C$_{13}$ methoxy group (Scheme 33). Dianion formation of **223** in neat TMEDA followed by reaction with diethyl oxalate provided a C$_{10}$-acylated product in 64% yield; no C$_{14}$ acylated product was observed. Transesterification afforded the

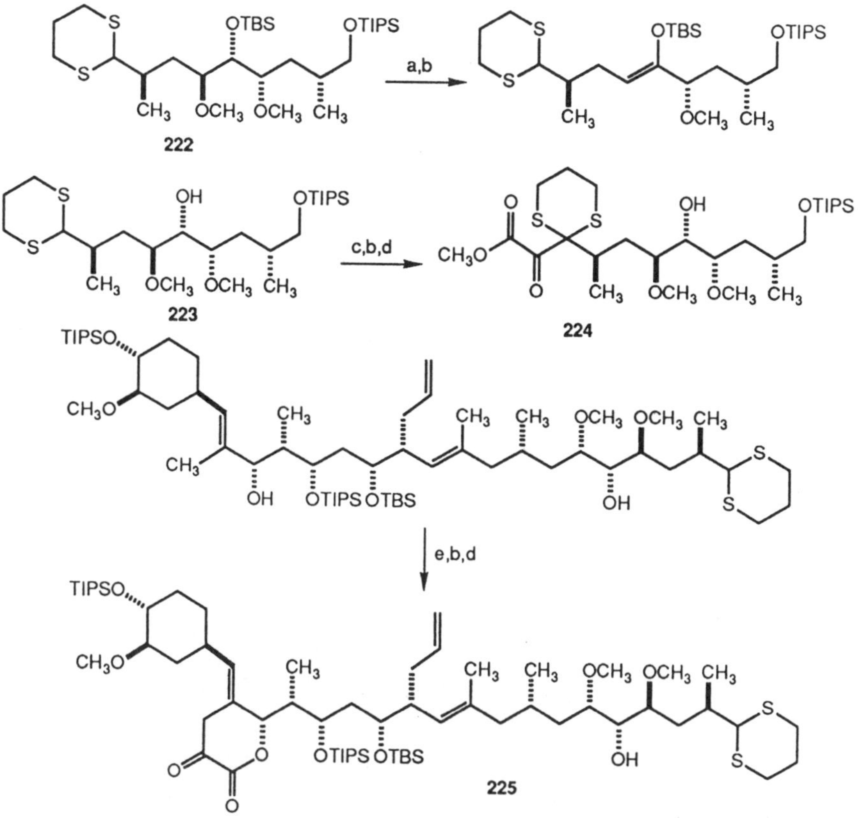

**Scheme 33.** (a) *n*-BuLi (1.5 equiv), DME-d$_{10}$, HMPA (2 equiv), −30°C; (b) (COOEt)$_2$, −30–0°C; (c) *n*-BuLi (4 equiv), neat TMEDA, −30°C; (d) NaOMe, MeOH; (e) *n*-BuLi (6 equiv), neat TMEDA, −30°C.

separable methyl ester **224**. Unfortunately, application of these conditions to 14,26-dihydroxy-**221** resulted primarily in metallation at the C$_{27}$ methyl to provide **225**; only a trace of the desired C$_{10}$ acylated product was found.

An alternate approach to the tricarbonyl system was investigated where the C$_{10}$ served as an electrophile, such as an aldehyde, rather than as a nucleophile. The plan was to add a protected α-hydroxyimide enolate to a C$_{10}$-aldehyde and oxidize to a tricarbonyl system. A model study of the aldol reaction of PMB imide **226** with C$_{10}$–C$_{18}$ aldehyde **227** gave aldol product **228** (Scheme 34).

The oxidation strategy with the methyl ester obtained from imide **228** by treatment with CH$_3$OMgBr was investigated next. Reductive removal of the PMB group followed by Swern oxidation of the resulting diol gave variable mixtures of the hydrate **229** and the free diketoester **230** in 88% yield (1-3:1). Removal of the

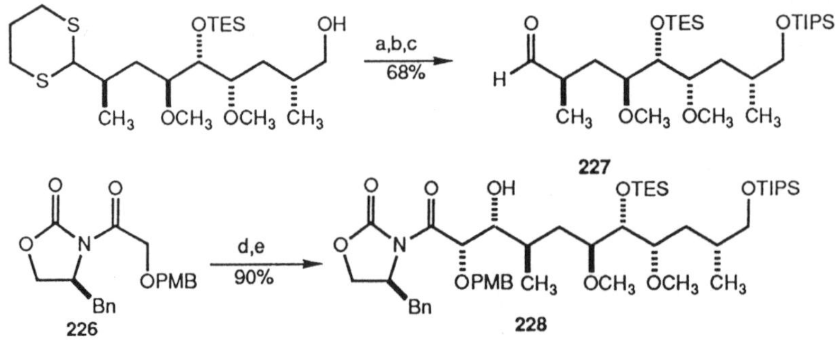

**Scheme 34.** (a) TIPSOTf, 2,6-lutidine; (b) NCS, AgNO₃, 2,6-lutidine, MeOH; (c) (HO)₂CHCOOH; (d) *n*-Bu₂BOTf, Et₃N, toluene, –50°C; (e) **227**, –30°C.

TES group at $C_{14}$ hydroxyl provided variable mixtures of two hemiketals, assigned to the 6- and 7-membered rings **231** and **232** (1–4:1, respectively) in 80% yield (Scheme 35). The center carbonyl ($C_9$=O) was effectively competing with the $C_{10}$ carbonyl for the internal nucleophile, the $C_{14}$ hydroxyl group.

As an earlier model study with the amide gave only the 6-membered ring, we reasoned that an imide or amide at $C_8$ might effectively screen $C_9$, thereby enhancing 6-membered ring selectivity. Swern oxidation of the diol **233** obtained from PMB deprotection of **228** gave tricarbonyl **234** which provided imide-hemiketal **235** in 66% overall yield without the formation of the undesired 7-membered ring hemiketal (Scheme 36).

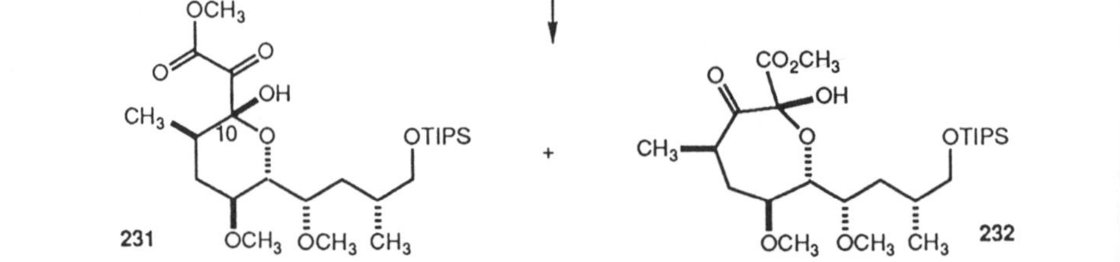

**Scheme 35.**

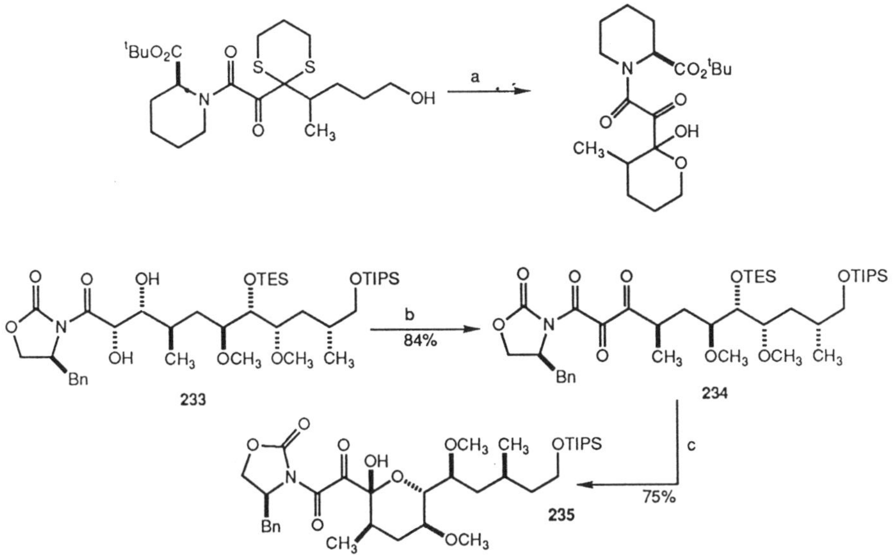

**Scheme 36.** (a) NCS, AgNO₃, lutidine, CH₃CN, H₂O; (b) (COCl)₂ DMSO, CH₂Cl₂, −78°C, Et₃N; (c) TFA, H₂O, THF.

Attempted hydrolysis of imide **235** with LiOOH resulted in oxidative cleavage of the $C_9$–$C_{10}$ bond. Hydrolysis with LiOH gave the carboxylic acid in low yield. Attempts to silylate **235** gave a 2:1 mixture of hemiketal and $C_{14}$ silylated hydroxy ketone. We therefore concluded that the hemiketal should be installed at a late stage in the synthesis. A final model study for the tricarbonyl system was performed to simulate a synthesis of FK-506 in which the tricarbonyl system is formed after macrocyclization.

The aldol adduct **236** derived from the $C_{14}$-TBS-protected aldehyde was hydrolyzed, silylated and condensed with *t*-butyl pipecolate to give amide **237** in 74% overall yield (Scheme 37). The PMB and TES groups were then removed to provide diol **238** in 79% yield from **237**. Swern oxidation provided the tricarbonyl **239** in 85% yield. Removal of silyl groups with HF/CH₃CN gave the desired hemiketal **240** in 80% yield; no 7-membered ring isomers were detected.

Selective removal of the TES group from **221** led to the $C_{26}$-hydroxy compound **241** in 93% yield (Scheme 38). Condensation with BOC-pipecolic acid gave the ester **242** in 98% yield. Conversion of the dithiane to the aldehyde **243** was accomplished via the dimethylacetal by a modification of Corey's method in 73% yield. An aldol reaction of the boron enolate of **226** with aldehyde **243** in toluene provided the desired adduct **244** in 88% yield. Adduct **244** was converted to the amino acid **245** by imide hydrolysis, silylation with concomitant deprotection of

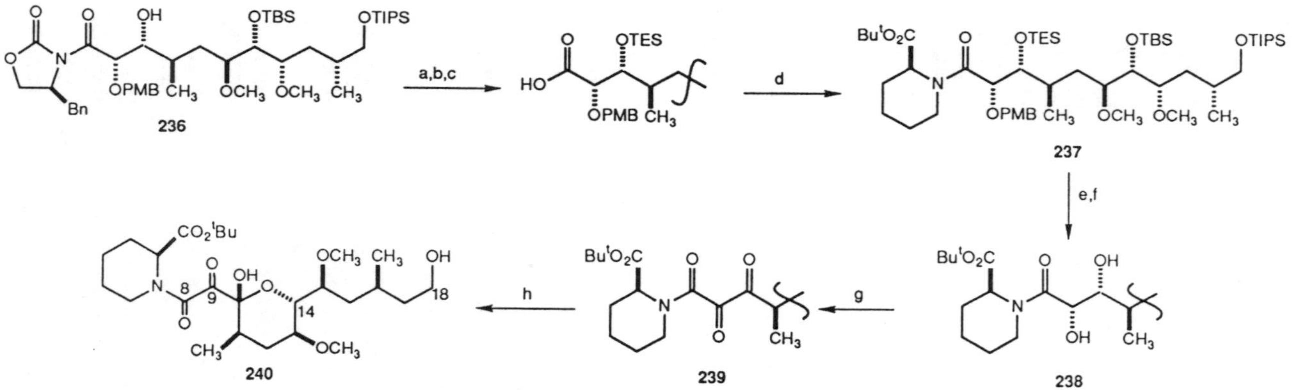

**Scheme 37.** (a) LiOH·H₂O, 30% H₂O₂, THF, H₂O, 0°C; (b) TESOTf (4 equiv), 2,6-lutidine (6 equiv), CH₂Cl₂, 0°C; (c) SiO₂; (d) *t*-butyl pipecolate, 2-chloro-1-methylpyridinium iodide, Et₃N, CH₂Cl₂; (e) DDQ, H₂O, CH₂Cl₂, 20°C; (f) TFA, H₂O, THF; (g) (COCl)₂, DMSO, CH₂Cl₂, −78°C; (h) aq. HF, MeCN.

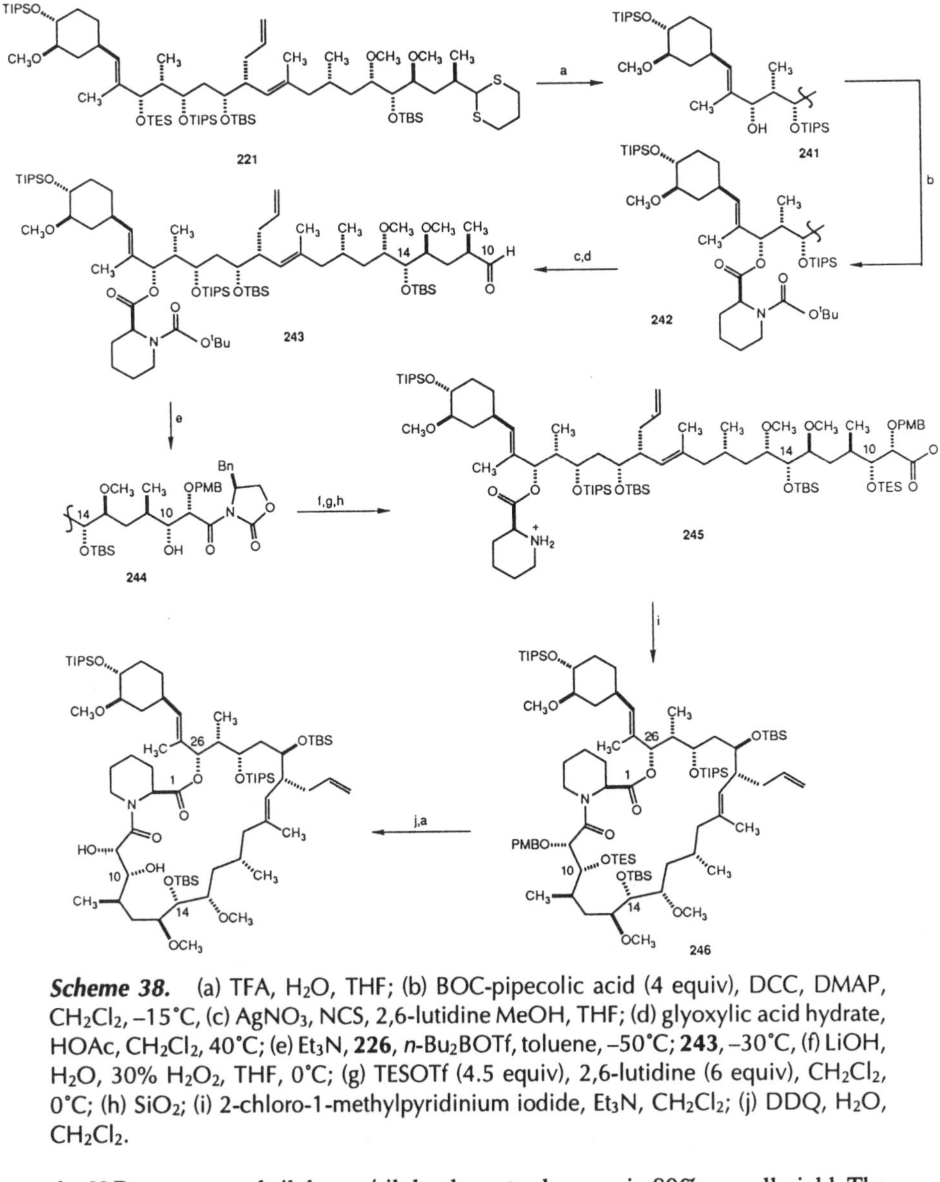

**Scheme 38.** (a) TFA, H$_2$O, THF; (b) BOC-pipecolic acid (4 equiv), DCC, DMAP, CH$_2$Cl$_2$, −15°C; (c) AgNO$_3$, NCS, 2,6-lutidine MeOH, THF; (d) glyoxylic acid hydrate, HOAc, CH$_2$Cl$_2$, 40°C; (e) Et$_3$N, **226**, n-Bu$_2$BOTf, toluene, −50°C; **243**, −30°C; (f) LiOH, H$_2$O, 30% H$_2$O$_2$, THF, 0°C; (g) TESOTf (4.5 equiv), 2,6-lutidine (6 equiv), CH$_2$Cl$_2$, 0°C; (h) SiO$_2$; (i) 2-chloro-1-methylpyridinium iodide, Et$_3$N, CH$_2$Cl$_2$; (j) DDQ, H$_2$O, CH$_2$Cl$_2$.

the N-Boc group and silyl ester/silylcarbamate cleavage in 80% overall yield. The unstable amino acid **245** was best treated immediately with Mukaiyama's chloropyridinium salt under high dilution conditions, which gave the macrolactam **246** in 85% yield. Deprotection of the PMB group with DDQ and treatment of the product with aqueous acid gave diol **247** in 74% overall yield.

Oxidation of diol **247** under double Swern conditions gave tricarbonyl **248** in 80% yield after one recycle of semi-oxidized products (Scheme 39). Interestingly,

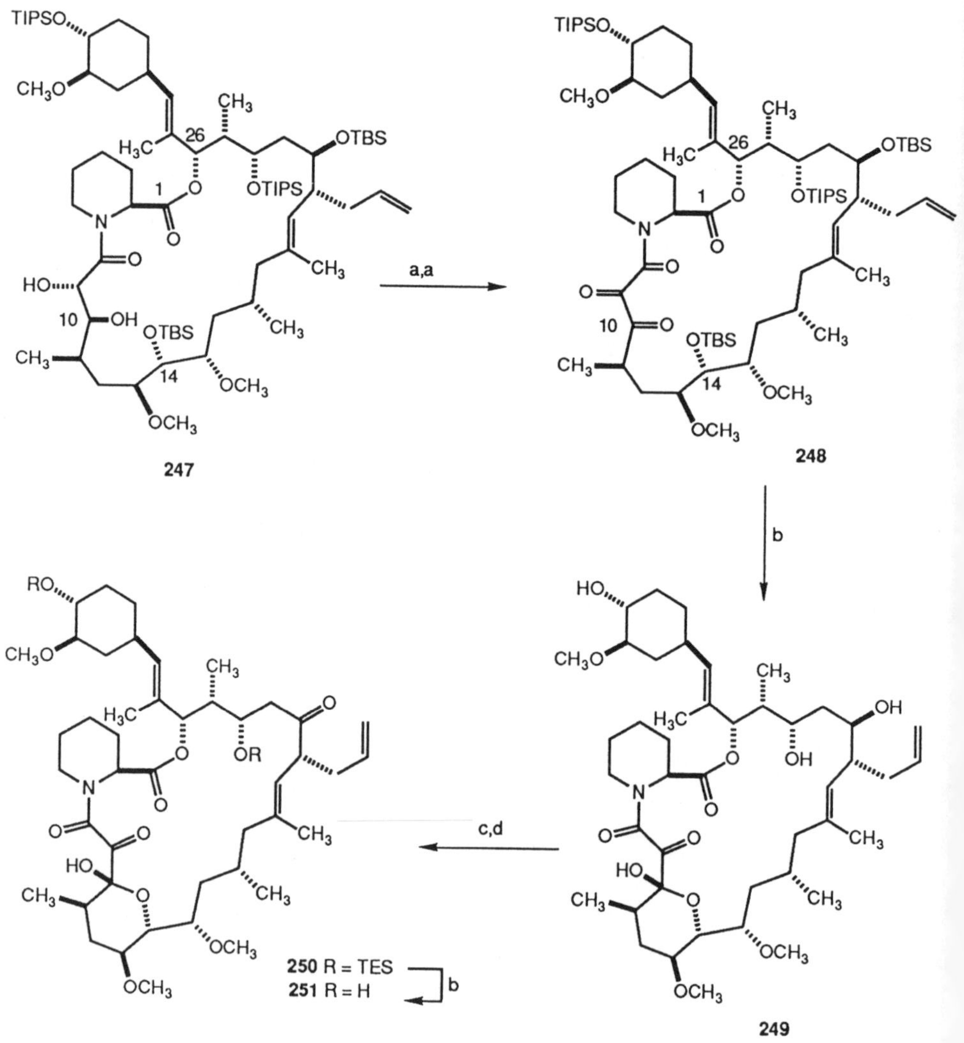

**Scheme 39.** (a) $(COCl)_2$, DMSO, $CH_2Cl_2$, $-78°C$; $Et_3N$; (b) aq. HF, MeCN; (c) TESCl, pyr, $0°C$; (d) Dess-Martin periodinane, pyridine, $CH_2Cl_2$.

treatment of the diol with the Dess-Martin periodinane resulted mostly in cleavage of the $C_9$–$C_{10}$ bond to give the dialdehyde. Treatment of the tricarbonyl **248** with aqueous $HF/CH_3CN$ gave 22-dihydro-FK-506 (**249**) in 82% yield. Treatment of **249** with TESCl/pyridine gave the desired 24,32-bis-TES-22-dihydro-FK-506 (**249**) in 70% yield. Oxidation with the Dess-Martin reagent now gave bis-TES-FK-506 (**250**) in 61% yield, and TES deprotection produced pure FK-506 (**251**) in

81% yield. This material was identical with natural material by $^1$H NMR, COSY-45, $^{13}$C NMR, optical rotation at 6 wavelengths, and TLC in solvent systems.

## F.  Chemistry of FK-506

Degradation work on FK-506 (**251**) was published[10] in 1987 by Tanaka and his co-workers at Fujisawa and Kyoto University. Ozonolysis of **251** followed by reductive and alkaline work-up gave cyclohexyl aldehyde (**252**, Fig. 4). The reduction of **252** produced an alcohol whose structure was confirmed by comparison of NMR data with that of an authentic sample. Hydrolysis of FK-506 with 1 $N$ NaOH produced α,β-unsaturated aldehyde **253**, while treatment with NaH under nonaqueous conditions gave dienol **254**. Ozonolysis of the dihydro derivative of FK-506, obtained by catalytic hydrogenation of the allyl group followed by reductive and alkaline workup, gave dihydropyrone **255**. This established the linkage of the original allyl group. From the above chemical evidence the partial structure **256** was deduced. The structure of the key α,β-diketo amide **257** was derived from the following data. Acid hydrolysis of FK-506 gave *L*-pipecolic acid. The direct ozonolysis of FK-506 also produced lactone **258** whose structure was established from NMR experiments. Baeyer-Villiger oxidation of **258** with MCPBA gave acetate **259**, confirming the structural assignment of **258**. After methylation and acetylation of alkaline treated FK-506, the reaction mixture was subjected to ozonolysis to give, after reductive workup, the intact tricarbonyl fragment **260**. The structural assignment of **260** was made by comparison of its NMR data with those of **258** and methyl *N*-acetylpipecolate. These data supported the partial structure **257**. The reasonable postulate that the partial structures **256** and **257** are connected through lactone and olefin linkages led to the full structure of FK-506.

Merck scientists later found that the structure of degradate **260**, proposed by Fujisawa, was incorrect, since treatment of FK-506 with aqueous hydroxide actually results in a benzilic acid rearrangement of the $C_8$–$C_{10}$ tricarbonyl portion of the molecule.[38] Exposure of bis-TIPS protected FK-506 ($C_{32}$-OH, $C_{24}$-OH) to lithium hydroxide followed by acidification produced hydroxy-acid (**261**, Scheme 40). The presence of a carboxylic acid group was indicated by the isolation of the methyl ester upon treatment with diazomethane. The characteristic $^{13}$C NMR resonance of $C_9$ at 196 ppm was absent, and a new carbonyl resonance appeared at 173.1 ppm. Additionally, the $C_{10}$ resonance at 97 ppm was shifted upfield to 82.4 ppm. Decarboxylation of **261** with lead tetraacetate followed by desilylation gave nor-$C_9$-FK-506 (**262**).

Repetition of the reported degradation protocol of FK-506 gave a methyl ester monoacetate species that was identical to the one described by Tanaka et al.[10] However, extensive spectroscopic investigations support a rearranged pyridooxazinedione structure **263** and not the originally proposed tricarbonyl

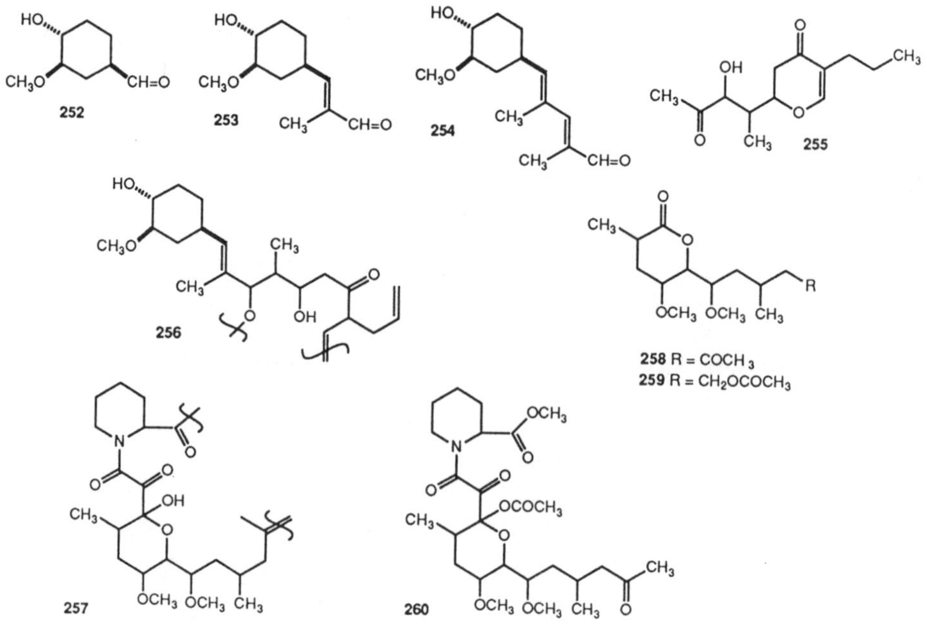

**Figure 4.** Degradation products of FK-506.

fragment **260**. Further confirmation of the inconsistency of structure **260** with spectral data was gained from synthetic fragment **240**.

Due to the extremely facile nature of the rearrangement, coupled with the intriguing possibility that the reactive tricarbonyl system is responsible for the remarkable biological activity of FK-506 (see Section 3), we investigated the nature of the bond reorganization.[39] Initial addition of LiOH to the highly electrophilic $C_9$ ketone was thought to allow ring-chain tautomerization of the hemiketal linkage, exposing the ketone at $C_{10}$ to produce **A** (Scheme 41). A 1,2-acyl shift of $C_8$ to $C_{10}$ would then afford the hydroxy-acid **B**. Alternatively, the alkoxide **A** generated at $C_{10}$ by LiOH would suffer a 1,2-ketol shift of $C_{11}$ to $C_9$, thereby affording an intermediate 7-membered lactone, which would presumably be saponified under the reaction conditions to **B**. If the rearrangement proceeds through the acyl shift mechanism "**a**", then decarboxylation of the intermediate hydroxy-acid **B** would result in the loss of the original $C_9$ carbonyl group as $CO_2$ to afford **C**. In order to differentiate these possibilities, the FK-506 derivative **251*** [13]C-labelled at $C_9$ was synthesized. Treatment of an admixture of 99 atom % **251*** (bis TES protected at $C_{32}$–OH and $C_{24}$–OH) and unlabelled **251** with aqueous LiOH gave, after workup, the crude rearranged hydroxy-acid. The isomer of **261*** bearing the label at the quaternary carbon could not be observed at this point. Decarboxylation gave nor-$C_9$ **262** bearing a small amount of the original label at the hemiketal carbon. Further

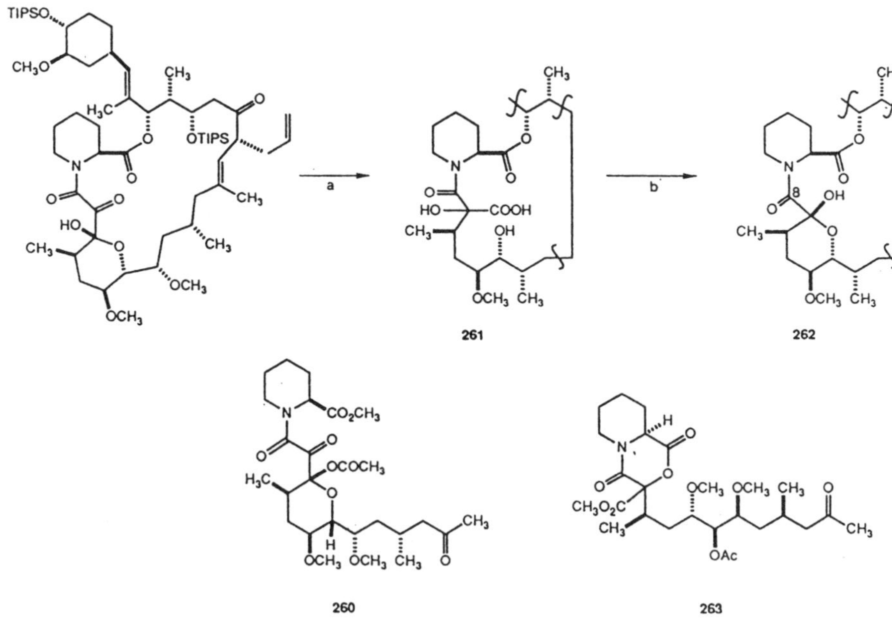

**Scheme 40.** (a) LiOH (1.03 equiv), H$_2$O, THF, 0–25°C; (b) Pb(OAc)$_4$, benzene, 25°C.

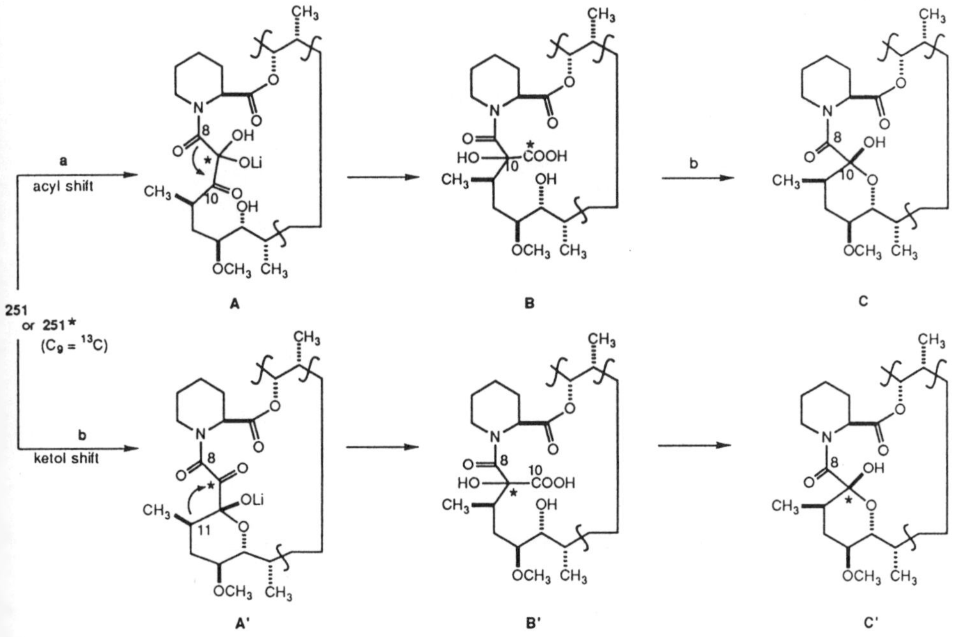

**Scheme 41.** (a) LiOH, H$_2$O, THF, 0°C; (b) Pb(OAc)$_4$, benzene, 25°C.

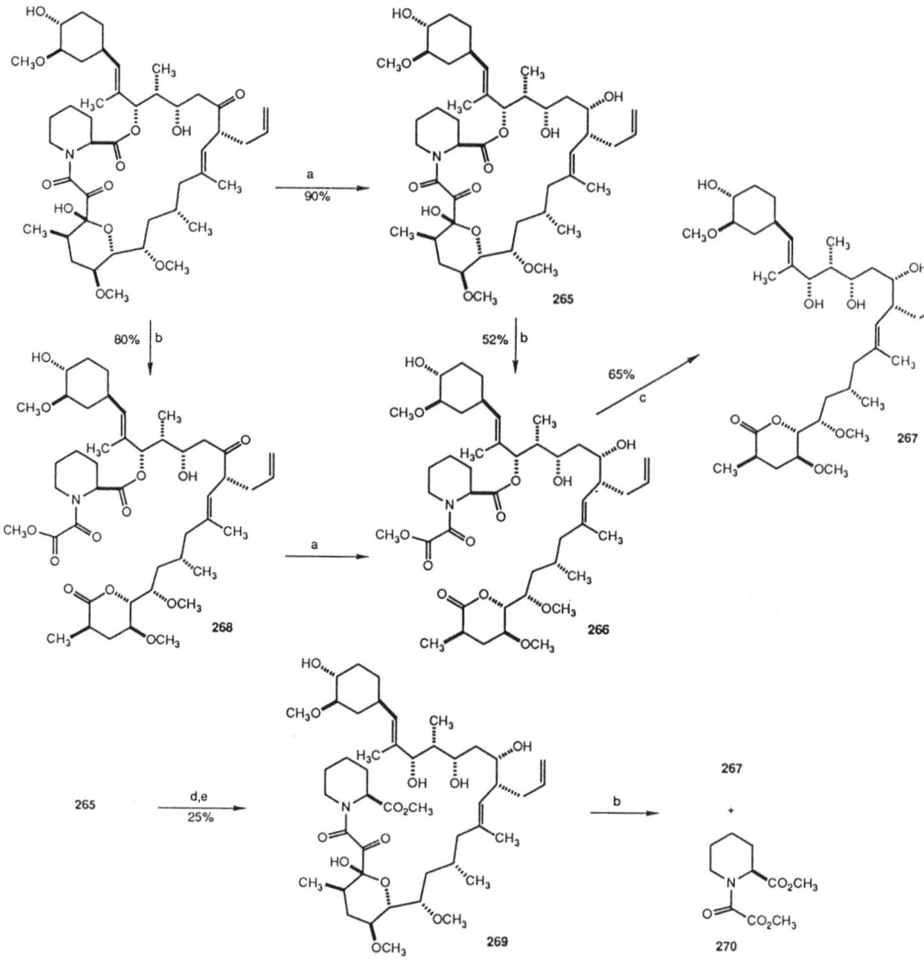

**Scheme 42.** (a) Li(s-Bu)$_3$BH, THF, −78°C; (b) Pb(OAc)$_4$, MeOH, 0–25°C; (c) LiOH (10 equiv), H$_2$O, THF; 10% HCl, pH =1; (d) LiOH (1.1 equiv), H$_2$O, THF; Dowex 15W, H$^+$ form; (e) CH$_2$N$_2$, ether, 0°C.

[13]C NMR work indicated that approximately 3% of the original label was present at the hemiketal carbon, implying 3% rearrangement by the ketol shift mechanism "**b**". Thus the major pathway of the rearrangement appears to be through the expected acyl shift mechanism "**a**".

Danishefsky found[40] that reduction of FK-506 with Li(s-Bu)$_3$BH occurred selectively at C$_{22}$ to afford a single alcohol **265** in 90% yield (Scheme 42). Examination of molecular models suggested that attack of hydride would occur

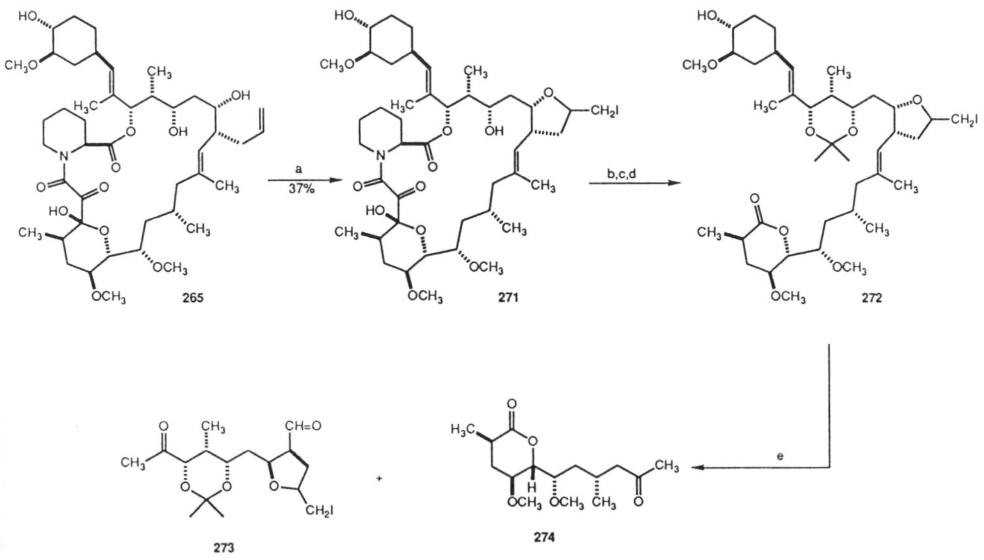

**Scheme 43.** (a) NIS (1.5–2.0 equiv), $CH_2Cl_2$, 2–4 days; (b) Pb(OAc)$_4$, MeOH, 52%; (c) LiOH, MeOH-THF-H$_2$O (2:3:1), 89%; (d) acetone, CuSO$_4$, CSA, 76%; (e) O$_3$, $CH_2Cl_2$, −78°C; Me$_2$S.

from the β-face to provide a product with the 22-*S* stereochemistry. Treatment of **265** with lead tetraacetate in methanol gave the amide diester **266**. Further hydrolysis produced the lactone tetraol **267** in 65% yield. Treatment of FK-506 with lead tetraacetate in methanol gave **269**, which, upon reduction with Li(*s*-Bu)$_3$BH, gave **266** as the sole product. Exposure of **265** to LiOH followed by acidification and esterification with diazomethane gave **269**, the 22-dihydro-*seco* acid corresponding FK-506. Treatment of **269** with lead tetraacetate in methanol afforded **267** and **270**.

Treatment of **265** with *N*-iodosuccinimide (NIS) slowly provided a single iodoether **271** in 37% yield with recovery (36%) of **265** (Scheme 43). Oxidation with lead tetraacetate, hydrolysis with LiOH, and acetalyzation of C$_{24}$ and C$_{26}$ hydroxyls afforded lactone **272**. Ozonolysis of **272** gave **273** and **274**.

# III.  BIOLOGY OF FK-506

## A.  In Vivo Pharmacology—Animal Studies

*Models of Transplantation*

FK-506 has been studied in a wide variety of animal models of experimental tissue and organ transplantation. In general, FK-506 is effective in blocking allograft rejection in a number of species and can be used at doses 10-100 times lower than the dose of CsA required in the same models.[41-46] In rats, FK-506 prolongs cardiac allograft survival at doses as low as 0.1 mg/kg/day when administered intramuscularly. As with CsA, short courses of treatment with FK-506 starting at the time of cardiac grafting can lead to indefinite allograft acceptance in this species, suggesting the induction of antigen-specific unresponsiveness (tolerance) in this model. In addition, FK-506 could inhibit ongoing cardiac allograft rejection when therapy was delayed until 5 days post-transplant, a phenomenon not observed in animals treated with CsA. In a similar fashion, FK-506 was effective in the treatment of small intestine allotransplantation and graft-versus-host disease in the rat, whereas CsA was not.[47] Tsuchimoto et al.[48] have reported that a 14-day treatment with FK-506 is effective in inducing indefinite graft survival in a model of rat orthotopic liver transplantation. In contrast to the other models discussed, the long-term survival of skin allografts in the rat appears to require maintenance therapy with FK-506.[41]

Transplantation experiments in the dog have been somewhat more controversial, since some groups suggest that dose-limiting toxicities (see below) compromise the effectiveness of therapy,[49] while other groups report much greater success.[43,45] Ochai et al.[43] reported that a dose of 0.16 mg/kg/day intramuscularly or 1 mg/kg/day orally was effective in prolonging kidney graft survival. Although side effects were observed, the authors concluded that FK-506 had an acceptable therapeutic index in this species. In contrast, dogs treated with similar doses of FK-506 by Thiru et al.[49] experienced severe side effects that prevented completion of the study.

Only modest prolongation of kidney allograft survival was noted in cynomolgus monkeys treated orally with 0.3 mg/kg/day of FK-506.[45] However, in baboons FK-506 given orally at up to 18 mg/kg/day for 90 days, produced significant prolongation of kidney allografts.[46,50] Although the doses required for a good therapeutic effect were 10 times greater in baboons than in dogs, FK-506 appeared to be relatively nontoxic in subhuman primates.

## Models of Autoimmunity

With respect to autoimmune diseases, researchers have observed beneficial effects of FK-506 have been observed in rat models of experimental autoimmune encephalomyelitis, experimental autoimmune uveoretinitis and collagen-induced arthritis.[51-53] Consistent with results previously obtained with CsA, FK-506 was only effective in these models when given during induction of the disease, but had no effect when given "therapeutically" after the onset of the disease. In two strains of spontaneously autoimmune mice, MRL/lpr and (NZB x NZW)$F_1$ mice, FK-506 prolongs lifespan, reduces proteinuria, and prevents the progression of nephropathy.[54,55] However, no differences in the levels of anti-dsDNA antibodies between treated and control animals were detected.

## Pharmacokinetics

Using a sensitive immunoassay, the pharmacokinetics of FK-506 have been measured in rats, dogs and baboons.[56] FK-506 is rapidly absorbed following oral administration, and when given in a solid dispersion formulation, bioavailability ranges from 9–30% in the dog and rat, respectively. It is extensively distributed in various tissues, and concentrations exceeding plasma levels are observed in lung, spleen, heart and kidney. The half-life of FK-506 in various species is approximately 9 hr. The drug is completely metabolized prior to elimination, with only a small proportion of the compound excreted in the urine. While the metabolic pathways of FK-506 have not been fully characterized, it does decrease the activity of hepatic drug metabolizing enzymes.

## Toxicity

The toxicity of FK-506 in dogs has, until quite recently, been the major impediment to the clinical development of this compound. While there are only modest clinical or pathological changes in rodents or primates treated with prolonged courses of FK-506 at therapeutic doses, dogs experience severe side effects at these dosage levels.[43,45,49] The most dramatic side effects in dogs relate to the gastrointestinal tract where emesis, weight loss and intussusception have been frequently observed. In addition, myocardial degeneration and arteritis, pancreatic acinar cell degeneration and degranulation, renal tubular cell vacuolation and centilobular hepatocyte swelling have been reported.[45] It is interesting to note that although intussusception has been considered a species-specific phenomenon, similar pathological findings have been noted in rats administered FK-506 intravenously at 18 mg/kg.[57]

In rats and baboons given FK-506 orally for 13 weeks, the principal histopathologic changes were seen in the kidneys (tubule basophilia and areas of interstitial inflammatory cells) and pancreata (islet cell vacuolation and angiectasis), which were confirmed by the serum biochemistries and urinalysis.[57] No evidence of vasculitis was noted, even following four week intravenous administration of the drug. Other signs of toxicity noted in baboons included lethargy, vomiting, weight loss and anorexia.

## B.   In Vivo Pharmacology—Human Studies

## Transplantation

Clinical trials began in February 1989 at the University of Pittsburgh, with a focus primarily on liver transplantation.[58] The rationale for this decision was twofold: first, the severe toxicity observed in dogs was not demonstrated in primates and, therefore, may not translate to man; second, the failure rate of liver allografts remains unacceptably high with conventional immunosuppressive therapy. The Pittsburgh group has reported remarkable success in their early trials.[58] FK-506 was used first in recipients under conventional immunosuppression who had rejection, nephrotoxicity, or both. This salvage therapy was successful in seven of the ten attempts. Subsequently, patients who underwent fresh orthotopic liver transplantation were given FK-506 plus low-dose steroids from the outset. None of these first six patients had rejection, and serious side effects were not encountered. Thus, it appears likely that clinical evaluation of FK-506 will be extended to other areas of organ transplantation, and it may be a significant improvement over CsA.

### Other Clinical Studies

Two reports from the Pittsburgh group illustrate the potential of FK-506 in other disease processes. One patient with CsA-induced hemolytic uremic syndrome was treated effectively with FK-506.[59] FK-506 was also efficacious in a child with steroid-resistant focal sclerosing glomerulonephritis and an adult with mesangial proliferative glomerulonephritis.[60] Since CsA has been shown to be beneficial in a number of autoimmune diseases in man (e.g., psoriasis, type 1 diabetes, uveitis),[2] investigation of FK-506 in these clinical situations will undoubtedly be initiated soon.

### Pharmacokinetics

The pharmacokinetic properties of FK-506 in man are similar to those observed in experimental animals.[56] The drug is incompletely absorbed after oral administration, is extensively distributed in the body and is eliminated by metabolism. Half-life of FK-506 in man averages 8.7 hr FK-506 increases CsA plasma concentrations and has the potential for inhibition of metabolism of other drugs.

### Toxicity

FK-506 administered at 0.15 mg/kg intravenously and orally has been remarkably free of serious toxicities in the early clinical trials of liver transplantation. Side effects observed in a significant number of patients treated with intravenous FK-506 include headache, nausea, vomiting, diarrhea, sensation of warmth/hyperesthesia,tremors, itching and flushing.[61] The incidence and severity of symptoms decreased markedly when the patients were switched to oral therapy. Evaluation of metabolic and hemodynamic parameters found to be affected by FK-506 in animal toxicology studies revealed no alterations in glucose metabolism, pancreatic function or cardiac performance.[62-64] Although moderate hyperkalemia and elevated serum creatinine is seen in FK-506-treated subjects, the renal toxicity of FK-506 appears to be less severe than that of CsA.[65]

## C.  Effects of FK-506 on Lymphocyte Activation

### Lymphokine Gene Expression

T-cell activation occurs following a specific recognition event via the T-cell receptor/CD-3 complex combined with additional signals provided by accessory cells and/or cytokines (for a recent review, see Krensky et al.[66]) A complex series of biochemical events ensues, including phosphoinositide turnover, activation of protein kinase C, a rise in intracellular $Ca^{++}$ and phosphorylation of proteins on serine, threonine and tyrosine residues. These processes lead to the coordinate expression of set of gene products critical for lymphocyte growth and differentiation. Although there is no evidence that CsA or FK-506 affect the early biochemical events associated with T-cell receptor recognition,[67-69] a great deal of attention has been focused on their effects on lymphocyte gene expression.

It is well known that one of the principal modes of CsA's immunosuppressive action is the inhibition of the expression of a discrete set of lymphokine genes at the level of transcriptional initiation. Recently, Tocci et al.[70] documented that FK-506 inhibits induction of the same set of lymphokine genes, including IL-2, IL-3, IL-4, GM-CSF, TNFα, and interferon-γ. Furthermore, nuclear run-off transcription studies reveal that both FK-506 and CsA directly inhibit transcription of the IL-2 gene. A more detailed molecular picture has emerged from the work of Crabtree and co-workers, who have identified several regions of the IL-2 promoter critical for IL-2 gene expression.[71] CsA was found to specifically inhibit the appearance of DNA binding activity (as assayed by gel retardation) of one such region, termed NF-AT, and abolished the ability of the NF-AT binding site to activate a linked promoter in transfected mitogen-stimulated T lymphocytes.[72] Recent experiments with FK-506 have revealed remarkably similar results (G. Crabtree, personal communication). Thus, at every level at which lymphokine gene regulation has been examined, the effects of FK-506 and CsA are virtually indistinguishable.

There are several experimental systems that can be used to document the selectivity of FK-506 and CsA with respect to gene transcription. First, FK-506, like CsA, does not affect expression of the mRNA's for IL-1α or IL-1β in LPS-activated human monocytes or of stromelysin, collagenase or TIMP in IL-1-treated synovial fibroblasts.[70] Second, transcription of other genes expressed following T-cell activation, such as c-fos, the transferrin receptor and, in particular, the IL-2 receptor α chain, are not inhibited by these compounds.[70] This selectivity is also reflected by the fact that inhibition of T-cell proliferation in cultures treated with CsA or FK-506 can be reversed by addition of recombinant IL-2.[40] The finding suggests that, with the exception of IL-2 or similar growth factors, all other gene products and pathways remain intact in cells treated with these immunosuppressive agents.

## Cellular Site of Action of FK-506

The data reviewed in the preceding section support the widely held notion that the principal site of action of FK-506 and CsA is at the level of lymphokine gene regulation. While it is reasonable to conclude that inhibition of lymphokine production is the critical event leading to these agents' immunosuppressive effects in vitro, it is unlikely they act directly at the transcriptional level. Instead, it is more appropriate to consider these compounds as selective for certain types of signal transduction pathways. Two areas of investigation strengthen this view: (1) the lack of effect of FK-506 and CsA on certain lymphocyte activation pathways; and (2) the effect of the immunosuppressants on processes that do not involve transcriptional regulation.

Several experimental systems have been identified in which the ability of FK-506 and CsA to inhibit lymphocyte signal transduction events is dependent on the mode of cellular activation. In general, only activation pathways that cause a measurable rise in intracellular $Ca^{++}$ were sensitive to CsA or FK-506. For example, CsA and FK-506 inhibit IL-2 production when human lymphocytes are activated via the CD3/T-cell receptor complex but not when the same cells are triggered via the CD28 pathway.[73,74] Thus, transcription of gene products that are typically sensitive to the immunosuppressive agents can be rendered resistant when an alternative activation pathway is utilized.

A second line of evidence that supports the hypothesis that FK-506 and CsA are selective for a subset of $Ca^{++}$-associated signal transduction pathway(s) in lymphocytes is their effect on events that are unlikely to involve transcriptional regulation. A good example of this phenomenon in T cells is the ability of CsA to inhibit granule exocytosis of cytotoxic lymphocytes triggered by antibodies to the T-cell receptor/CD-3 complex.[75,76] A similar observation has recently been made in neutrophils, where degranulation induced by a $Ca^{++}$ ionophore is blocked by FK-506 and CsA (M. J. Forrest and N. H. S., unpublished observations). Thus, the data described in this section suggest that the site of action of these immunosuppressive agents is likely to be distal to receptor/membrane-associated events but proximal to the transcriptional factors per se.

## D.   Biochemical Site of Action of FK-506

## Identification of a Major Cytosolic Receptor

A number of proteins have been hypothesized to mediate the immunosuppressive activity of CsA. Cyclophilin, discovered by Handschumaker et al.[77] in 1984, is a ubiquitous and abundant cytosolic protein that is highly conserved among eukaryotic organisms,[78] and is responsible for the uptake of CsA into the cell. Several observations have strengthened the hypothesis that cyclophilin is involved in CsA's mechanism of action. First, the specificity of cyclophilin binding was examined with a limited set of cyclosporin analogues, and, within this narrow series, immunosuppressive activity was found to correlate with cyclophilin binding.[79] Second, cyclophilin was recently discovered to catalyze the *cis-trans* isomerization of peptidyl-proline bonds, and CsA was shown to inhibit this enzymatic activity.[80,81] The discovery that the *Drosophila ninaA* mutant, which is defective in the conversion of opsin to rhodopsin, has a mutation in a cyclophilin-like gene[82,83] illustrates how this molecule may be involved in a signal transduction pathway.

The remarkable similarities between FK-506 and CsA carry over to the biochemical level. Siekierka et al.[84] first showed that a radiolabeled analogue of FK-506 readily accumulated in the cytosol of the T-cell lymphoma line JURKAT in a temperature-dependent manner. The protein responsible for this accumulation was similar to cyclophilin in cytosolic location and abundance, but was slightly smaller in molecular weight (11-12 kD versus 18 kD for cyclophilin), was heat stable and did not bind CsA. Subsequently, both Siekierka et al.[85] and Harding et al.[86] purified this protein to homogeneity. Like cyclophilin, the FK-506 binding protein (FKBP) possesses peptidyl-prolyl *cis-trans* isomerase (PPIase) activity, which is inhibited by its own immunosuppressive ligand but not by CsA. Furthermore, FK-506 binds to FKBP with a very potent $K_d$ of 0.4–0.8 nM, as estimated from Scatchard analysis of direct binding or competition binding data.[85] The affinity measurements are consistent with FK-506's potency in cellular assays and support FKBP as a mediator of the immunosuppressive activity of FK-506.

Studies on the enzymatic mechanism of proline isomerization by FKBP and cyclophilin have been recently carried out.[87,88] Two mechanistic alternatives have been proposed for this class of enzymes: (1) nucleophilic addition of an active site thiol to the carbonyl carbon of the -Xaa-Pro- to form a tetrahedral, hemithioorthoamide intermediate; or (2) catalysis by distortion in which the PPIase binds and stabilizes a transition state that is characterized by partial rotation about the C–N amide bond. Recent experiments using determinations of secondary deuterium isotope effect[87] and [13]C NMR-based investigations[88] strongly support the latter mechanism in which the enzymes use free energy released from favorable, non-covalent interactions with the substrate to stabilize the transition state. Indeed, it has been hypothesized that the α-keto amide of FK-506 serves as a surrogate for the twisted amide of a bound peptide substrate.[88] This suggestion has gained support from studies comparing the substrate specificities of FKBP and cyclophilin. Using substrates of the general structure Suc-Ala-Xaa-Pro-Phe-*p*-nitroanilide, cyclophilin-catalyzed isomerization showed little dependence on Xaa, while FKBP enzymatic rates varied markedly. FKBP showed a strong preference for Leu or Phe in the Xaa position.[89,90] These observations are consistent with the notion that cyclophilin and FKBP are members of a family of PPIases that may possess distinct substrate specificities with the cell and may play diverse physiologic roles.

The finding that FKBP, like cyclophilin, possesses PPIase activity provides evidence that these enzymes play a critical role in lymphocyte signal transduction. While the molecular mechanism associated with such isomerases in lymphocyte activation remains to be discovered, the fact that both FKBP and cyclophilin are PPIases (inhibited by their respective immunosuppressive ligands) must be viewed as more than a fortuitous coincidence. We and others have speculated that FKBP and cyclophilin catalyze a critical proline isomerization step(s) in a component(s) of the $Ca^{++}$-associated signal transduction pathways described in the preceding section. In this model, the activity of signal transduction elements, such as transcriptional regulatory factors, protein kinases or ion channels, could be "conformationally" regulated by peptidyl-prolyl bond isomerization. FK-506 and CsA, as potent inhibitors of their respective PPIases, would selectively block events catalyzed by these proteins.

## IV.  CONCLUDING OBSERVATIONS

Since the discovery of FK-506 in 1985, information concerning its mechanism of action at the cellular, molecular and biochemical levels has grown enormously. FK-506 has been shown to be remarkably similar to CsA in its site of action with respect to its ability to inhibit lymphokine production at the level of transcriptional initiation and in its selectivity for a subset of signal transduction pathways that are associated with a rise in intracellular $Ca^{++}$. In addition, FK-506 and CsA interact with distinct, abundant, highly-conserved, cytosolic proteins, both of which possess related PPIase activities. It is anticipated that, through medicinal chemistry, these insights will lead to the rational design of novel immunosuppressive drugs that would extend therapeutic potential of such agents.

## ABBREVIATIONS

| | |
|---|---|
| AIBN | Azobisisobutyronitrile |
| BINAP | 2,2'-Bis(diphenylphosphino)-1,1'-binaphthyl |
| Bn | Benzyl |
| BX | *t*-Butoxycarbonyl |
| Bz | Benzoyl |
| CSA | Camphorsulfonic acid |
| DBU | 1,8-Diazabicyclo[5.4.0]undec-7-ene |
| DCC | 1,3-Dicyclohexylcarbodiimide |
| DDQ | 2,3-Dichloro-5,6-dicyano-1,4-benzoquinone |
| DET | Diethyl tartrate |

DIBAL-H   Diisobutylaluminum hydride
DIEA      N,N-Diisopropylethylamine
DIPHOS-4  1,4-Bis(Diphenylphosphino)butane
DIPT      Diisopropyl tartrate
DMAP      4-Dimethylaminopyridine
DME       Ethylene glycol dimethyl ether
DMPU      1,3-Dimethyl-3,4,5,6-tetrahydro-2(1H)-pyrimidinone
DMS       Dimethyl sulfide
HMPA      Hexamethylphosphoramide
KHMDS     Potassium hexamethyldisilazide
LDA       Lithium diisopropylamide
MCPBA     3-Chloroperoxybenzoic acid
MEM       2-Methoxyethoxymethyl
MOM       Methoxymethyl
MOMCl     Chloromethylmethyl ether
NBD       Bicyclo[2.2.1]hepta-2,5-diene
NBS       N-Bromosuccinimide
NCS       N-Chlorosuccinimide
NIS       N-Iodosuccinimide
PDC       Pyridinium dichromate
Piv       Pivaloyl, Trimethylacetyl
PMB       *p*-Methoxybenzyl
PMP       *p*-Methoxyphenyl
PPTS      Pyridinium p-toluensulfonate
TBAF      Tetrabutylammonium fluoride
TBAI      Tetrabutylammonium iodide
TBDPS     *t*-Butyldiphenylsilyl
TBS       *t*-Butyldimethylsilyl
TEA       Triethylamine
TES       Triethylsilyl
TFA       Trifluoroacetic acid
TFAA      Trifluoroacetic anhydride
TIPS      Triisopropylsilyl
TMEDA     N,N,N',N'-Tetramethylethylenediamine
TMS       Trimethylsilyl
TMSI      Iodotrimethylsilane

# REFERENCES

1. Cohen, D.J.; Loertscher, R.; Rubin, M.; Tilney, N.L.; Carpenter, C.B.; Strom, T.B. *Ann. Int. Medicine* **1984**, *101*, 667.
2. Bach, J.-F. *Transplant. Proc.* **1989**, *21 (supp 1)*, 97.
3. Goto, T.; Kino, T.; Hatanaka, H.; Nishiyama, M.; Okuhara, M.; Kohsaka, M.; Aoki, H.; Imanaka,

H. *Transplant. Proc.* **1987**, *19*, 4; Kino, T.; Hatanaka, H.; Goto, T.; Okuhara, M. J. *Agricultural Chem. Soc. Japan* **1989**, *63*, 224; Pento, J.T. *Drugs of the Future* **1989**, *14*, 751; Tanaka, H.; Kuroda, A.; Marusawa, H.; Hatanaka, H.; Kino, T.; Goto, T.; Hashimoto, M.; Taga, T. *J. Am. Chem. Soc.* **1987**, *109*, 5031.

4. Kino, T.; Hatanaka, H.; Miyata, S.; Inamura, N.; Nishiyama, M.; Yajima, T.; Goto, T.; Okuhara, M.; Kohsaka, M.; Aoki, H.; Ochiai, T. *J. Antibiot. (Tokyo)* **1987**, *40*, 1256.

5. Sawada, S.; Suzuki, G.; Kawase, Y.; Takaku, F. *J. Immunol.* **1987**, *139*, 1797.

6. Woo, J.; Stephen, M.; Thomson, A.W. *Immunology* **1988**, *65*, 153.

7. Kay, J.E.; Doe, S.E.; Benzie, C.R. *Cell. Immunol.* **1989**, *124*, 175.

8. Dumont, F.J.; Staruch, M.J.; Koprak, S.L.; Melino, M.R.; Sigal, N.H. *J. Immunol.* **1990**, *144*, 251.

9. Kino, T.; Hatanaka, H.; Hashimoto, M.; Nishiyama, M.; Goto, T.; Okuhara, M.; Kohsaka, M.; Aoki, H.; Imanaka, H. *J. Antibiot.* **1987**, *40*, 1249.

10. Tanaka, H.; Kuroda, A.; Marusawa, H.; Hatanaka, H.; Kino, T.; Goto, T.; Hashimoto, M. *J. Am. Chem. Soc.* **1987**, *109*, 5031.

11. Tanaka, H.; Kuroda, A.; Marusawa, H.; Hashimoto, M.; Hatanaka, H.; Kino, T.; Goto, T.; Okuhara, M. *Transplant. Proc.* **1987**, *19*, 11.

12. Taga, T.; Tanaka, H.; Goto, T.; Tada, S. **1987**, *Acta Cryst.* **1987**, *43*, 571.

13. Hatanaka, H.; Kino, T.; Miyata, S.; Inamura, N.; Kuroda, A.; Goto, T.; Tanaka, H.; Okuhara, M. *J. Antibiot.* **1988**, *41*, 1592.

14. Hatanaka, H.; Iwami, M.; Kino, T.; Goto, T.; Okuhara, M. *J. Antibiot.* **1988**, *41*, 1586.

15. Hatanaka, H.; Kino, T.; Asano, M.; Goto, T.; Tanaka, H.; Okuhara, M. *J. Antibiot.* **1989**, *42*, 620.

16. Williams, D.R.; Benbow, J.W. *J. Org. Chem.* **1988**, *53*, 4643.

17. Kocienski, P.; Stocks, M. *Tetrahedron Lett.* **1988**, *29*, 4481.

18. Egbertson, M.; Danishefsky, S.J. *J. Org. Chem.* **1989**, *54*, 11.

19. Villalobos, A.; Danishefsky, S.J. *J. Org. Chem.* **1989**, *54*, 12.

20. Ireland, R.E.; Wipf, P. *Tetrahedron Lett.* **1989**, *30*, 919.

21. Sitahama, H.; Mikami, A.; Morimoto, Y.; Japanese Chem. Society Meeting, 21L39, April 1989.

22. Wasserman, H.H.; Rotello, V.M.; Williams, D.R.; Benbow, J.W. *J. Org. Chem.* **1989**, *54*, 2785.

23. Rao, A.V.R.; Chakraborty, T.K.; Reddy, K.L. *Tetrahedron Lett.* **1990**, *31*, 1439.

24. Askin, D.; Volante, R.P.; Reamer, R.A.; Ryan, K.M.; Shinkai, I. *Tetrahedron Lett.* **1988**, *29*, 227.

25. Schreiber, S.L.; Sammakia, T.; Uehling, D.E. *J. Org. Chem.* **1989**, *54*, 15.

26. Askin, D.; Volante, R.P.; Ryan, K.M.; Reamer, R.A.; Shinkai, I. *Tetrahedron Lett.* **1988**, *29*, 4245.

27. Smith III, A.B.; Hale, K.J. *Tetrahedron Lett.* **1989**, *30*, 1037.

28. Wang, Z. *Tetrahedron Lett.* **1989**, *30*, 6611.

29. Stocks, M.; Kocienski, P. *Tetrahedron Lett.* **1990**, *31*, 1640.

30. Mills, S.; Desmond, R.; Reamer, R.A.; Volante, R.P.; Shinkai, I. *Tetrahedron Lett.* **1988**, *29*, 281.

31. Desmond, R.; Mills, S.G.; Volante, R.P.; Shinkai, I. *Tetrahedron Lett.* **1988**, *29*, 3895.

32. Schreiber, S.L.; Smith, D.B. *J. Org. Chem.* **1989**, *54*, 9.

33. Jones, A.B.; Yamaguchi, M.; Patten, A.; Danishefsky, S.J.; Ragan, J.A.; Smith, D.B.; Schreiber, S.L. *J. Org. Chem.* **1989**, *54*, 19.

34. Corey, E.J.; Huang, H-C. *Tetrahedron Lett.* **1989**, *30*, 5235.

35. Smith III, A.B.; Hale, K.J.; Laakso, L.M.; Chen, K.; Riera, A. *Tetrahedron Lett.* **1989**, *30*, 6963.

36. Rao, A.V.R.; Chakraborty, T.K.; Purandare, A.V. *Tetrahedron Lett.* **1990**, *31*, 1443.

37. Jones, T.K.; Mills, S.G.; Reamer, R.A.; Askin, D.; Desmond, R.; Volante, R.P.; Shinkai, I. *J. Amer. Chem. Soc.* **1989**, *111*, 1157. Recent total synthesis (a) Jones, A.B.; Villalobos, A.; Linde, R.G.; Danishefsky, S.J. *J. Org. Chem.* **1990**, *55*, 2786. (b) Nakasuka, M.; Ragan, J.A.; Sammakia, T.; Smith, D.B.; Uehling, D.E.; Schreiber, S.L. *J. Am. Chem. Soc.* **1990**, *112*, 5583

38. Askin, D.; Reamer, R.A.; Jones, T.K.; Volante, R.P.; Shinkai, I. *Tetrahedron Lett.* **1989**, *30*, 671.

39. Askin, D.; Reamer, R.A.; Joe, D.; Volante, R.P.; Shinkai, I.; *Tetrahedron Lett.* **1989**, *30*, 6121.

40. Coleman, R.S.; Danishefsky, S.J. *Heterocycles* **1989**, *28*, 157.
41. Inamura, N.; Nakahara, K.; Kino, T.; Goto, T.; Aoki, H.; Yamaguchi, I.; Kohsaka, M.; Ochiai, T. *Transplantation* **1988**, *45*, 206.
42. Nakajima, K.; Sakamoto, K.; Ochiai, T.; Nagata, M.; Asano, T.; Isono, K. *Transplantation* **1988**, *45*, 1146.
43. Ochiai, T.; Nagata, M.; Nakajima, K.; Suzuki, T.; Sakamoto, K.; Enomoto, K.; Gunji, Y.; Uematsu, T.; Goto, T.; Hori, S.; Kenmochi, T.; Nakagouri, T.; Asano, K.; Isono, K.; Hamaguchi, H.; Tsuchida, K.; Nakahara, N.; Inamura, N.; Goto, T. *Transplantation* **1987**, *44*, 729.
44. Ochiai, T.; Sakamoto, K.; Gunji, Y.; Hamaguchi, K.; Isegawa, N.; Suzuki, T.; Shimada, H.; Hayashi, H.; Yasumoto, A.; Asano, T.; Isono, K. *Transplantation* **1989**, *48*, 193.
45. Todo, S.; Ueda, Y.; Demetris, J.A.; Imventarza, O.; Nalesnik, M.; Venkataramanan, R.; Makowka, L.; Starzl, T.E. *Surgery* **1988**, *104*, 239.
46. Todo, S.; Demetris, A.; Ueda, Y.; Imventarza, O.; Cadoff, E.; Zeevi, A.; Starzl, T.E. *Surgery* **1989**, *106*, 444.
47. Hoffman, A.L.; Makowka, L.; Banner, B.; Cai, X.; Cramer, D.V.; Pascualone, A.; Todo, S.; Starzl, T.E. *Transplantation* **1990**, *49*, 483.
48. Tsuchimoto, S.; Kusumoto, K.; Nakajima, Y.; Kakita, A.; Uchino, J.; Natori, T.; Aizawa, M. *Transplant. Proc.* **1989**, *21*, 1064.
49. Thiru, S.; Collier, D.; Calne, R.Y. *Transplant. Proc.* **19 (suppl. 6)**, 98 (1987).
50. Imventarza, O.; Todo, S.; Eiras, G.; Ueda, Y.; Furukawa, H.; Wu, Y.M.; Zhu, Y.; Oks, A.; Demetris, J.; Starzl, T.E. *Transplant. Proc.* **1990**, *22*, 64.
51. Takagishi, K.; Yamamoto, M.; Nishimura, A.; Yamasaki, G.; Kanazawa, N.; Hotokebuchi, T.; Kaibara, N. *Transplant. Proc.* **1989**, *21*, 1056.
52. Inamura, N.; Hashimoto, M.; Nakahara, K.; Aoki, H.; Yamaguchi, I.; Kohsaka, M. *Clin. Immunol. Immunopathol.* **1988**, *46*, 82.
53. Inamura, N.; Hashimoto, M.; Nakahara, K.; Nakajima, Y.; Nishio, M.; Aoki, H.; Yamaguchi, I.; Kohsaka, M. *Int. J. Immunopharmacol.* **1988**, *10*, 991.
54. Yamamoto, K.; Mori, A.; Nakahama, T.; Ito, M.; Okudaira, H.; Miyamoto, T. *Immunology* **1990**, *69*, 222.
55. Takabayashi, K.; Koike, T.; Kurasawa, K.; Matsumura, R.; Sato, T.; Tomioka, H.; Ito, I.; Yoshiki, T.; Yoshida, S. *Clin. Immunol. Immunopathol.* **1989**, *51*, 110.
56. Venkataramanan, R.; Jain, A.; Cadoff, E.; Warty, V.; Iwasaki, K.; Nagase, K.; Krajack, A.; Imventarza, O.; Todo, S.; Fung, J.J.; Starzl, T.E. *Transplant. Proc.* **1990**, *22*, 52.
57. Ohara, K.; Billington, R.; James, R.W.; Dean, G.A.; Nishiyama, M.; Noguchi, H. *Transplant. Proc.* **1990**, *22*, 83.
58. Starzl, T.E.; Todo, S.; Fung, J.; Demetris, A.J.; Venkataramman, R.; Jain, A. *Lancet* **1989**, *2*, 1000.
59. McCauley, J.; Bronsther, O.; Fung, J.; Todo, S.; Starzl, T.E. *Lancet* **1989**, *2*, 1516.
60. McCauley, J.; Tzakis, A.G.; Fung, J.J.; Todo, S.; Starzl, T.E. *Lancet* **1990**, *335*, 674.
61. Shapiro, R.; Fung, J.J.; Jain, A.B.; Parks, P.; Todo, S.; Starzl, T.E. *Transplant. Proc.* **1990**, *22*, 35.
62. Kang, Y.; Mazer, M.A.; Dewolf, A.M.; Fung, J.J.; Gasior, T.; Venkataramanan, R.; Starzl, T.E. *Transplant. Proc.* **1990**, *22*, 21.
63. Vanthiel, D.H.; Iqbal, M.; Jain, A.; Fung, J.; Todo, S.; Starzl, T.E. *Transplantation Proceedings* Feb **1990**, *22*, 37.
64. Mieles, L.; Todo, S.; Fung, J.J.; Jain, A.; Furukawa, H.; Susuki, M.; Starzl, T.E. *Transplant. Proc.* **1990**, *22*, 41.
65. McCauley, J.; Fung, J.; Jain, A.; Todo, S.; Starzl, T.E. *Transplant. Proc.* **1990**, *22*, 17.
66. Krensky, A.M.; Weiss, A.; Crabtree, G.; Davis, M.M.; Parham, P. *New Eng. J. Med.* **1990**, *322*, 510.
67. Fidelus, R.K.; Laughter, A.H. *Transplantation* **1986**, *41*, 187.

68. Bijsterbosch, M.K.; Klaus, G.G.B. *Immunology* **1985**, *56*, 435.
69. Fujii, Y.; Fujii, S.; Kaneko, T. *Transplantation* **1989**, *47*, 1081.
70. Tocci, M.J.; Matkovich, D.A.; Collier, K.A.; Kwok, P.; Dumont, F.; Lin, S.; Degudicibus, S.; Siekierka, J.J.; Chin, J.; Hutchinson, N.I. *J. Immunol.* **1989**, *143*, 718.
71. Crabtree, G.R. *Science* **1989**, *243*, 355.
72. Emmel, E.A.; Verweij, C.L.; Durand, D.B.; Higgins, K.M.; Lacy, E. Crabtree, G.R. *Science* **1989**, *246*, 1617.
73. June, C.H.; Ledbetter, J.A.; Gillespie, M.M.; Lindsten, T.; Thompson, C.B. *Molec. Cell. Biol.* **1987**, *7*, 4472.
74. Kay, J.E.; Benzie, C.B. *Immunol. Lett.* **1990**, in press.
75. Lancki, D.; Kaper, B.P.; Fitch, F.W. *J. Immunol.* **1989**, *142*, 416.
76. Trenn, G.; Taffa, R.; Hohman, R.; Kincade, R.; Shevach, E.M. Sitkovsky, M. *J. Immunol.* **1989**, *42*, 3796.
77. Handschumacher, R.E.; Harding, M.W.; Rice, J.; Druggs, R.J.; Speicher, D.W. *Science* **1984**, *226*, 544.
78. Koletsky, A.J.; Harding, M.W.; Handschumacher, R.E. *J. Immunol.* **1986**, *137*, 1054.
79. Quesniaux, V.F.; Schieier, J.M.; Wenger, R.M.; Hiestand, P.C.; Harding, M.W.; Van Regenmortal, M.H.V. *Eur. J. Immunol.* **1987**, *17*, 1359.
80. Takahashi, N.; Hayano, T.; Suzuki, M. *Nature* **1989**, *337*, 473.
81. Fischer, G.; Wittmann, L.B.; Lang, K.; Kiefhaber, T.; Schmid, F.X.; *Nature* **1989**, *337*, 476.
82. Schneuwly, S.; Shortridge, R.D.; Larrivee, D.C.; Ono, T.; Ozaki, M.; Pak, W.L. *Proc. Natl. Acad. Sci. USA* **1989**, *86*, 5390.
83. Shieh, B.H.; Stamnes, M.A.; Seavello, S.; Harris, G.L.; Zuker, C.S. *Nature* **1989**, *338*, 67.
84. Siekierka, J.J.; Staruch, M.J.; Hung, S.H.; Sigal, N.H. *J. Immunol.* **1989**, *143*, 1580.
85. Siekierka, J.J.; Hung, S.H.; Poe, M.; Lin, C.S.; Sigal, N.H. *Nature* **1989**, *341*, 755.
86. Harding, M.W.; Galat, A.; Uehling, D.E.; Schreiber, S.L. *Nature* **1989**, *341*, 758.
87. Harrison, R.K.; Stein, R.L. *Biochemistry* **1990**, *29*, 1684.
88. Rosen, M.K.; Standaert, R.F.; Galat, A.; Nakatsuka, M.; Schreiber, S.L. *Science* **1990**, *248*, 863.
89. Harrison, R.K.; Stein, R.L. *Biochemistry* **1990**, *29*, 3813.
90. Albers, M.W.; Walsh, C.T.; Schreiber, S.L. *J. Org. Chem.* **1990**, *55*, 4984.

# THE DEVELOPMENT OF KETOROLAC:
## IMPACT ON PYRROLE CHEMISTRY AND
## ON PAIN THERAPY*

Joseph M. Muchowski

*Contribution no. 810 from the Syntex Institute of Organic Chemistry.

**Advances in Medicinal Chemistry,**
**Volume 1, pages 109–135.**

# ABSTRACT

5-Aroyl-1,2-dihydro-3$H$-pyrrolo[1,2-$a$]pyrrole-1-carboxylic acids (**4**, $n = 2$, Scheme 1) comprise a series of bicyclic arylacetic acid derivatives, with powerful antiinflammatory and analgesic activity, which were rationally designed on the basis of literature leads. Efficient linear and convergent syntheses of these compounds were devised.

Ketorolac **49**, one member of the above group of compounds, showed modest antiinflammatory activity in acute and chronic models of inflammation, e.g., in the rat carrageenan edema (55 × phenylbutazone), rat cotton pellet granuloma (0.5 × indomethacin) and rat adjuvant arthritis (2 × naproxen) assays. In contrast, it elicited extraordinarily strong analgesic activity in the mouse phenylquinone writhing assay (347 × aspirin). This observation and a remarkably low propensity to cause gastrointestinal erosion on chronic administration to rats (7 days) were instrumental in the selection of this agent for development as an analgesic in humans.

Upon resolution of ketorolac, it was found that essentially all of the biological activity resides in the (–) enantiomer which has the ($S$)-absolute configuration. This stereochemistry is identical to that found in the active isomers of other α-substituted arylacetic acid antiinflammatory/analgesic agents such as naproxen and clindanac.

In humans, ketorolac can be administered orally, intravenously, intramuscularly, and topically (in solution or as a gel). Orally, a 10-mg dose is about as effective as 10 mg of i.m. morphine sulfate, for the relief of moderate to severe postoperative pain; while 10 mg and 30 mg intramuscularly are at least equivalent to 12 mg of morphine, with the 30 mg dose having a significantly longer duration of action. Ketorolac does not cause respiratory depression, has a lesser effect on bowel function than morphine, and shows no evidence of withdrawal symptoms or tachyphylaxis after prolonged use.

Oral ketorolac (10 mg) is equiefficacious with aspirin (650 mg) for the relief of postpartum pain. For pain requiring long-term treatment, it is more effective (10 mg, q.i.d.) than aspirin (650 mg, q.i.d.) with the incidence of side effects being similar for the two drugs. Successive administration for five days by oral or parental routes (10–30 mg, q.i.d.) to normal volunteers caused significantly less irritation to gastric and duodenal mucosa than did aspirin (650 mg q.i.d.). Oral ketorolac (10 or 20 mg) was as effective as or superior to acetaminophen (1000 mg) or to an acetaminophen (600 mg) codeine (60 mg) combination in the relief of post-surgical pain.

Ketorolac has been released for marketing in 18 countries worldwide and is marketed in 9 countries as of December 1991.

# I. INTRODUCTION

In the early part of the 1970s, our research groups in Palo Alto and Mexico City were charged with the mission of developing new, high potency ($\geq$ indomethacin) non-steroidal antiinflammatory agents in which the central and peripheral side effects were markedly less than those of agents in use at that time. A particularly troublesome side effect associated with the chronic use of most, if not all, of the potent non-steroidal antiinflammatory drugs (NSAID's) is the propensity to cause gastrointestinal irritation.[1] It was considered of paramount importance that this side effect, especially, be minimized in any new agent. Although the primary mission was not achieved, an appreciable number of the compounds synthesized in our studies, were found through classical methods of evaluation in animal models, to be much more effective in the attenuation of pain than of other inflammatory sequelae. Extensive evaluation of several such compounds in animals led to the selection of two of them for human clinical trials, one of which (ketorolac) has been recently approved for use as an analgesic in humans in several countries. This article gives an account of the development of ketorolac, the impact that this agent has had on the chemistry of pyrroles, and its potential role in pain therapy.

Around the time that our studies were initiated, tolmetin (1, Scheme 1) and zomepirac (2) were two pyrroleacetic acid derivatives used in the treatment of rheumatoid arthritis[2] and pain.[3-5] These drugs were selected as one class of arylacetic acids, among several others,[6-8] as lead compounds for structural manipulation. In this regard, Carson and Wong[9] reported that methylation of zomepirac in the acetic acid side chain, e.g., as in 3 resulted in a fourfold increase in the antiiflammatory potency as measured by the rat paw kaolin edema assay. It therefore appeared to be of interest to determine what effect constraining the carbon atoms of the N- and C- methyl groups in a bicyclic framework, as found in 4, would have on the antiinflammatory activity in animal models.[10]

The first attempt to synthesize bicyclic compounds of structural type 4 was made by Iriarte and Garcia[12] who condensed 5-(2-thienoyl)pyrrole-2-acetic acid ethyl ester (5, Scheme 2) with 1,2-dibromoethane in the presence of potassium carbonate. Under optimized conditions, the required compound 6 was obtained as one com-

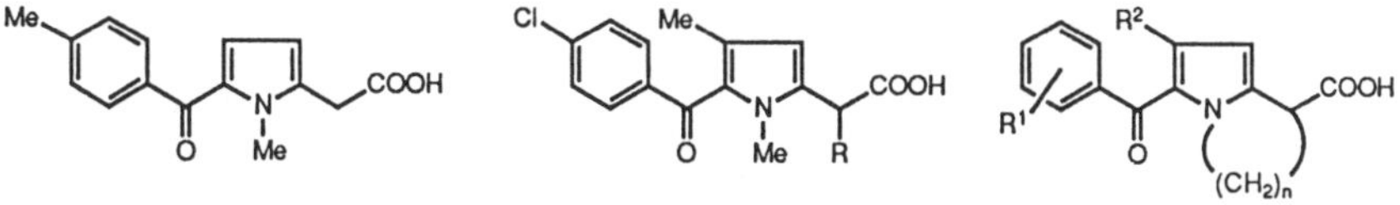

|  |  |  |
|---|---|---|
| 1 | 2, R = H | 4 |
|  | 3, R = Me |  |

*Scheme 1.*

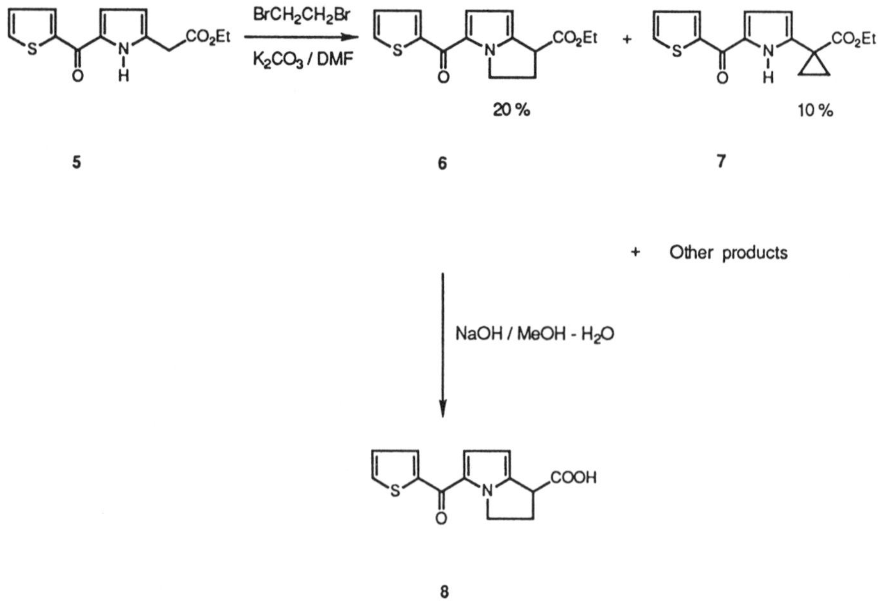

**Scheme 2.**

ponent of a complex mixture, which also contained the structurally isomeric cyclopropane derivative **7**, in about 20% yield. Saponification of **6** gave the corresponding carboxylic acid **8** (RS-63058) which, in animal models, was shown not only to have quite respectable antiinflammatory activity (70 × phenylbutazone, rat carrageenan paw assay) but also to be an extraordinarily powerful analgesic agent (350 × aspirin, mouse phenylquinone writhing assay). This latter observation engendered much excitement in our group which was, however, tempered by a certain degree of unease because the process shown in Scheme 2 was clearly unsuitable for the synthesis of the large number of derivatives (of **8**) usually required for structure activity optimization purposes. Nevertheless, an all-out effort by the Mexico group soon resulted in the development of several closely related synthetic processes, which provided access to many of the congeners of **8** that were eventually synthesized.

## II.  CLASSICAL SYNTHESES OF 5-AROYL-1,2-DIHYDRO-3H-PYRROLO[1,2-a]PYRROLE-1-CARBOXYLIC ACIDS

In the early phases of the above studies, Carpio and Valdes[13] demonstrated that the cyclization of ethyl N-(2-iodoethyl)pyrrole-2-acetate (**9**, Scheme 3) to ethyl 1,2-

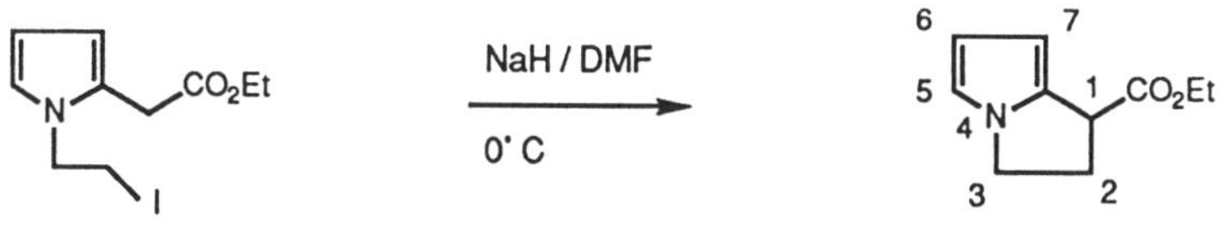

**Scheme 3.**

dihydro-3*H* pyrrolo[1,2-*a*]pyrrole-1-carboxylate (**10**), could be effected efficiently at 0°C with sodium hydride in dimethylformamide (DMF). Since the aroylation of **10** was expected,[9,14] and subsequently shown, to provide analogs of **6**, synthetic efforts were focused on efficient means of generation of **10** or derivatives thereof.

The first successful approach to the desired bicyclic system[15] was based on a Hantzsch pyrrole synthesis.[16] Reaction of freshly generated bromoacetaldehyde with the enamine **11** (Scheme 4) derived from ethanolamine and dimethyl acetonedicarboxylate, in boiling acetonitrile, gave the pyrrole derivative **12** in modest yield. This simple molecule contains all of the atoms needed for the generation of the required bicyclic system. It was converted into the methanesulfonate which, upon reaction with excess sodium iodide in hot acetonitrile, gave the iodide **13** in 90% yield from the alcohol **12**. As expected, cyclization of **13** to the bicyclic diester **14** occurred rapidly and efficiently under mild conditions, with sodium hydride in DMF solution. Removal of the extra carboxyl group from **14** was accomplished by a three-step sequence based on the different steric environment of the two ester moieties. Thus, saponification of **14** with potassium hydroxide gave the dicarboxylic acid **15**, the least hindered carboxyl group of which was selectively esterified, in high yield, with 2-propanol under Fischer conditions.[17] As anticipated for a pyrrole-3-carboxylic acid derivative (ref. 16, pp. 329–335), thermal (270°C) decarboxylation of **16** gave the bicyclic mono ester **17** in 18–20% overall yield.[18,19]

An important aspect of the above process is that it has considerable flexibility. For example, substitution of propanolamine for ethanolamine in the Hantzsch condensation step gave the 1-(3-hydroxypropyl) diester **18** (Scheme 5) which could be transformed into the 5,6,7,8-tetrahydropyrrolo[1,2-*a*]pyridine-1-carboxylic acid derivative **19** (Scheme 5).[19] Furthermore, when the α-halocarbonyl component of the Hantzsch reaction was a bromomethylketone such as bromoacetone, 1-bromo-2-butanone, etc., the condensation products derived therefrom were readily converted into the 6-substituted congeners of **17, 20a–20d**.[19]

A different synthetic approach, which will not be described here, but which

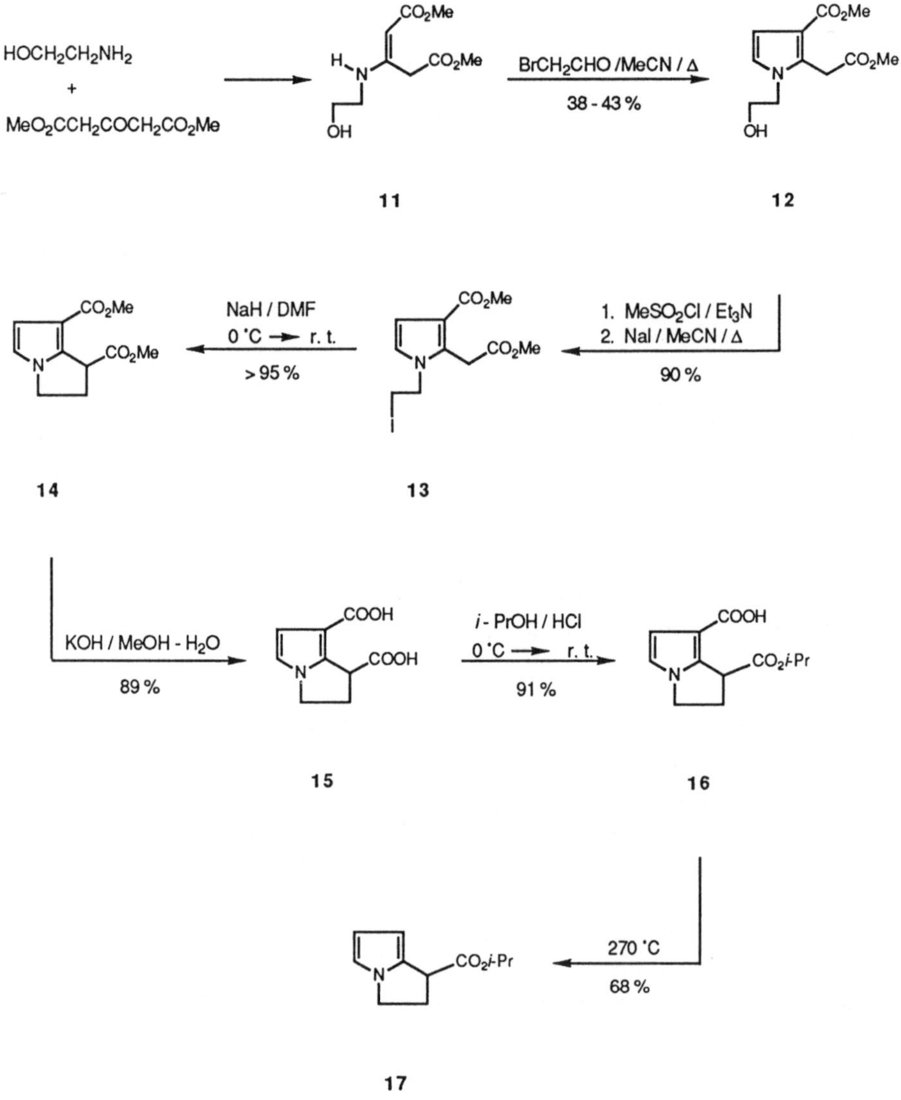

**Scheme 4.**

involved cyclization of intermediates analogous to **9**, provided the 6-halo sub-stituted bicyclic esters **20e** and **20f** (Scheme 5).[19.]

Another synthetic route to the desired bicyclic systems, tactically different from that described above, was devised by P. Gallegra of the Process Development group in Palo Alto. In this case, the 2-carbon side chain was attached to C-2 of the pyrrole

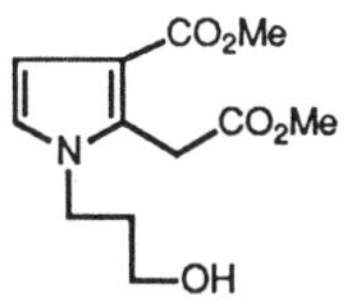

## Scheme 5

**18**

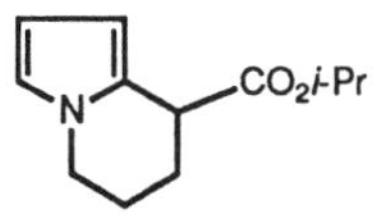

**19**

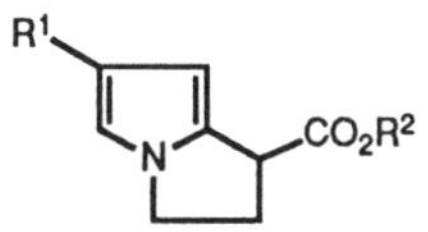

**20** a, $R^1$ = Me, $R^2$ = $i$-Pr
b, $R^1$ = Et, $R^2$ = $i$-Pr
c, $R^1$ = MeO(CH$_2$)$_2$, $R^2$ = $i$-Pr
d, $R^1$ = MeO(CH$_2$)$_3$, $R^2$ = $i$-Pr
e, $R^1$ = Cl, $R^2$ = Me
f, $R^1$ = Br, $R^2$ = Me
g, $R^1$ = MeS, $R^2$ = Me

*Scheme 5.*

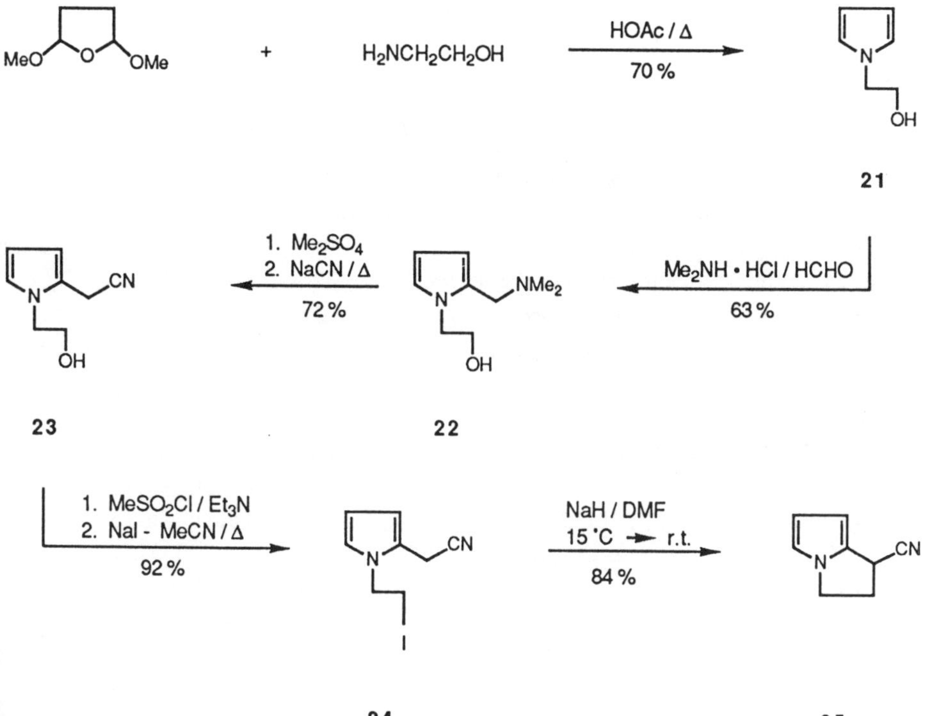

MeO—O—OMe + H$_2$NCH$_2$CH$_2$OH →[HOAc / Δ] [70 %] **21**

**23** ←[1. Me$_2$SO$_4$ / 2. NaCN / Δ] [72 %] **22** ←[Me$_2$NH • HCl / HCHO] [63 %]

**24** ←[1. MeSO$_2$Cl / Et$_3$N / 2. NaI - MeCN / Δ] [92 %]

→[NaH / DMF / 15 °C → r.t.] [84 %] **25**

*Scheme 6.*

115

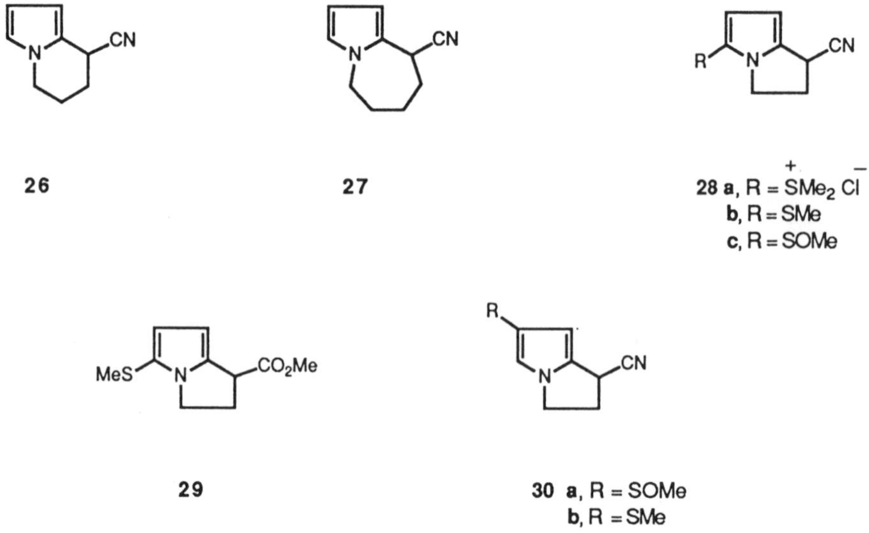

26                                27                          28 a, R = SMe$_2$ Cl
                                                               b, R = SMe
                                                               c, R = SOMe

29                                                30 a, R = SOMe
                                                     b, R = SMe

*Scheme 7.*

nucleus in a three-step sequence beginning with a Mannich reaction on 1-(2-hydroxyethyl)pyrrole (**21**, Scheme 6), which was obtained by a Paal-Knorr type synthesis (ref 16, pp. 77–81) from 2,5-dimethoxytetrahydrofuran and ethanolamine. Quaternization of the Mannich base **22** with dimethyl sulfate followed by reaction with hot aqueous sodium cyanide, in the same pot, gave the acetonitrile derivative **23**. The cyclization of **23** was accomplished in a manner analogous to that described above, the bicyclic nitrile **25** being obtained in 24.5% overall yield.[19]

Utilization of reaction sequences entirely analogous to that of Scheme 6, but commencing with 3-amino-1-propanol or 4-amino-1-butanol, provided the 6- and 7-membered systems **26** and **27** (Scheme 7) corresponding to the nitrile **25**.[19]

The nitrile **25** served as the progenitor of the bicyclic 6-methylthio compounds **20g** (Scheme 4) and **30b**. These entities were both derived from **28b** (Scheme 7) which was easily prepared, in over 90% yield, by the low temperature generation of the sulfonium salt **28a** (*N*-chlorosuccinimide/dimethylsulfide/–55 °C) and subsequent pyrolysis thereof in boiling toluene.[20] The sulfide **28b** was converted by standard procedures into the ester **29**, which underwent trifluoroacetic acid-induced isomerization to **20g**, a reaction typical of α-alkylthiopyrroles.[21] Similar acid-induced isomerization of the sulfoxide **28c** to the thermodynamically more

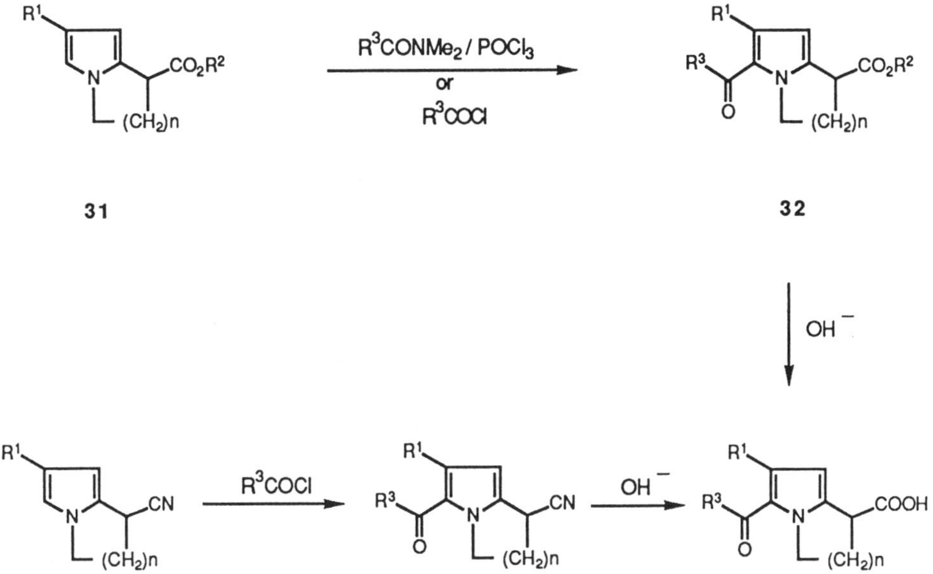

**Scheme 8.**

stable isomer **30a**[21] followed by reduction with the triphenylphosphine-iodine-sodium iodide system of Olah et al.[22] provided the sulfide **30b**.

With reliable supplies of the various bicyclic esters and nitriles in hand, it became a simple matter to convert them into analogs of compound **6** (RS-63058). These carboxylic acids **35** (Scheme 8) were prepared by alkaline hydrolysis of the acylated esters **32** or nitriles **34**. The esters **32** were generated by acylation of **31** with the Vilsmeier-Haack reagent derived from an *N,N*-dimethylcarboxamide and phosphorus oxychloride in boiling 1,2-dichloroethane[9] or with a carboxylic acid chloride in an inert, high boiling solvent.[23] This latter process, which occurs in the absence of a Friedel-Crafts catalyst, was the preferred method for the preparation of the nitriles **34**. Furthermore, unlike the Vilsmeier-Haack method, sterically encumbered acyl groups could be introduced with ease and acid-catalyzed migration of the acyl moiety from the α- to the β-position of the pyrrole nucleus was obviated.[24,25]

## III.  STRUCTURE-ACTIVITY RELATIONSHIP OF 5-SUBSTITUTED 1,2-DIHYDRO[3*H*]PYRROLO[1,2-*a*]PYRROLE-1-CARBOXYLIC ACIDS AND RELATED COMPOUNDS

Non-steroidal antiinflammatory agents are generally considered to exert their effects by virtue of the inhibition of cyclooxygenase, the enzyme responsible for the in vivo transformation of arachidonic acid into PGG, the progenitor of the natural prostaglandins.[26] The ability of such agents to block the synthesis of prostaglandins accounts, in large measure, for their beneficial effects on inflammation and analgesia, as well as for their injurious effects on the gastrointestinal tract of animals and man.[27]

### A.  Acute Assays

The carboxylic acids of structural type **35** were initially evaluated for antiinflammatory and analgesic activity in the carrageenan rat paw edema assay and the mouse phenylquinone writhing assay, respectively (see ref. 25 for a description of these assays). The following general comments can be made with regard to the structural features required for high activity in both assays in those congeners of **35** not substituted in the pyrrole nucleus (Table 1):

1.  Aroyl or heteroaroyl group at C-5 (*n* = 1). Alkanoyl (e.g., benzyl, entry 4, log P = 2.52) or cycloalkanoyl (e.g., cyclobutyl, entry 5, log P = 2.41) groups at this position gave rise to compounds which were weakly active in both assays even though the log P values of these compounds were close to that of the highly active parent compound (entry 1, log P = 2.72).
2.  Carboxyl group at C-1. All compounds wherein the carboxylic acid moiety was located elsewhere (e.g., at C-2 or C-7 in **35** (*n* = 1), data not tabulated) were essentially inactive.
3.  Saturated ring, 5-membered. Larger ring sizes (entries 2 and 3) resulted in the virtual abolition of activity. These results are consistent with the known sensitivity to steric bulk of that region of the cyclooxygenase active site which is occupied by the methyl group of aryl propionic acid antiinflammatory agents.[28] Indeed, the low degree of activity associated with compounds of type **35** (*n* = 1) methylated α- to the carboxylic acid group (≤ 10 × the standards in both assays) is readily explicable in terms of a further exacerbation of an already sterically demanding situation at this region of the active site.

A detailed study of the quantitative structure-activity relationships of ca. forty 5-(substituted)benzoyl-1,2-dihydro-3*H*-pyrrolo[1,2-a]pyrrole-1-carboxylic acids for each assay revealed that the potencies of these compounds could not be

**Table 1.** Antiinflammatory and Analgesic Activities of 5-Substituted 1,2-Dihydro-3H-pyrrolo[1,2-a]pyrrole-1-carboxylic Acids and Related Compounds

| Entry | System | R | n | Rat Paw (PB = 1)[a] | Mouse Writhing (aspirin = 1)[b] |
|-------|--------|---|---|---------------------|--------------------------------|
| 1 | A | H | 1 | $55^c$ | 347 |
| 2 | A | H | 2 | $<1^d$ | 9 |
| 3 | A | H | 3 | ~0.4 | ~3 |
| 4 | B | PhCH$_2$ | — | 3 | ≤14 |
| 5 | B | c-C$_4$H$_7$ | — | ~1.5 | 30 |
| 6 | A | 2-Me | 1 | ~1.5 | 35 |
| 7 | A | 2-Cl | 1 | ~4 | 35 |
| 8 | A | 2-I | 1 | 1.5 | <1 |
| 9 | A | 2-OH | 1 | ~4 | ~100 |
| 10 | A | 3-Me | 1 | ≤0.4 | ~7 |
| 11 | A | 3-F | 1 | 6 | ~20 |
| 12 | A | 3-Cl | 1 | ≤4 | 35 |
| 13 | A | 3-OMe | 1 | <1 | ~4 |
| 14 | A | 3-OH | 1 | <1 | <0.7 |
| 15 | A | 3-NO$_2$ | 1 | <1 | <1 |
| 16 | A | 4-Me | 1 | $39^c$ | 165 |
| 17 | A | 4-F | 1 | $61^c$ | 183 |
| 18 | A | 4-Cl | 1 | $21^c$ | 83 |
| 19 | A | 4-MeO | 1 | $83^c$ | 117 |
| 20 | A | 4-OH | 1 | <10 | <3 |
| 21 | A | 4-SO$_2$Me | 1 | <0.5 | <0.6 |
| 22 | A | 4-CH$_2$=CH | 1 | $127^c$ | ~200 |
| 23 | B | 2-thenyl | — | $70^e$ | 350 |
| 24 | B | 3-thenyl | — | 53 | 180 |
| 25 | B | 2-furyl | — | 4 | 65 |
| 26 | B | N-Me-2-pyrrolyl | — | 5 | 142 |

[a] ED$_{30}$ for phenylbutazone = 15 mg/kg.
[b] ED$_{50}$ for aspirin = 70 mg/kg.
[c] Data taken from ref. 25, Table IV.
[d] Data for entries 2–15, 20, 21 are historical values taken from Table III, ref 25.
[e] Data for entries 23–26, unpublished results.

correlated with log P nor with the electronic substituent constants $f$F and $r$R.[29] The analgesic and antiinflammatory activities were, however, acceptably correlated with precise steric and hydrogen bonding properties of the substituents (see ref. 25, equations 7 and 10). Thus, the ortho position is tolerant of moderately bulky substituents (e.g., Cl, entry 7; OH, entry 9; Table 1) whereas the meta position is not (entries 10–15). Hydrogen bond acceptor substituents are not tolerated at the meta or the para positions (e.g., entries 15 and 21) and the para position has specific shape requirements. On the basis of the above correlations, the 4-vinylbenzoyl compound (entry 22) was predicted to be highly active and this prediction was confirmed when the compound was synthesized and submitted to the assays.

In connection with their studies on compounds related to tolmetin, Carson and Wong[9] reported that the introduction of a methyl group into the 4-position of the pyrrole nucleus (as in 2, for example) caused at least a twofold increase in the antiinflammatory potency over compounds without this substituent, as measured by the rat paw kaolin edema assay. It was therefore of considerable interest to determine the effect of comparable substitution in the pyrrolopyrrole series of compounds. Since the data cited above, and that in Table 1, clearly showed that the most active entities were those in which the benzoyl group was unsubstituted or monosubstituted in the para position, no ortho- or meta- substituted benzoyl congeners were synthesized in this series of compounds.[20,30]

More often than not, the introduction of a methyl or a chloro group at C-6 increased the antiinflammatory potency relative to the unsubstituted compounds (entries 1, 16–19 in Table 1 and entries 1–10, Table, 2) and for some compounds this increase was quite significant (4- to 6-fold, entries 4 and 9, Table 2). At the same time there was a trend toward a reduction in the analgesic potency of these compounds. Thus, the net result of the 6-substituent was to decrease the analgesic: antiinflammatory ratio over that found in the 6-unsubstituted congeners, although this effect was not a strong one. Surprisingly, however, a 6-methylthio moiety, though nearly isolipophilic ($\pi = 0.61$) with a methyl group ($\pi = 0.56$), substantially increased the analgesic: antiinflammatory ratio (entries 17–21). In fact, except for the $p$-chloro compound (entry 20) these entities were nearly devoid of antiinflammatory activity.

The above results clearly indicate that the region of the cyclooxygenase active site occupied by the 6-substituent can tolerate a moderate degree of steric bulk. It may account for the lack of activity found for compounds with a 2-methoxyethyl, 3-methoxypropyl or a methylsulfinyl group at this position (entries 15, 16 and 22) although other explanations such as an inappropriate lipophilicity, adverse hydrogen bonding effects, etc. are also possible.

One of the consequences of a substituent at C-6 is that it is expected to force the phenyl group out of the CO-pyrrolyl plane, a conformation which is likely for ketorolac in the solid state at least (see below) and one that is probably of significance at the cyclooxygenase active site.[26,28,31] The NMR spectral data provide strong evidence for the predominance of such a conformation in solution

**Table 2.** Antiinflammatory and Analgesic Activities of 6- or
7-Substituted-5-aroyl-1,2-dihydro-3*H*-pyrrolo[1,2-a]pyrrole-1-carboxylic Acids

| Entry | $R^1$ | $R^2$ | $R^3$ | Rat Paw (PB = 1)[a] | Mouse Writhing (aspirin = 1)[b] |
|---|---|---|---|---|---|
| 1 | H | Me | H | 43[c] | 250 |
| 2 | Me | Me | H | 43 | 250 |
| 3 | F | Me | H | 103 | 302 |
| 4 | Cl | Me | H | 122 | 167 |
| 5 | OMe | Me | H | 73 | 130 |
| 6 | H | Cl | H | 10 | 240 |
| 7 | Me | Cl | H | 32 | 158 |
| 8 | F | Cl | H | 139 | 151 |
| 9 | Cl | Cl | H | 88 | 167 |
| 10 | OMe | Cl | H | 55 | 201 |
| 11 | H | Br | H | 73 | 314 |
| 12 | F | Br | H | 96 | 120 |
| 13 | H | Et | H | 26 | 108 |
| 14 | F | Et | H | 46 | 121 |
| 15 | H | (CH$_2$)$_2$OMe | H | ≤0.3 | 7.6 |
| 16 | H | (CH$_2$)$_3$OMe | H | <10 | ≤3 |
| 17 | H | SMe | H | 2[d] | 78 |
| 18 | Me | SMe | H | 3.4 | 301 |
| 19 | F | SMe | H | ≤10 | 60 |
| 20 | Cl | SMe | H | 35 | 130 |
| 21 | OMe | SMe | H | <10 | 50 |
| 22 | Br | SOMe | H | 0.5 | <3 |
| 23 | H | H | Me | 18 | ~20 |
| 24 | H | H | Cl | 12 | ~35 |
| 25 | H | H | Br | ~2.5 | ~10 |

[a]ED$_{30}$ for phenylbutazone = 15 mg/kg.
[b]ED$_{50}$ for aspirin = 70 mg/kg.
[c]Data for entries 1–14 taken from ref 30.
[d]Data for entries 15–20 taken from ref. 20.
[e]Data for entries 21–23, unpublished results.

for the 6-methyl compounds. Thus, the methyl absorption for the bicyclic ester **20a** (Scheme 5) is found at δ 2.03, whereas in the corresponding 5-aroylated carboxylic acids (entries 1–5, Table 2) and the isopropyl esters thereof, it occurs at δ1.86 ± 0.04 and 1.85 ± 0.03, respectively. This upfield shift is consistent with a conformation in which the methyl group occupies the shielding region above the plane of the aryl moiety.[32]

Whereas a substituent at C-6 augmented the antiinflammatory potency, one at C-7 caused a significant reduction in potency in both assays (entries 23–25) relative

**Table 3.** Antinflammatory Profile of Selected
5-Aroyl-1,2-dihydro-3H-pyrrolo[1,2-a]pyrrole-1-carboxylic Acids

| Entry | $R^1$ | $R^2$ | Rat Paw (PB = 1) | Mouse Writing (aspirin = 1) | Cotton Pellet (indomethacin = 1) | Adjuvant Arthritis (naproxen= 1)[b] | GI Erosion $MED^c$,mg/kg |
|-------|-------|-------|------|------|------|------|------|
| 1 | H | H | 55 | 347 | 0.5 | 2 | 5 |
| 2 | Me | H | 39 | 165 | <0.3 | 0.6 | 3 |
| 3 | F | H | 61 | 183 | 1–2 | 1–3 | 5 |
| 4 | Cl | H | 21 | 83 | 1–3 | 4 | 3 |
| 5 | MeO | H | 83 | 117 | 0.6 | <0.3 | 7 |
| 6 | CH$_2$=CH | H | 127 | ~200 | 1.0 | 7 | 1 |
| 7 | H | Me | 43 | 250 | 1 | 1 | ~1.7 |
| 8 | Me | Me | 43 | 250 | 0.7 | 3 | 2.5 |
| 9 | F | Me | 103 | 302 | ~1 | ~3 | 1.2–1.7 |
| 10 | Cl | Me | 122 | 167 | ~2 | 3.5 | ~0.6 |
| 11 | H | Cl | 10 | 246 | 0.6 | ≤8 | ~1.7 |
| 12 | Me | Cl | 32 | 158 | 0.2 | 11 | 5 |
| 13 | F | Cl | 139 | 151 | 1 | 13 | ~0.3 |
| 14 | Cl | Cl | 88 | 167 | 3 | 40 | ~0.2 |
| 15 | MeO | Cl | 55 | 201 | 0.7 | 4 | ~1.3 |
| 16 | Me | MeS | 3.4 | 301 | 0.2 | 0.2 | 19 |
| zomepirac | — | — | 26 | 36 | — | — | 10 |
| naproxen | — | — | 11 | 7 | 0.2 | 1 | 30 |
| indomethacin | — | — | 16 | ~60 | 1 | 7 | — |

$^a$ED$_{30}$ = 3 mg/kg.
$^b$ED$_{50}$ = 1.7 mg/kg.
$^c$Gastrointestinal erosion, 7-day chronic assay; minimum effective dose.
$^d$MED GI erosion/ED$_{30}$ rat paw.

to the corresponding unsubstituted compound.[33] This result may be a consequence of an unfavorable steric interaction between this substituent and the carboxylic acid moiety at C-1.

## B. Chronic Assays

The compounds which showed high potency in the acute screens were selected for study in chronic models of inflammation, which included the cotton pellet granuloma and adjuvant-induced arthritis assays. In addition, these compounds were evaluated for gastrointestinal erosive potential in a 7-day assay in rats (for a description of these assays, see ref. 25). The minimum effective dose (MED) that caused gastrointestinal erosion was then used to calculate a therapeutic ratio (MED divided by $ED_{30}$ for the rat carrageenan paw assay). From the data in Table 3 it is apparent that many of the compounds are powerful antiinflammatory agents, several of them being equipotent with or more potent than indomethacin (entries 3,4,6,7,9,10,13,14). These compounds also showed the highest degree of gastrointestinal erosive activity although two of the para-fluoro compounds (entries 3 and 9) were exceptions in this regard. The majority of the compounds shown in Table 3 were subjected to extensive pharmacological evaluation in animals as possible antiinflammatory and analgesic agents. Without exception, the compounds which were powerful antiinflammatory agents in the chronic assays were eliminated as clinical candidates because of their liability to precipitate gastrointestinal irritation and/or because of other deleterious effects. Two compounds did, however, survive the selection criteria of high potency, good therapeutic ratio, synthetic accessibility, etc., and they (ketorolac and anirolac, entries 1 and 5) were chosen for evaluation as analgesic agents in humans.

# IV. ABSOLUTE CONFIGURATION OF KETOROLAC

Ketorolac and congeners thereof contain a single asymmetric center and it was therefore expected that, like other chiral cyclooxygenase inhibitors,[35] the pharmacological activity would reside in one of the enantiomers. The resolution of ketorolac was effected by fractional crystallization of the cinchonidine and cinchonine salts, from which the pure (−) and (+) forms, respectively, were obtained upon acidification.[38] Essentially all of the antiinflammatory and analgesic activity, as measured by the carrageenan rat paw edema and the mouse phenylquinone writhing assays, was found in the levorotatory isomer, which, was about twice as active as the racemate. The absolute configuration of the inactive (+) isomer was deduced to be (R) by a single crystal X-ray analysis of the amide derived from (+)-(R)-1-(1-naphthyl)ethylamine (Fig. 1). Thus (−)-ketorolac must have the (S) absolute stereochemistry, identical to that found for other α-substituted arylacetic antiinflammatory/analgesic agents such as naproxen[39] and clindanac.[40]

The computer generated perspective drawing of the R,R amide depicted in Figure 1 reveals a most interesting conformational phenomenon. The ketone carbonyl group O-1–C-9 and the pyrrole nucleus are effectively coplanar [signed torsion angle O-1–C-9–C-5–N-4 = −3.5°(1)] and the phenyl group is twisted out of the plane by ca. 56 ° [signed torsion angle C-15–C-10–C-9–O-1 = + 55.8°(1)]. This situation is dramatically highlighted by the very different C-9–C-10 and C-5–C-9 bond distances of 1.516 and 1.418 Å, respectively. The C-9–C-10 distance is close to that expected for a normal Ar-R carbon-carbon bond (1.505 Å) and longer than a typical Ar-CO bond (1.47 Å) in aralkylketones.[41a] The reduced C-5– C-9 bond length is indicative of substantial overlap of the pyrrole and ketone carbonyl group π systems. Because of this strong overlap, the phenyl ring is free to adopt a conformation in which the non-bonded interactions between the hydrogen atoms

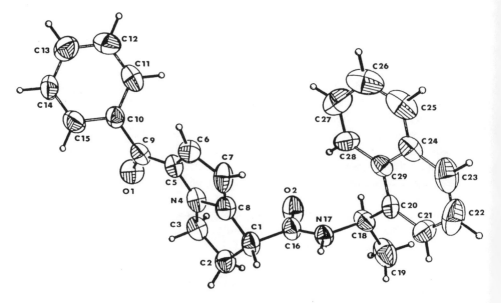

**Figure 1.** Computer generated perspective drawing of amide of (+)-(*R*)-1-(1-naph-thyl)ethylamine and (+)-(*R*)-5-benzoyl-1,2-dihydro-3*H*-pyrrolo[1,2-*a*]-1-carboxylic acid.

ortho to the carbonyl group and H-6 of the pyrrolopyrrole nucleus are minimized. This conformation is likely to be found in ketorolac itself and may well be of significance at the level of the cyclooxygenase active site. In addition, the short C-5–C-9 bond distance is a clear-cut rationalization of the long known (ref. 16, p. 283) low reactivity of pyrrolyl aldehydes and ketones, a reactivity entirely consistent with that expected for vinylogous amides.

It is noteworthy that although the solid state conformation of benzophenone is not dissimilar to that of the *R,R* amide, little or no shortening of the Ar–CO bonds is observed in benzophenone.[41b]

# V.  CLINICAL EFFICACY OF KETOROLAC AND ANIROLAC

## A.  Ketorolac

Ketorolac, as the tris-(hydroxymethyl)methylammonium salt (tromethamine salt) has been under clinical investigation on a worldwide basis for about 10 years, and two monographs reviewing the results of various studies have now been published.[42,43] It can be administered orally, intravenously, intramuscularly and even topically as an ophthalmic solution or as a gel.[44] A brief account of some of the studies is given below.

A single oral 10-mg dose of ketorolac was found to be equivalent to 10-mg of intramuscular (i.m.) morphine sulfate for the relief of moderate to severe postoperative pain[45] and ketorolac had a plasma half life of 5.4 hr.[46] In other studies, in post-surgical patients,[47,48] 10-mg and 30-mg i.m. doses of ketorolac were at least as effective as 12-mg of i.m. morphine and the 30-mg dose was significantly longer acting. No respiratory depression was observed at the 30-mg dose level and the recovery of bowel function was earlier as compared to 6 mg or 12 mg of morphine.[49] Furthermore, subjects showed no withdrawal symptoms on being deprived of the drug following five days of dosing (30 mg, q.i.d.). Also, inasmuch as ketorolac does not bind to $\mu$, $\kappa$ or $\delta$ opiate receptor subtypes, it is unlikely that it will cause physical dependence or abuse of the type associated with opioid analgesics.[50] With regard to other opiates, ketorolac (30 mg, i.m.) was found to be more potent and longer acting and to give significantly better pain relief than 50 or 100 mg of i.m. meperidine (pethidine) against postoperative pain[51] and against pain after wisdom tooth extraction.[52] Similarly, a 30-mg i.m. dose was superior in efficacy to 30 mg of i.m. pentazocine in patients with moderate to very severe pain after major surgery and was as well tolerated as placebo.[53]

Oral ketorolac (10 mg) had an efficacy similar to aspirin (650 mg) in the relief of moderate to severe postpartum uterine pain.[54] In a multicenter study in patients with pain requiring long-term treatment, ketorolac (10 mg, q.i.d.) had increased efficacy without increased side effects as compared to 650 mg q.i.d. of oral aspirin.[55] Normal volunteers given therapeutic doses (10–30 mg, q.i.d.) for five days by oral or parenteral routes showed significantly less irritation to the gastric and duodenal mucosa than did individuals given aspirin (650 mg, q.i.d.) as determined by endoscopic examination.[56]

A single 10- or 20-mg dose of ketorolac was comparable in efficacy to 1000 mg of acetaminophen (paracetamol) in the relief of postoperative orthopedic pain and gave a longer lasting peak analgetic effect at the higher dose.[57] In outpatients with pain after wisdom tooth extraction,[58] 10- or 20-mg doses of ketorolac were significantly superior to aspirin (650 mg) or to an acetaminophen (600 mg)/codeine (60 mg) combination. In another study, orally administered ketorolac (10–40 mg/day) and paracetamol (1000–4000 mg/day)/codeine (60–240 mg/day) were of comparable efficacy for the treatment of moderate to severe pain after gynecological surgery[59] and had a similar incidence of side effects.

Finally, ketorolac ophthalmic solution appears to be a useful alternative to corticosteroid therapy in reducing postoperative inflammation after cataract extraction and intraocular lens implantation.[60]

## B. Anirolac

Far fewer clinical trials have been undertaken with anirolac than with ketorolac. These studies indicate that for the relief of moderate to severe postoperative pain,

100 mg of oral anirolac is more potent and longer lasting than 650 mg of aspirin[61] or 550 mg of naproxen sodium.[62] Similar results were observed for the relief of postpartum pain.[63] The side effects elicited by anirolac were minimal in all of the reported studies.

## VI. CONVERGENT SYNTHESIS OF KETOROLAC AND CONGENERS THEREOF

A significant advantage of the syntheses of ketorolac and derivatives described above was broad applicability, a feature that was particularly important with regard to the structure-activity studies. The multistep and linear nature of these processes were, however, not desirable for large-scale synthetic purposes. Of several strategies considered, the highly convergent process depicted retrosynthetically in Scheme 9 was especially attractive because good literature precedents existed for each step. Thus, the generation of the bicyclic compound **39** by the intramolecular nucleophilic displacement of a leaving group X had precedent in the studies of Vecchieti et al.[64] for a closely related system. Second, the single-step attachment of the four carbon unit (**37** to **38**) required for the above cyclization was grounded on the facile nucleophilic cleavage of spiro activated cyclopropane-1,1-dicarboxy-

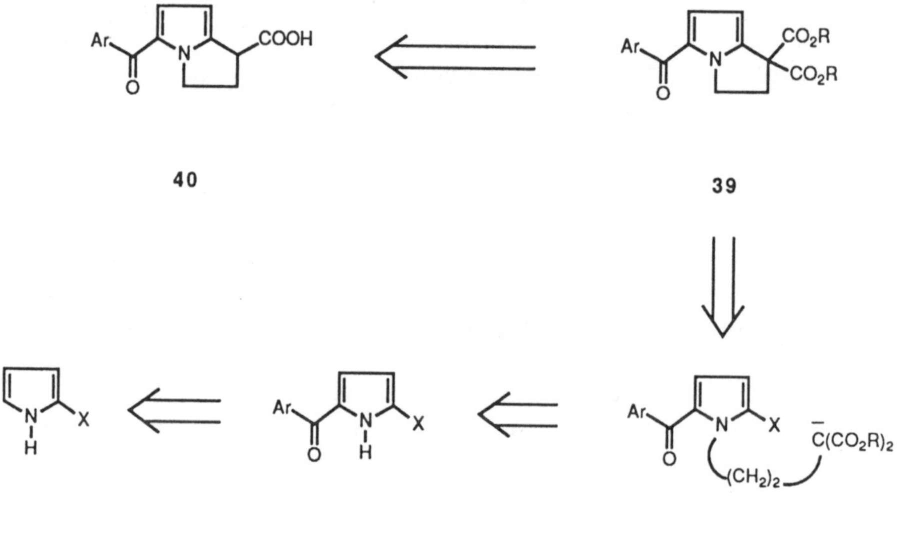

*Scheme 9.*

lates, a concept developed by Danishefsky and co-workers.[65] An account of the successful implementation of the synthetic strategy shown in Scheme 9 is given below.

Low temperature reaction of pyrrole with the *N*-chlorosuccinimide dimethylsulfide adduct gave the sulfonium salt **41** [Scheme 10,[66]] which on thermolysis, either neat or in boiling 1,2-dichloroethane, gave 2-methylthiopyrrole **42** in ca. 60% overall yield. This synthesis of **42**, based on observations of Tomita et al.[67] in the indole area, yielded material free of disubstituted contaminants because of the deactivating effect of the dimethylsulfonium group. Compound **42** was selected as the starting material on the expectation that the methylthio group would serve both to direct aroylation to C-5 and, after oxidation to the sulfone, to function as a leaving group [i.e., as methyl sulfinate[68] in the intramolecular displacement process. As predicted, acylation of **42** with *N,N*-dimethylbenzamide, under Vilsmeier-Haack conditions, gave the 5-benzoyl compound **43** exclusively. Reaction of the sodium salt thereof (generated with sodium hydride) with an equimolar amount of the spirocyclopropane **44**, in DMF solution at 55 °C, gave the Meldrum's acid derivative **45** which, on oxidation with *m*-chloroperbenzoic acid, was converted into the sulfone **46**. The stage was now set to examine the ring closure reaction. Surprisingly, the sodium salt of **46** was recovered unchanged after heating in DMF at 90°C, a result which may be explicable in terms of insufficient anion nucleophilicity and/or excessive steric hindrance to cyclization. This problem was resolved by conversion of **46** into the dimethyl ester **47**. The sodium salt of this compound was converted into a mixture of the expected bicyclic malonate **48a** (83%) and a lesser amount of the monomethyl ester **48b**. Saponification of the mixture and subsequent acidification gave ketorolac (**49**) in ca. 35% overall yield from 2-methylthiopyrrole. This process, though unoptimized, was nearly twice as efficient as the Hantzsch-based synthesis (19% overall) of ketorolac.

This reaction sequence could, of course, be used to prepare substituted aryl congeners of ketorolac and, with appropriate modifications, derivatives containing a substituent in the pyrrole nucleus.[66] For example, chlorination of **41** with sulfuryl chloride at –76 °C gave a new sulfonium salt **50** (Scheme 11), which on heating in boiling xylene was converted into the very air sensitive chlorosulfide **51**. Acylation of this compound, in situ, with *N,N*-dimethyl-4-chlorobenzamide under Vilsmeier-Haack conditions, gave the trisubstituted pyrrole derivative **52** (46% yield from **50**) which was then transformed into 5-(4-chlorobenzoyl)-6-chloro-1,2-dihydro-3*H*-pyrrolo[1,2-*a*]pyrrole carboxylic acid (**53**) via the reaction sequence described above.[66]

In the above processes, an extra step is always needed to activate the methylthio moiety for displacement. This would be unnecessary for a substituent which is activating in the resonance sense but, which nevertheless, can serve as a leaving group (e.g., Cl, Br, etc.). To test this concept,[66] 2-benzoylpyrrole (**54**) was converted into the dibromide **55** (Scheme 12) with excess bromine[69] and the sodium salt thereof was reacted with the cyclopropane derivative **44** at 80°C in DMF. After

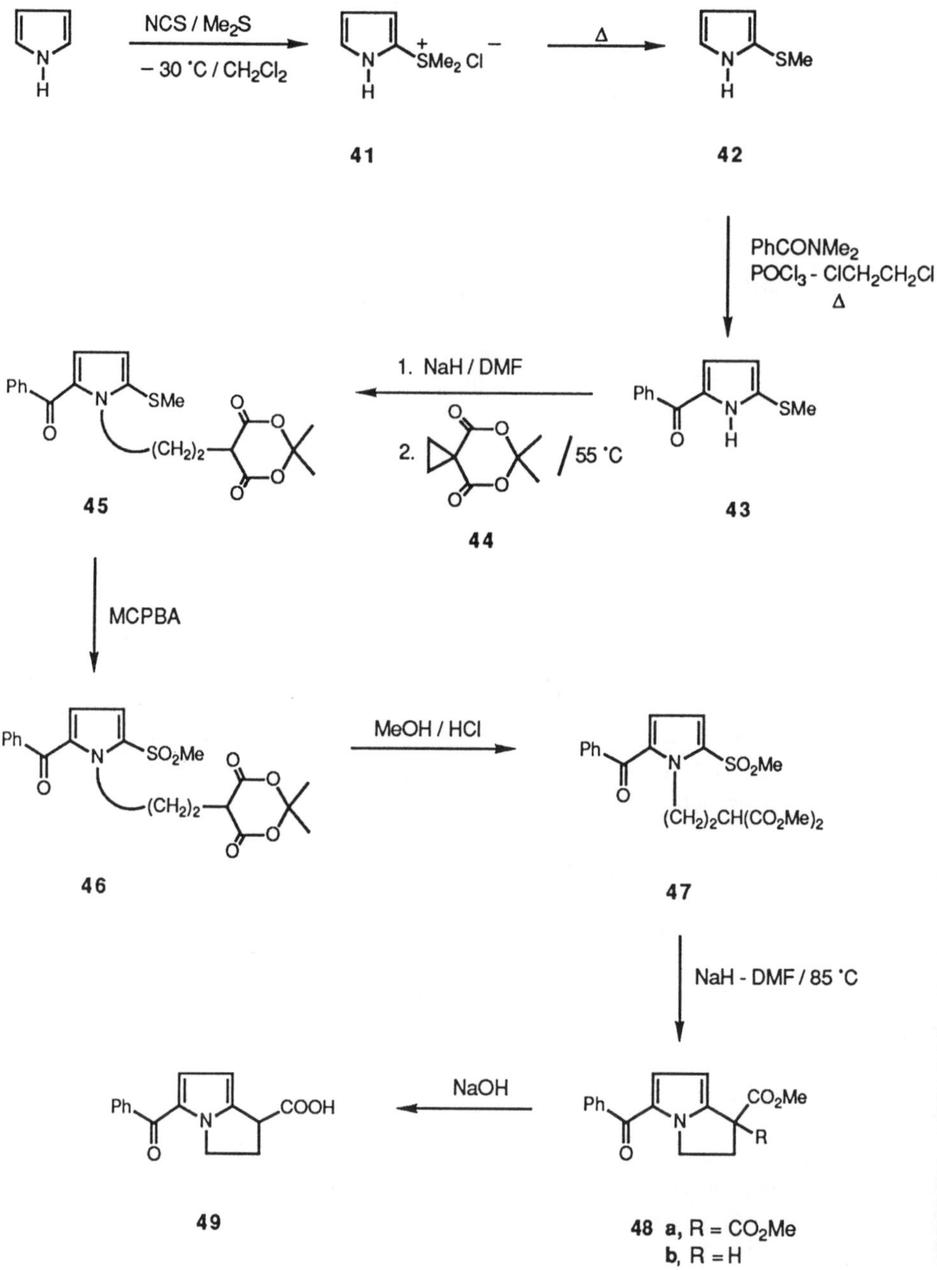

*Scheme 10.*

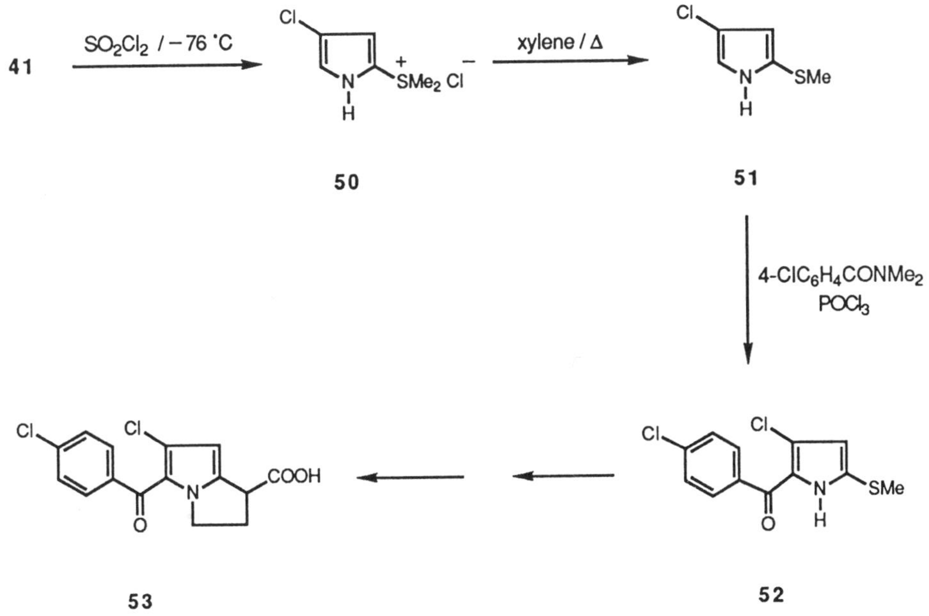

**Scheme 11.**

4 hours, the sodium salt of the required adduct **56** could be isolated, in about 95% yield, merely by dilution with ether and filtration. Application of the usual reaction sequence to this substance resulted in the highly efficient conversion thereof into a 1:6 mixture of the bicyclic mono- and diesters **57a** and **57b**. Sequential saponification, hydrogenolytic debromination and acidification, without purification of intermediates, gave ketorolac in 47% overall yield based on 2-benzoylpyrrole.

Even though this synthetic sequence (not optimized) is considerably more efficient than the sulfinate displacement route, it is obviously limited to those congeners of ketorolac that are not susceptible to the hydrogenolysis conditions. Furthermore, utilization of the dibromo compound **55** as the starting material is not reagent-sparing and the number of synthetic steps is unchanged from the sulfinate process because of the need to remove the unrequired bromine atom. The obvious solution to both problems was a good synthesis of 2-halo-5-acylpyrroles. In fact, 2-chloro-5-aroylpyrroles were shown to be readily accessible by the Vilsmeier-Haack acylation of 2-chloropyrrole.[70,71] The usual synthetic sequence was applicable to these compounds with the intramolecular displacement of chloride occurring readily at 100 °C. 2-Halopyrroles, however, are notoriously unstable entities,[73] the use of which on a large scale could never be seriously entertained. A very simple solution to this quandary was devised.[74] Reaction of the crude iminium salt **58** (Scheme 13), generated from pyrrole under Vilsmeier-Haack conditions,

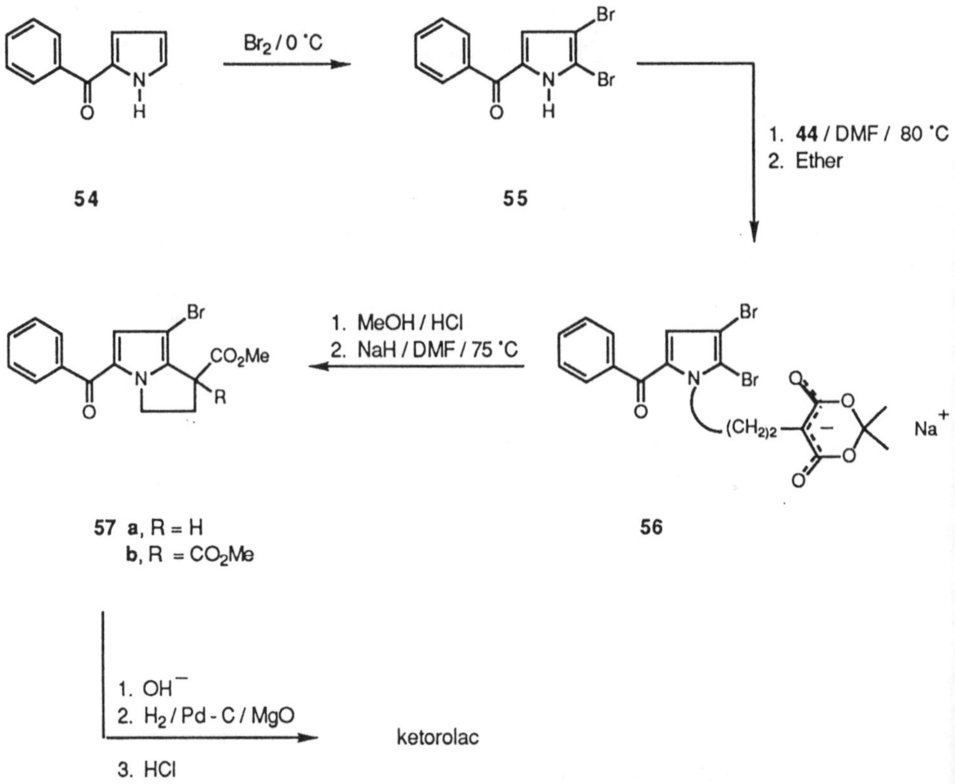

*Scheme 12.*

with excess potassium cyanide in acetonitrile, gave the "Strecker-type" nitrile **59** which was brominated exclusively at the vacant α-position with N-bromosuccinimide in THF at −78 °C. The unstable bromo compound thus obtained was then hydrolyzed, via the stable, isolable azafulvene **60**, to 2-bromo-5-benzoylpyrrole (**60**). The entire process could be carried out, in high yield, without purification of any of the intermediates. Compound **60** could be used as a source of ketorolac with an efficiency quite comparable to that observed for the dibromide **55**.[75]

## VII. IMPACT OF KETOROLAC ON PYRROLE CHEMISTRY AND ON PAIN THERAPY

### A. Pyrrole Chemistry

During the studies involving the preparation of ketorolac and congeners, several synthetic problems associated with pyrrole chemistry itself were encountered.

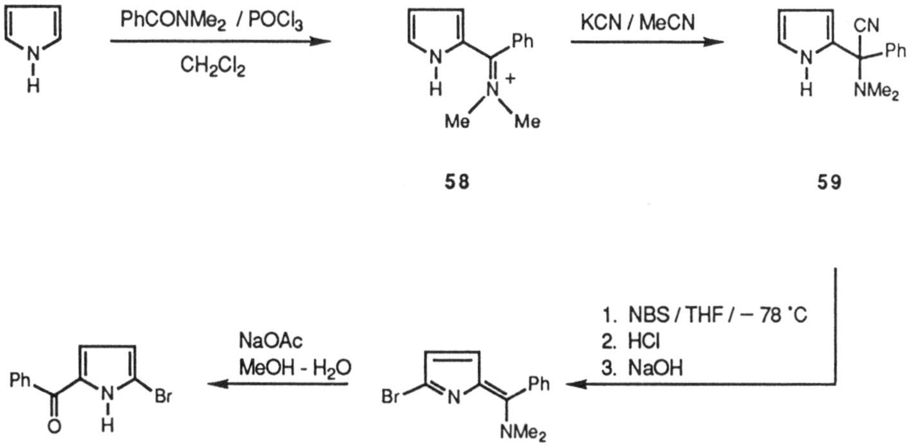

**Scheme 13.**

Some of these were of a fundamental nature and the investigation thereof continued even after the specific problems related to ketorolac were solved. Thus, there have been both direct and indirect consequences of the development of this entity on the chemistry of pyrroles. Specifically, a general route to β-substituted pyrroles has been devised based on the highly selective kinetic electrophilic substitution of *N*-triisopropylsilylpyrrole.[76] 3-Bromo-1-triisopropylsilylpyrrole, generated in this way, is easily converted by halogen-metal interchange into the 3-lithio compound and this formal 3-lithiopyrrole provides access to numerous other 3-substituted pyrroles.[76] It has also been established by us and others[21] that the acid-induced isomerization of the readily available α-substituted pyrroles frequently is a preparatively useful source of the β-isomers. In addition, we developed new routes to pyrrole-2-carboxaldehydes[77,78] and acylpyrroles,[79] and mild processes for reduction of acylpyrroles to alkylpyrroles[80] and acylpyrroles to acylpyrrolidines;[81] also the first examples of intramolecular carbenoid addition to the pyrrole nucleus were reported.[82] Therefore, the impact of the development of ketorolac on pyrrole chemistry has been significant. Not only is the methodology cited above being utilized by others (see refs. 83 and 84 for example), but we have used this information in developing new potential clinical entities in therapeutic areas quite unrelated to that of ketorolac.

## B.  Pain Therapy

The injectable formulation of ketorolac has recently been released for marketing in the U.S. (1989) New Zealand (1990) and Italy (1990), while permission to market the oral dosage form has so far been granted in Italy (1990) and Denmark (1990). As a consequence, the place of ketorolac in pain therapy still remains to be established. In view of its potency and relatively low incidence of side effects, its major initial use is likely to be as an alternative to morphine and other opiates for the relief of moderate to severe postoperative pain. The success of ketorolac in the marketplace for this and other indications will depend on several factors not the least of which will be the willingness of medical practitioners to prescribe what they may view as just another non-steroidal antiinflammatory/analgesic agent.[43]

## NOTE ADDED IN PROOF

Ketorolac in various dosage forms has now been approved for use in 18 countries worldwide, e.g., intramuscular (16), intravenous (2), oral (6), and opthalmic (5). The intramuscular (IM) and oral formulations are being marketed in Canada, Italy, Israel, Korea, and Spain while only the IM formulation has reached the marketplace in Sweden, the United Kingdom, and the United States. In the United States ketorolac has already made very significant inroads into those areas of pain therapy traditionally occupied by the opiates (e.g., emergency wards, postoperative pain, etc.).

## REFERENCES AND NOTES

1.  Winter, C.A. in Progress in Drug Research, Vol. 10, Ed by Jucker, E. Birkhauser-Verlag, Basel. 1966. pp. 163-166.
2.  Berkowitz, S.A.; Bernhard, G.; Bilka, P.J.; Blechman, W.J.; Marchesano, J.M.; Rosenthal, M.; Wortham, G.F. *Curr. Ther. Res.* **1974**, *16*, 442.
3.  "Zomepirac: A New Non-Narcotic Analgesic"; Proceedings of the Symposium on Zomepirac; Atlanta, GA, 1979: *J. Clin. Pharmacol.* **1980**, *20*, 213.
4.  Whereas tolmetin continues to play a significant role in the therapy of inflammation, zomepirac has been withdrawn from the marketplace because of a low incidence (< 0.01 %) of anaphylactic side effects.[5.]
5.  *F.D.C. Reports* **1983**, *45*(10), 19; **1983**, *45*(11), 3,4.
6.  Dunn, J.P.; Green, D.M.; Nelson, P.H.; Rooks, W.H.; Tomolonis, A. Untch; K.G. *J. Med. Chem.* **1977**, *20*, 1557.
7.  Ackrell, J.; Antonio, Y.; Franco, F.; Landeros, R.; Leon, A.; Muchowski, J.M.; Maddox, M.L.; Nelson, P.H.; Rooks, W.H.; Roszkowski, A.P.; Wallach, M.B. *J. Med. Chem.* **1978**, *21*, 1035.
8.  Dunn, J.P.; Muchowski, J.M.; Nelson, P.H. *J. Med. Chem.* **1981**, *24*, 1097.
9.  Carson, J.R.; Wong, S. *J. Med. Chem.* **1973**, *16*, 172.
10. The origin of this simple but logical concept was, in fact, somewhat different. During the course of a discussion of the structure-activity relationship of compounds such as **1-3** at a consultant's meeting, Dr. Arthur Kluge wondered if similar compounds incorporating the 1,2-dihydro-3*H*-pyrrolo[1,2-*a*]pyrrole system, as in **4** (*n* = 2), might be worth considering. This suggestion was prompted by his interest in the pheromones of certain tropical butterflies, one component of which was known[11] to contain the above mentioned bicyclic system. The relationship between inflammation and butterfly pheromones is not patently obvious!

11. Meinwald, J.; Meinwald, Y.E. *J. Am. Chem. Soc.* **1966**, *88*, 1305.

12. Iriarte, J.; Garcia, I. Unpublished data, Division de Investigacion, Syntex, S.A. de C.V., Mexico.

13. Carpio, H. Valdes, D. Unpublished data, Division de Investigacion, Syntex, S.A. de C.V., Mexico.

14. Carson, J.R.; McKinstry, D.N.; Wong, S. *J. Med. Chem.* **1971**, *14*, 646.

15. This process was devised by J. Ackrell and F. Franco, Division de Investigacion, Syntex, S.A. de C.V., Mexico.

16. Jones, R.A.; Bean, G.P. *The Chemistry of Pyrroles*. Academic Press, London. 1977. pp. 57-63.

17. A similarly selective esterification of **15** can be effected with methanol but careful monitoring of the progress of the reaction is necessary to minimize the formation of **14**.

18. This process has been optimized and can be conducted on a multi-kilo scale.

19. Carpio H.; Galeazzi, E.; Greenhouse, R.; Guzman, A.; Velarde, E.; Antonio, Y.; Franco, F.; Leon, A.; Perez, V.; Salas, R.; Valdes, D.; Ackrell, J.; Cho, D.; Gallegra, P.; Halpern, O.; Koehler, R.; Maddox, M.L.; Muchowski, J.M.; Prince, A.; Tegg, D.; Thurber, T.C. Van Horn; A.R.; Wren, W. *Can. J. Chem.* **1982**, *60*, 2295.

20. Muchowski, J.M.; Galeazzi, E.; Greenhouse, R.; Guzman A.; Perez, V.; Ackennan, N.; Ballaron, S.A.; Rovito, J.R.; Tomolonis, A.J.; Young, J.M.; Rooks, W.H. *J. Med. Chem.* **1989**, *32*, 1202.

21. DeSales, J.; Greenhouse, R.; Muchowski, J.M. *J. Org. Chem.* **1982**, *47*, 3668. Carmona, O.; Greenhouse, R.; Landeros, R.; Muchowski, J.M. *J. Org. Chem.* **1980**, *45*, 5336.

22. Olah, G.A.; Gupta, B.G.B.; Narang, S. *Synthesis* **1978**, 137.

23. Carson, J.R. U.S. Patent 3 998 844, 1976.

24. Carson, J.R.; Davis, N.M. *J. Org. Chem.* **1981**, *46*, 839.

25. Muchowski, J.M.; Unger, S.H.; Ackrell, J.; Cheung, P.; Cooper, G.F.; Cook, J.; Gallegra, P.; Halpern, O.; Koehler, R.; Kluge, A.F.; Van Horn, A.R.; Antonio, Y.; Carpio, H.; Franco, F.; Galeazzi, E.; Garcia, I.; Greenhouse, R.; Guzman, A.; Iriarte, J.; Leon, A.; Pena, A.; Perez, V.; Valdes, D.; Ackerman, N.; Ballaron, S.A.; Krishna Murthy, D.V.; Rovito, J.R.; Tomolonis, A.J; Young, J.M.; Rooks, W.H. *J. Med. Chem.* **1985**, *28*, 1037.

26. Nelson, P.H. in *CRC Handbook of Eicosanoids: Prostaglandins and Lipids. Vol. 11. Drugs Acting via the Eicosanoids*. Ed. by Willis, A.L. CRC Press Inc., Boca Raton, Florida. 1989. pp. 59-64.

27 Robert, A. *Gastroenterology* **1979**, *77*, 761.

28. Gund, P.; Shen, T.Y. *J. Med. Chem.* **1977**, *20*, 1146.

29. Unger, S.H. *Drug Design* **1980**, *9*, 48.

30. Muchowski, J.M.; Cooper, G.F.; Halpern, O.; Koehler, R.; Kluge, A.F.; Simon, R.L.; Unger, S.H.; Van Horn, A.R.; Wren, D.L.; Ackrell, J.; Antonio, Y.; Franco, F.; Greenhouse, R.; Guzman, A.; Leon, A.; Ackerman, N.; Ballaron, S.A.; Krishna Murthy, D.V.; Rovito, J.R.; Tomolonis, A.J.; Young, J.M.; Rooks, W.H. *J. Med. Chem.* **1987**, *30*, 820.

31. Sankawa, U.; Shibuya, M.; Ebizuka, Y.; Noguchi, H.; Kinoshita, T.; Iitaka, Y. *Prostaglandins* **1982**, *24*, 21.

32. Jackman, L.M. *Applications of NMR Spectroscopy in Organic Chemistry*. Pergamon Press; London, 1959. p. 125.

33. The 7-methyl compound (entry 23) was synthesized by A.R. Van Horn, unpublished data. For the synthesis of the 7-halo compounds (entries 24, 25) see ref. 34.

34. Muchowski, J.M.; Greenhouse, R. U.S. Patent 4,456,759, 1984.

35. Ketorolac was 2.8- and 5.8-fold more potent than indomethacin with regard to the inhibition of the prostaglandin synthetases from bovine seminal vesicles[36] and human platelet microsomes,[37] respectively.[25]

36. Tomlinson, R.V.; Ringold, H.J.; Qureshi, M.C.; Forchielli, E. *Biochem. Biophys. Res. Commun.* **1972**, *46*, 552.

37. Hammerstrom, S.; Falardeau, P. *Proc. Natl. Acad. Sci. USA* **1977**, *74*, 3691. Flower, R.J.; Cheung, H.S.; Cushman, D.W. *Prostaglandins* **1973**, *4*, 325.

38. Guzman, A.; Yuste, F.; Toscano, R.A.; Young, J.M.; Van Horn, A.R.; Muchowski, J.M. *J. Med. Chem.* **1986**, *29*, 589.

39. Riegl, J.; Maddox, M.L.; Harrison, I.T. *J. Med. Chem.* **1974**, *17*, 377.

40. Kamiya, K.; Wada, Y.; Nishikawa, M. *Chem. Pharm. Bull.* **1975**, *23*, 1589.

41. (a) Gordon, A.J.; Ford, R.A. *The Chemists Companion.* Wiley, New York. 1972. pp. 107–108.
    (b) Fleischer, E.B.; Sung, N.; Hawkinson, S. *J. Phys. Chem.* **1968**, *72*, 4311.

42  *Drugs of the Future* **1988**, *13*, 976.

43. Buckley, M. M-T.; Brogden, R.N. *Drugs* **1990**, *39*, 86.

44. Diebschlag, W.; Nocker, W.; Bullingham. *J. Clin. Pharmacol.* **1990**, *30*, 82.

45. Yee, J.; Brown, C.R.; Sevelius, H.; Wild, V. *Clin. Pharmacol. Ther.* **1984**, *35*, 285.

46. Mroszczak, E.J.; Ling, T.; Yee, J.; Massey, I.; Phil, D.; Sevelins, H. *Clin. Pharmacol. Ther.* **1985**, *37*, 215.

47. Yee, J.P.; Koshiver, J.E.; Allbon, C.; Brown, C.R. *Pharmacotherapy* **1986**, *6*, 253.

48. O'Hara, D.A.; Fragen, R.J.; Kinzer, M.; Pemberton, D. *Clin. Pharmacol. Ther.* **1987**, *41*, 556.

49. Rubin, P.; Yee, J.P.; Murthy, V.S.; Seavey, W. *Clin. Pharmacol. Ther.* **1987**, *41*, 182.

50. Lopez, M.; Waterbury, D.L.; Michel, A.; Seavey W.; Yee, J. *Pharmacologist* **1987**, *29*, 136.

51. Yee, J., Bradley, R., Stanski, D., Cherry, C. *Clin. Pharmacol. Ther.* **1986**, *39*, 237.

52. Fricke, J.; Angelocci, D. *Clin. Pharmacol. Ther.* **1987**, *41*, 181.

53. Estenne, B.; Julien, M.; Charleux, H.; Arsac, M. Arvis, G.; Loygue, J. *Curr. Ther. Res.* **1988**, *43*, 1173.

54. Bloomfield, S.S.; Mitchell, J.; Cissell, G.B.; Barden, T.P.; Yee, J.P. *Pharmacotherapy* **1986**, *6*, 247.

55. Rubin, P.; Murthy, V.S.; Yee, J. *Clin. Pharmacol. Ther.* **1987**, *41*, 229.

56. Lanza, F.L.; Karlin, D.A.; Yee, J.P. *Am. J. Gastroenterol.* **1987**, *82*, 939.

57. McQuay, H.J.; Poppleton, P.; Carroll, D.; Summerfield, R.J.; Bullingham, R.E.S.; Moore, R.A. *Clin. Pharmacol. Ther.* **1986**, *39*, 89.

58. Forbes, J.A.; Butterworth, G.A.; Kehm, C.K.; Grodin, C.D.; Yee, J.P.; Beaver, W.T. *Clin. Pharmacol. Ther.* **1987**, *41*, 162.

59. Vangen, O.; Doessland, S.; Lindbaek, E. *J. Int. Med. Res.* **1988**, *16*, 443.

60. Flach, A.J.; Lavelle, C.J.; Olander, K.W.; Retzlaff, J.A.; Sorenson, L.W. *Opthamology* **1988**, *95*, 1279.

61. Kantor, T.G.; Hopper, M.; Peterson, K.; Koshiver, J.; Schwartz, K. *Clin. Pharmacol. Ther.* **1986**, *39*, 202.

62. Brown, C.; Wild, V.; Peterson, K.; Koshiver, J.; Schwartz, K. *Clin. Pharmacol. Ther.* **1986**, *39*, 183.

63. Bloomfield, S.S.; Cissell, G.B.; Peters, N.M.; Mitchell, J.; Nelson, E.D.; Barden, T.P. *Clin. Pharmacol. Ther.* **1987**, *42*, 89.

64. Vecchieti, V.; Dradi, E.; Lauria, F., *J. Chem. Soc. (C)* **1971**, 2554.

65. Danishefsky, S. *Acc. Chem. Res.* **1979**, *12*, 66.

66. Franco, F.; Greenhouse, R.; Muchowski, J.M. *J. Org. Chem.* **1982**, *47*, 1682.

67. Tomita, K.; Terada, A.; Tachikawa, R. *Heterocycles* **1976**, *4*, 729.

68. Berkoz, B.; Lewis, B.; Muchowski, J.M. U.S. Patent 4,064,135, 1977; *Chem. Abstr.* **1978**, *88*, 170130 b.

69. Kinetic monobromination of **54** produces 2-benzoyl-4-bromopyrrole.

70. Muchowski, J.M.; Greenhouse, R. U.S. Patent 4,873,340, 1989.

71. Vilsmeier-Haack aroylation of 2-bromopyrrole gave 2-chloro-5-aroylpyrroles.[72]

72. Franco, F.; Greenhouse, R. Unpublished observations, Division de Investigacion, Syntex, S.A. de C.V., Mexico.

73. Cordell, G.A. *J. Org. Chem.* **1975**, *40*, 3161; Gilow, H.M.; Burton, D.E. *J. Org. Chem.* **1981**, *46*, 2221.

74. Bray, B.L.; Muchowski, J.M. *Can. J. Chem.* **1990**, *68*, 1305.

75. Bray, B.L. Unpublished observations, Syntex Research, Institute of Organic Chemistry, Palo Alto, CA.

76. Bray, B.L.; Mathies, P.H.; Naef, R.; Solas, D.R.; Tidwell, T.T.; Artis, D.R.; Muchowski, J.M.; *J. Org. Chem.* **1990**, *55*, 6317.

77. Muchowski, J.M.; Hess, P. *Tetrahedron Lett.* **1988**, *29*, 777; Muchowski, J.M.; Hess, P. *Tetrahedron Lett.* **1988**, *29*, 3215.

78. Bray, B.L.; Hess, P.; Muchowski, J.M.; Scheller, M.E. *Helv. Chim. Acta.* **1988**, *71*, 2053.

79. Bray, B.L.; Muchowski, J.M. *J. Org. Chem.* **1988**, *53*, 6115.

80. Greenhouse, R.; Ramirez, C.; Muchowski, J.M. *J. Org. Chem.* **1985**, *50*, 2961.

81. Kaiser, H-P.; Muchowski, J.M. *J. Org. Chem.* **1984**, *49*, 4203.

82. Galeazzi, E.; Guzman, A.; Pinedo, A.; Saldana, A.; Torre, D.; Muchowski, J.M. *Can. J. Chem.* **1983**, *61*, 454.

83. Jefford, C.W.; Johncock, W. *Helv. Chim. Acta* **1984**, *66*, 2666; Jefford, C.W.; Tang, Q.; Zaslona, A. *Helv. Chim. Acta* **1989**, *72*, 1749.

84. Ksander, K.; Bold, G.; Lattmann, R.; Lehmann, C.; Fruh, T.; Xiang, Y-B.; Inomata, K.; Buser, H-P.; Schreiber, J.; Zass, E.; Eschenmoser, A. *Helv. Chim. Acta* **1987**, *70*, 1115.

# APPLICATION OF SILICON CHEMISTRY IN THE CORTICOSTEROID FIELD

Douglas A. Livingston

**Advances in Medicinal Chemistry,**
**Volume 1, pages 137–174.**
**Copyright © 1992 by JAI Press Inc.**
**All rights of reproduction in any form reserved.**
**ISBN: 1-55938-170-1**

# I. INTRODUCTION

In this chapter, I would like to describe the development of new, silicon-based methodology[1] for the extension of cyanohydrins, which has commercial potential in its application to the synthesis of steroids of the pregnane family, especially the corticosteroids.[2] These hormones and hormone mimics have great therapeutic importance, with uses ranging from antiinflammatory drugs to birth control agents. Our work in this area was part of a larger program at the Upjohn Company, termed the "sitosterols utilization project", which has been active in the Chemical Process Research and Development Unit since the early 1970s.

In order for you to fully appreciate the need for new methodology in this area, it is important that I place this project in its proper historical context. Thus, I will first present some key background material, along with a brief review of methods for the synthesis of corticosteroids, developed both at Upjohn and at other companies.

# II. HISTORY[3]

The production of corticosteroids by Upjohn technically can be traced back to 1935, when a standardized, epinephrine-free adrenal cortex extract was offered by the company. This extract was used primarily in the treatment of Addison's disease and other disorders of the adrenal cortex, which constituted a very limited market. Hormone research at Upjohn, and at many other institutions, proceeded on a relatively even front during the following decade, with the active components of the adrenal cortex being isolated and characterized one by one.[4] However, no truly significant products appeared during this period.

It is well known that synthetic efforts towards the corticosteroids were pioneered by researchers at Merck during the Second World War, culminating in Sarett's famous synthesis of cortisone (11) in 1946 (Fig. 1).[5] In 1948, Philip S. Hench of the Mayo Clinic used synthetic cortisone from Merck to demonstrate its antiarthritic properties, eventually treating 14 severely crippled patients.[6] When this work was reported at a rheumatological meeting in New York in April of 1949, the commercial significance of corticosteroids was instantly recognized by many drug companies, including Upjohn.

The cost of the original Merck cortisone has been estimated at $200/gram in 1949 dollars.[7] This material was prepared from bile acid, a mixture of cholic (1) and desoxycholic (2) acids, obtained in 5–6% yield by the hydrolysis of bovine bile. The original synthesis required 33 steps, the majority of which involved walking the 12α-hydroxyl to the 11-position and manipulating the side chain (Fig. 1). The overall yield was about 3%. Improvements to this process eventually allowed the production of over 3000 kilograms in 1952, when the price dropped to around $10/gram.

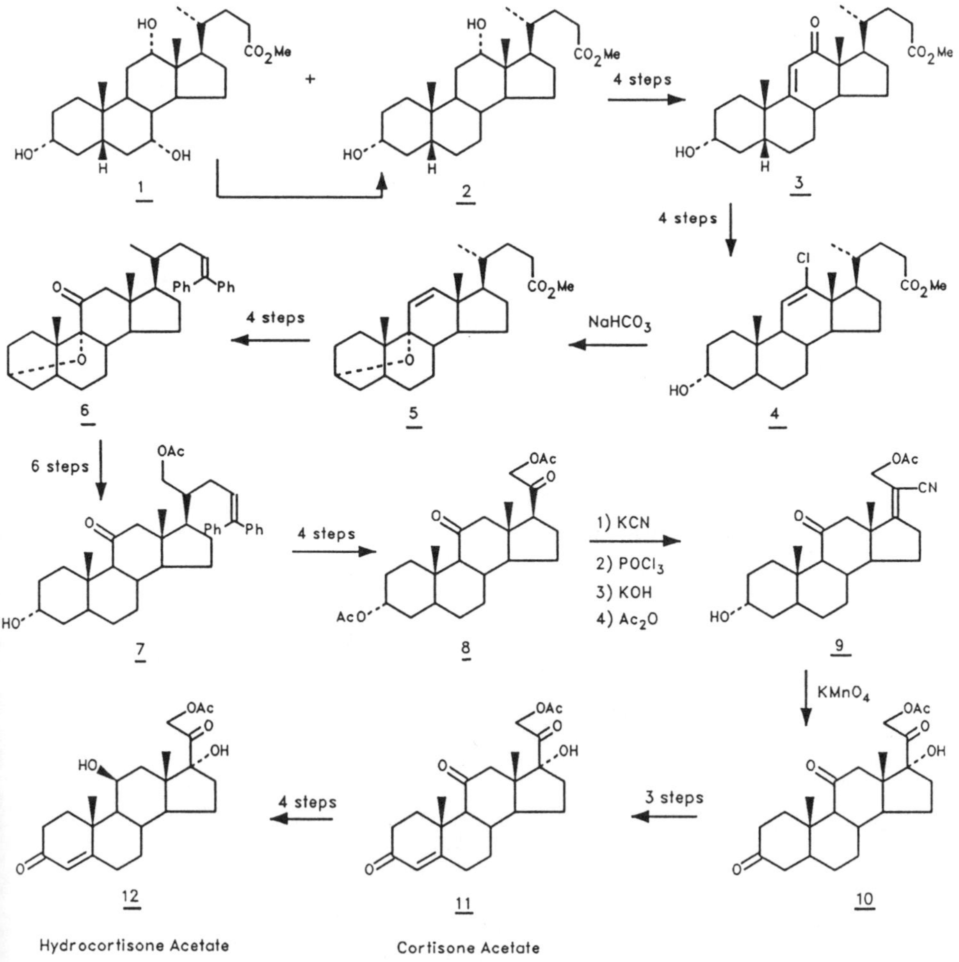

*Figure 1.*

Upjohn entered the synthetic cortisone market with a more efficient process that same year, and began marketing hydrocortisone (cortisol) in 1953. In the subsequent years of 1955 to 1958, the price of these corticosteroids gradually declined from $3.50/gram to $1.90/gram. The current price of hydrocortisone acetate, which is now the market leader in volume, is about $0.67/gram, or just $0.19/gram in 1958 dollars.[8] This impressive reduction in price is a reflection of several factors, not the least of which is steady improvements in synthetic methodology. Currently, the total world market for corticosteroid pharmaceuticals is over $1 billion, and Upjohn is the largest producer of bulk drug substance.

## III.  THE ORIGIN OF THE "SITOSTEROL PILE"

Two factors were primarily responsible for the efficiency of the Upjohn corticosteroid process. First, there was the use of progesterone as a key intermediate,
which was obtained from a unique source—the soy sterols. Second, there was the
use of certain mold cultures to introduce an 11α-hydroxyl group into progesterone
directly, a method first demonstrated by D.H. Peterson, H.C. Murray and co-
workers (see below).

The company had desired to get into the synthetic progesterone market following World War II, but found that the routes from cholesterol, a logical steroid
starting material, were blocked by European patents. A research program aimed at
identifying other steroid precursors eventually resulted in the processes of Heyl
and Herr[9] for the production of progesterone from stigmasterol.

Stigmasterol (**13**) represents about 20% of the "soy sterols", a mixture of sterols
and sterol esters found in soybean oil (Fig. 2). These sterols are concentrated in a
distillate fraction during routine purification of the oil. Part of the cost of processing
this distillate is recovered by the sale of tocopherols (vitamin E), also found in the
fraction.

Oxidation of the 3-hydroxyl and concomitant migration of the double bond to
give stigmastadienone (**14**), followed by ozonolysis of the 22–23 side-chain double
bond, leads to 3-keto-bis-nor-4-cholenaldehyde (BNA; **16**, Fig. 3). Of the various

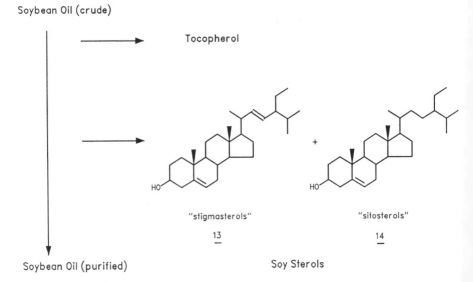

*Figure 2.*

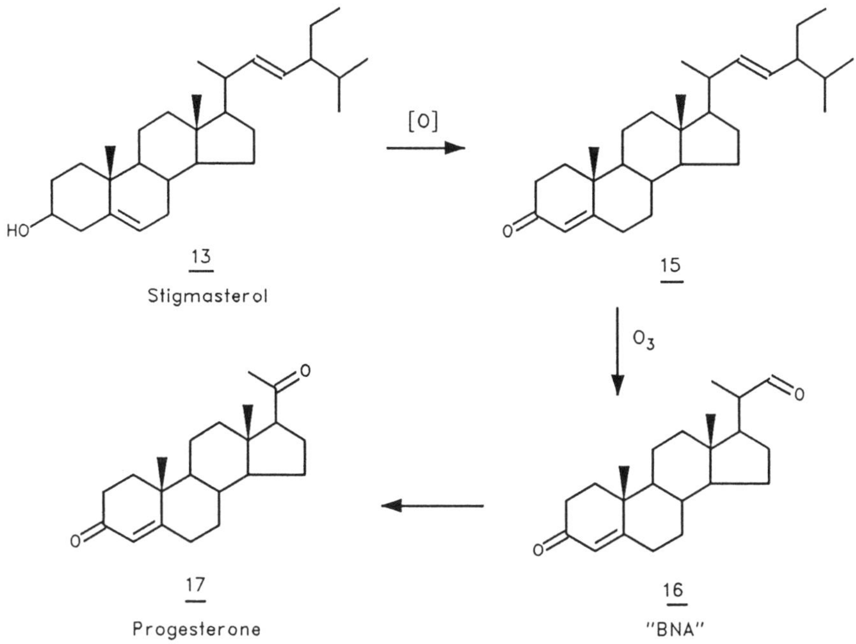

**13**

Stigmasterol

[O]

**15**

O₃

**17**

Progesterone

**16**

"BNA"

**Figure 3.**

methods that exist for oxidative excision of carbon-22 to give progesterone, two prominent ones are ozonolysis of the enol acetate[9b] or oxidation of the enamine.[9c,10]

With the ready availability of large amounts of progesterone, this compound became a logical substrate for the bioconversion studies of Murray and Peterson. They began a program of searching for microorganisms capable of introducing the key 11-oxy functionality present in the biologically active corticosteroids, efficiently screening for activity with the then-new technique of filter paper chromatography. Ultimately, a mold culture garnered from a windowsill in their laboratory successfully performed the 11α-hydroxylation of progesterone.[11] Following up on this discovery, they eventually identified a strain of *Rhizopus nigricans* that could accomplish this remarkable transformation in 90% yield (Fig. 4)![12]

The 11α-hydroxyl can be easily oxidized to the 11-ketone, found in cortisone (**11**, Fig. 1), which can then be reduced by hydride to the 11β-hydroxyl, the correct orientation found in hydrocortisone (**12**, Fig. 1), at an appropriate point in the sequence. Oxidation of the progesterone side chain to the dioxyacetone side chain found in the corticosteroids can be accomplished via the Favorskii sequence depicted in Figure 5.[13]

The efficiency of this progesterone-based process was further enhanced by the

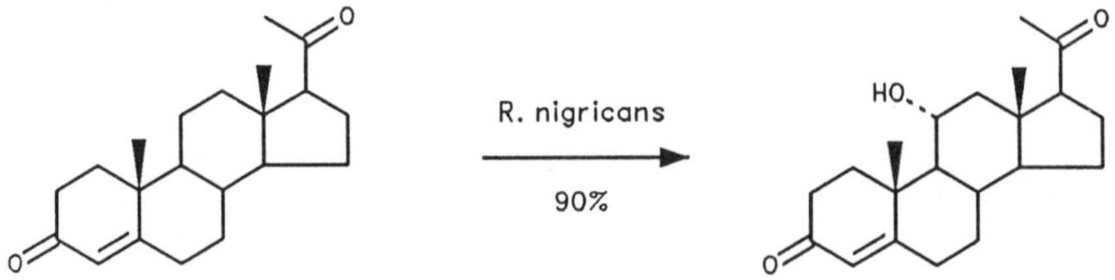

Figure 4.

work of W. Greiner and G. Fevig, who devised a countercurrent crystallization method for isolating the stigmasterol fraction from the soy sterols.[14] The residue from this separation, composed of approximately 80% of the original soy sterol mixture, was stockpiled in the back of the Upjohn plant in the form of 3000-pound "pigs". Of course, it was hoped that someday a method for the utilization of the sitosterol fraction might be developed to capitalize on this by-product of a by-product. As Upjohn grew to become the world's largest producer of corticosteroids, this sterol accumulation eventually took on the dimensions of a small mountain, a particularly prominent landmark in the surrounding flat Michigan landscape (Fig. 6).

The key to sitosterol utilization required the efficient degradation of the unfunctionalized C-17 side chain. Chemical means of accomplishing this are still not good enough; however, microbiologists were able to develop microorganisms that are capable of selectively degrading the side chain.[15]

Fortunately, microbes capable of consuming plant sterols are ubiquitous; otherwise, the world would be dotted with steroid mountains like the one in Kalamazoo. The pathway for microbial degradation has been well worked out.[16] The side chain is oxidatively degraded from the terminus, but progress is blocked at the 17-position. Simultaneously, degradation of the ring structure begins with oxidation at C-3

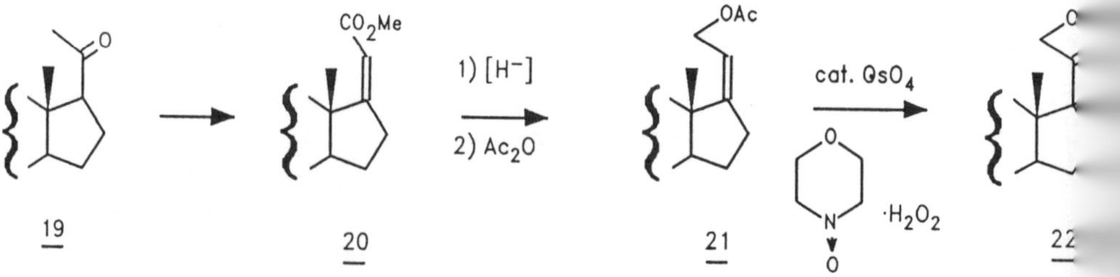

Figure 5.

**Figure 6.** The "Sitosterol Pile" (with the author skiing in Michigan)

and proceeds through introduction of $\Delta^{1,2}$, and 9α-hydroxylation to the unstable 9α-hydroxy-1,4-dienone (Fig. 7). Spontaneous fragmentation to the B-seco compound is followed by further oxidation to the 9,17-diketo-5-carboxylic acid, and eventually to carbon dioxide and water.

Critical intermediates in this pathway include the useful 17-ketones, androst-4-ene-3,17-dione (AD, **24**), androsta-1,4-diene-3,17-dione (ADD, **25**), and 9α-hydroxyandrost-4-ene-3,17-dione (9-OH-AD, **23**). In the early seventies, W. Marsheck and co-workers at Searle,[17] and later M. Wovcha and co-workers at Upjohn,[18] were successful in developing microbes that are inherently blocked from continuing the biodegradation beyond these points. Their strategy was to mutate particularly effective sterol degrading species of *Mycobacteria*, which are capable of surviving on sitosterol as their sole carbon source. The mutant colonies were then individually transferred to a medium containing a 17-keto steroid such as AD, **24**, as the only source of carbon. Those colonies that were incapable of surviving on the 17-keto steroid were cultured as candidates for full-scale production.

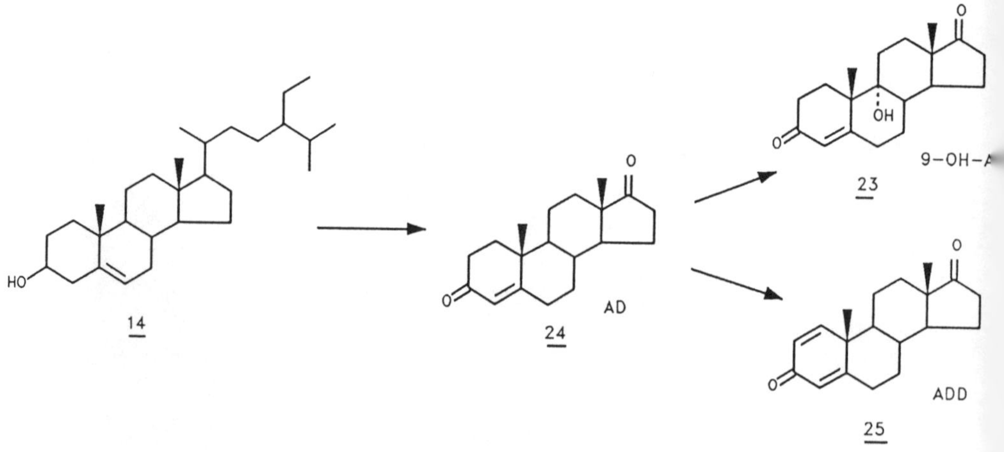

**Figure 7.**

Note that oxidation at the 3-position and migration of the double bond occur early in the degradation, giving the required 3-keto-4-ene "for free". Through selective protection of the 3-ketone and addition of nucleophiles from the less hindered α-face, AD is an ideal precursor for testosterone (**27**) various androgens [such as methyltestosterone (**28**) and testosterone esters], and the progestational agent, ethisterone (**29**, Figure 8).

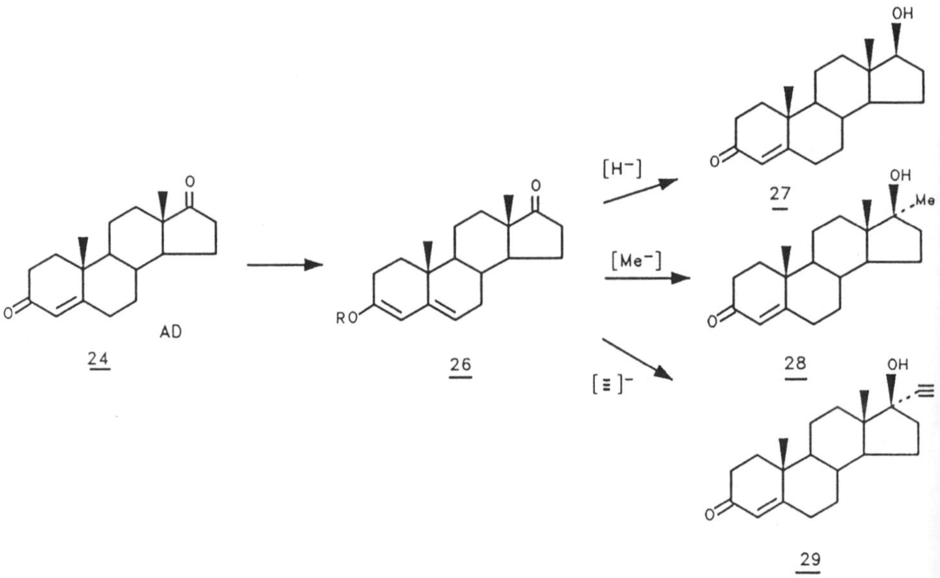

**Figure 8.**

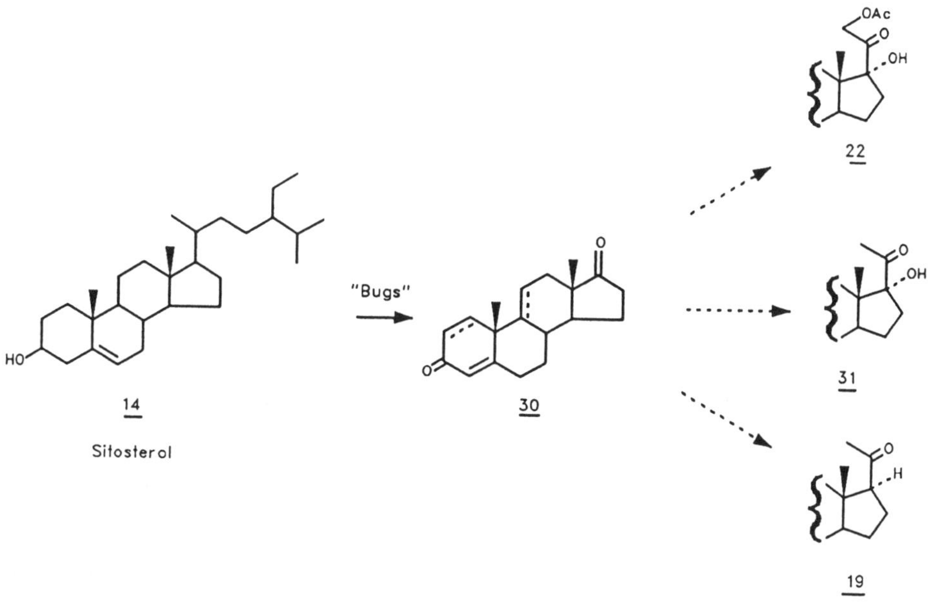

**Figure 9.**

The production of these hormones has provided an outlet for some sitosterol, but the market is small compared to the corticosteroids and related pregnanes. Upjohn currently sells about 34 different compounds containing the corticosteroid side chain (22), seven 17α-hydroxy progesterones (31), and two progesterones (19, Fig. 9).

Considerable technology exists for the elaboration of the A,B, and C rings from the above fermentation-derived substitution patterns. The real challenge has classically been in the efficient construction of the three required side chains from the 17-ketone.

The criterion for synthetic elegance is clear in this case: the methodology must accomplish the assembly of the carbon–carbon bond and add the oxygen substituents, and the stereocenter, at lower cost than the oxidative degradation through progesterone.

Although only one carbon–carbon bond technically needs to be made, the challenges are formidable. Beyond the tendency of the hindered, 5-membered ring ketone to enolize, nucleophiles that add successfully always do so from the α-face, leading to the incorrect orientation of the C-17 hydroxyl in kinetic additions. In the case of the corticosteroid side chain, the high oxidation state requires that either a complex nucleophile be added, or that subsequent elaboration be carried out.

Given the potential for profit, a wide variety of methods of varying efficiency have been developed over the years to solve these problems. We can divide these into two broad classes on the basis of their solution to the stereochemical problem. There are methods that put in the correct configuration of the 17-hydroxyl by initially forming a trigonal C-17, followed by kinetic oxidation from the α-face. Then, there are methods that rely on inversion of a 17β-hydroxyl, produced by kinetic attack of the nucleophile from the back (α) face. Since a complete review of this prior methodology is beyond the scope of this chapter, the interested reader is directed to a collection of excellent reviews on the subject.[19] However, to give a feeling for what has been accomplished in the field, a brief summary of some recent methodology follows.

## IV. PRIOR METHODS FOR SIDE-CHAIN INTRODUCTION

An example of the trigonal C-17 approach is the so-called chloroaldehyde chemistry, due to Hessler and Van Rheenen (Upjohn),[20] with an extension by Walker (Fig. 10).[21]

In this method, vinyllithium reagent (**33**) is added to the 17-ketosteroid (**32**). Hydrolysis on work-up results in elimination of the 17-hydroxyl, leading to

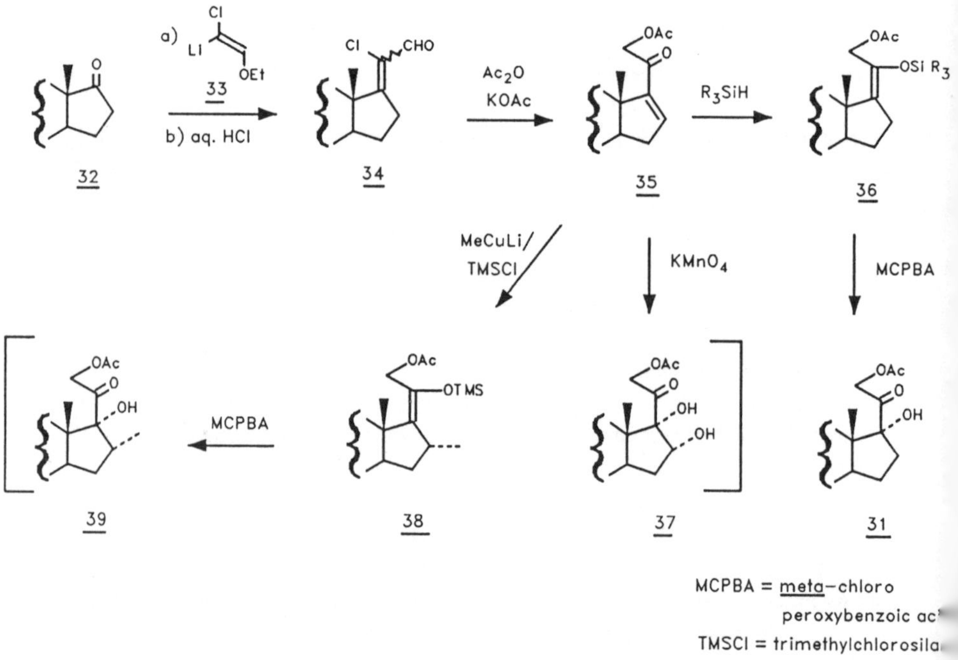
a)
33
b) aq. HCl
Ac2O
KOAc
R3SiH
MeCuLi/
TMSCl
KMnO4
MCPBA
MCPBA
32
34
35
36
39
38
37
31
MCPBA = meta-chloro
peroxybenzoic ac
TMSCl = trimethylchlorosila

*Figure 10.*

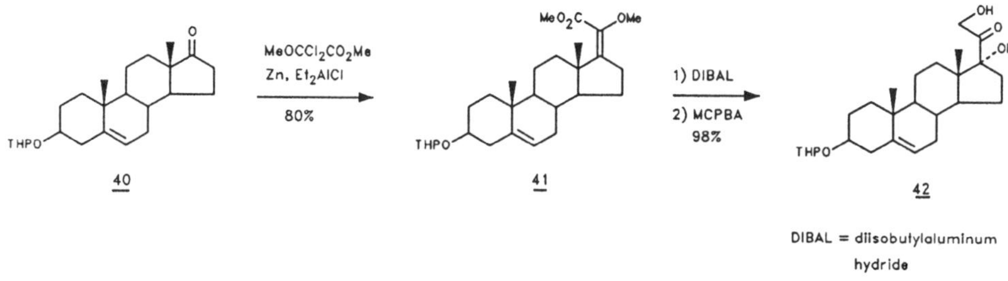

DIBAL = diisobutylaluminum hydride

**Figure 11.**

α,β-unsaturated, α-chloroaldehyde (**34**). Subsequent reaction with potassium acetate and acetic anhydride initiates a series of acetate migrations, displacement of the α-chloro group, which ultimately results in the $\Delta^{16}$ corticoid (**35**). Noble metal-catalyzed hydrosilylation gives the intermediate silyl enol ether (**36**) and subsequent oxidation produces the parent corticosteroid structure (**22**).[22]

Intermediate **35** is also an excellent precursor to 16-substituted corticosteroids, members of the "specialty steroid" class. These highly functionalized corticosteroids were introduced more recently than the "base" corticosteroids that have been discussed up until now. The volume of bulk drug for these specialties is not generally large, but the value is high. Triamcinolone derivatives (cf. **37**), can be prepared by direct hydroxylation of the 16- and 17-positions from the α-face of **35**.[23] Alternatively, nucleophiles such as methyl cuprate can be added conjugatively to the 16-position, and the resulting enolate or silyl enol ether (**38**) can be subsequently epoxidized and hydrolyzed to give 16α-methyl, "dexamethasone" derivatives (cf. **39**).[24]

An interesting case wherein the double bond is oriented in the opposite direction, toward C-20 in the key intermediate **41**, was reported by Daniewski and Wojciechowska.[25] This is formed by the condensation of 17-ketone (**40**) with the Takai reagent derived from trichloroacetic ester, and then displacement with methoxide. More directly, this can be obtained by condensation with dichloromethoxyacetic ester, as is depicted in Figure 11.

Reduction of the ester, oxidation of the enol ether, and hydrolysis gives the corticosteroid, (**42**). The high yields and inexpensive reagents make this route look attractive. A number of analogous approaches exist with other hetero groups at the pro-20 position, which also give the 20-ketone on oxidation.[26] These include isonitriles,[27] heterocycles,[28] and nitriles,[29] the latter being conceptually similar to the original Sarett synthesis (see intermediate **9** in Fig. 1).[30]

An elegant example of the hydroxyl inversion approach is contained in the chemistry of Shephard and Van Rheenen (Upjohn), the so-called allene-sulfoxide method.[31] A synthesis of hydrocortisone acetate from 9-OH-AD (**23**) by use of this

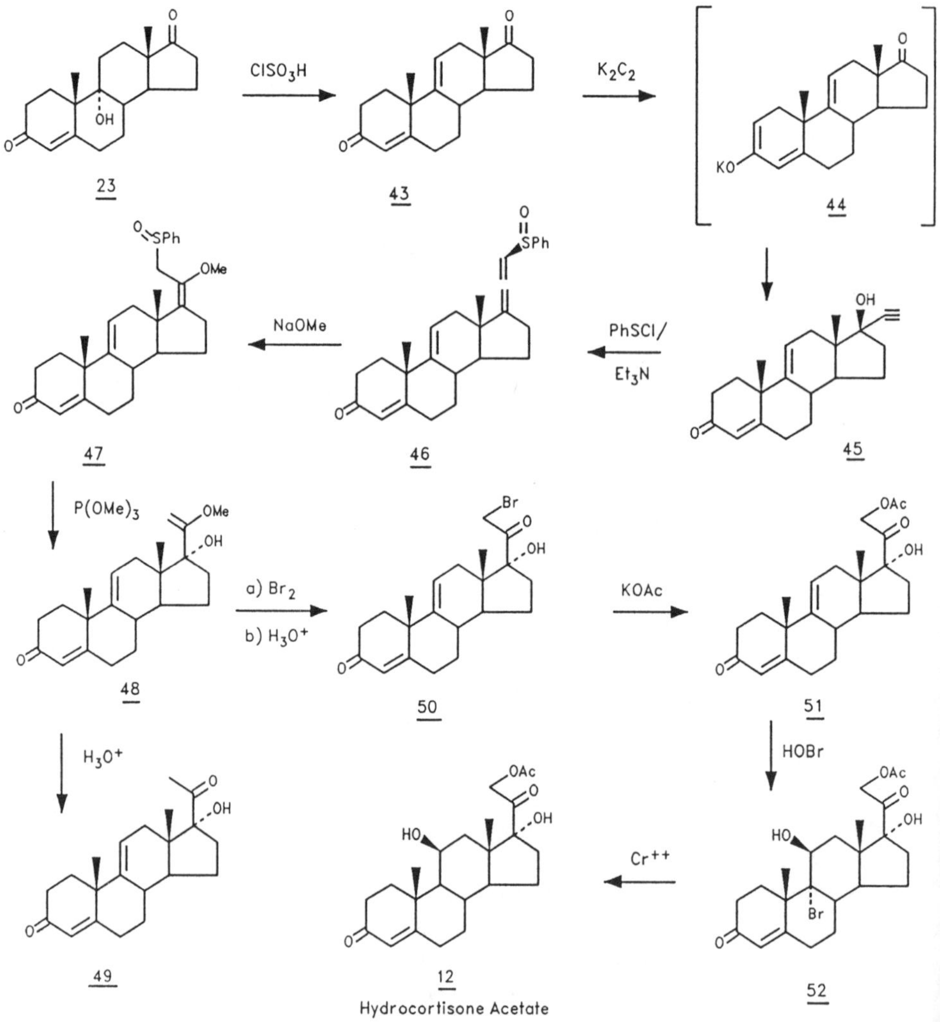

*Figure 12.*

chemistry, and incorporation of an 11β-hydroxyl as originally developed by Barton's RIMAC group,[32] is depicted in Figure 12.

The 9α-hydroxyl can be efficiently eliminated specifically to the $\Delta^{9,11}$-position (structure **43**) with strong acid,[33] or via the sulfinate ester.[34] The corresponding ethisterone derivative (**45**) is prepared by the selective addition of acetylene to the 17-ketone.[35] This selectivity actually results from prior, selective enolization of the 3-ketone by the reagent **44**. Condensation of the product propargyl alcohol (**45**)

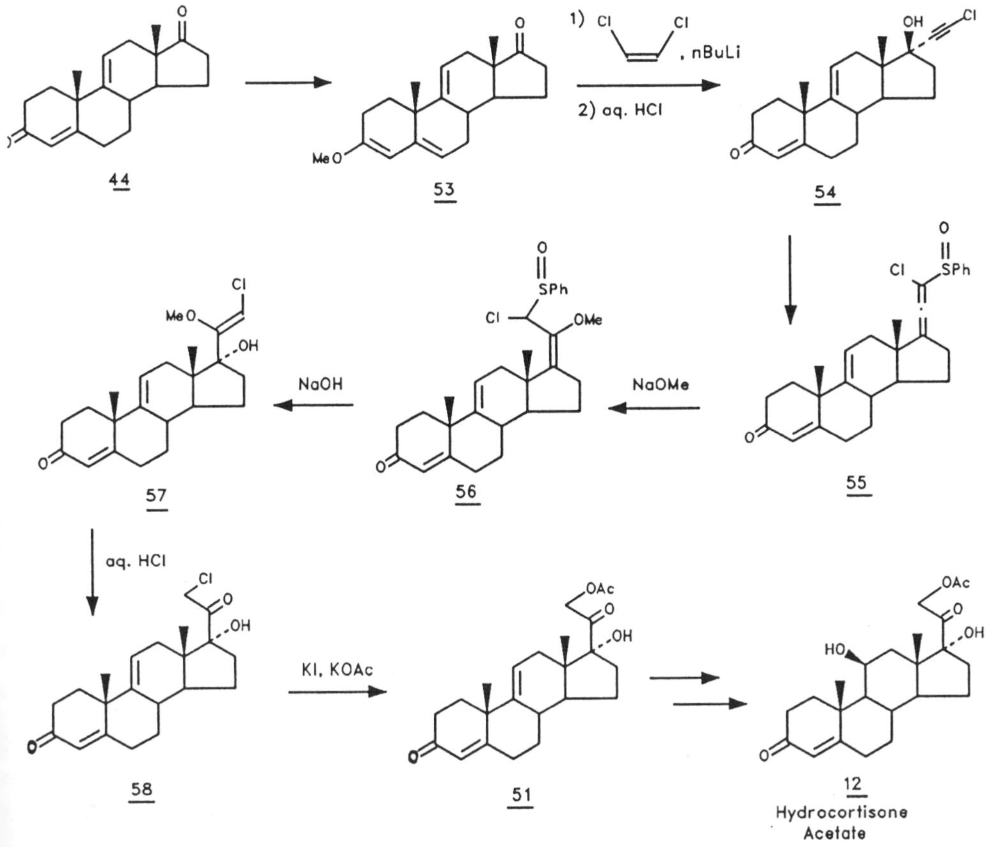

*Figure 13.*

with phenylsulfenyl chloride initiates rearrangement to the allene sulfoxide (**46**). This is treated (in situ) with sodium methoxide, which conjugatly adds to the pro-20 position of **46**. Warming sets up a mobile allyl sulfoxide-sulfenate equilibrium via a [2,3] sigmatropic shift. Addition of a thiophile, such as trimethyl phosphite, results in desulfurization from the sulfenate stage of the equilibrium, resulting in a net overall inversion of the 17-hydroxyl, as is found in $\Delta^{20,21}$ methyl enol ether (**48**). Hydrolysis affords $\Delta^{9,11}$-17$\alpha$-hydroxyprogesterone (**49**), which is a cor-ticosteroid precursor via Stork iodination (see Fig. 20, page 153). Oxidation of the enol ether directly to the corticosteroid can be accomplished via epoxidation and hydrolysis,[31,36] but it is more efficient to first brominate and hydrolyze to the 21-bromide (**50**), and then displace with potassium acetate to yield the $\Delta^{9,11}$ corticosteroid (**51**). Trans-diaxial addition of hypobromous acid across the 9,11

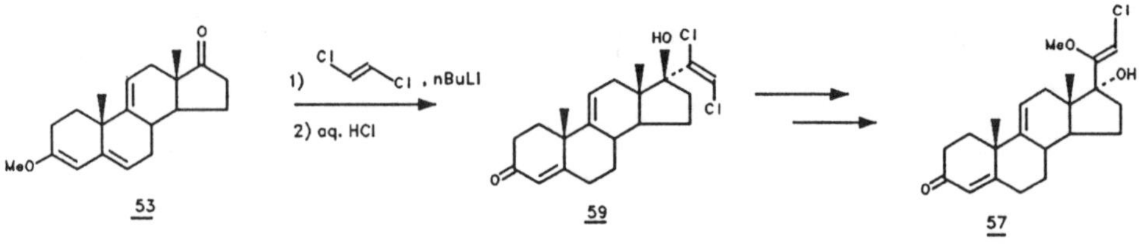

**Figure 14.**

double bond, followed by a free radical debromination yields hydrocortisone acetate (**12**).

An analogous method, wherein the 21-halide is introduced via chloroacetylene (**54**), and then carried through the process, was later developed by Walker at Upjohn[37] (Fig. 13).

In addition to avoiding the need for a separate oxidation step, desulfurization of the 17-sulfenate derived by rearrangement of allyl sulfoxide (**56**) occurs without requiring an added thiophile. However, protection of the 3-ketone against nucleophilic attack by the reagent is required with this methodology.

An analogous sequence from lithio *trans*-dichloroethylene has also been reported (Fig. 14).[38]

An example of a more direct inversion of the ethisterone derivative, via the nitrate ester (**61**), was recently developed by Nitta et al. at Mitsubishi (Fig. 15).[39] These workers reported that solvolysis of the nitrate ester was "preferable" among various esters that were studied. Acetylation of the 17-alcohol was necessary for the bromination/hydrolysis from **62**, presumably because participation by the 17-acetate was needed to stabilize an intermediate carbocation. However, with the 17-acetate in place, the 21-bromide in **63** is very difficult to displace; hydrolysis of the hindered 17-acetate is almost impossible without prior migration to a free 21-hydroxyl. Structures such as **63** thus represent something of a synthetic dead end.

Hofmeister and co-workers of Schering AG[40] recently demonstrated an alterna-

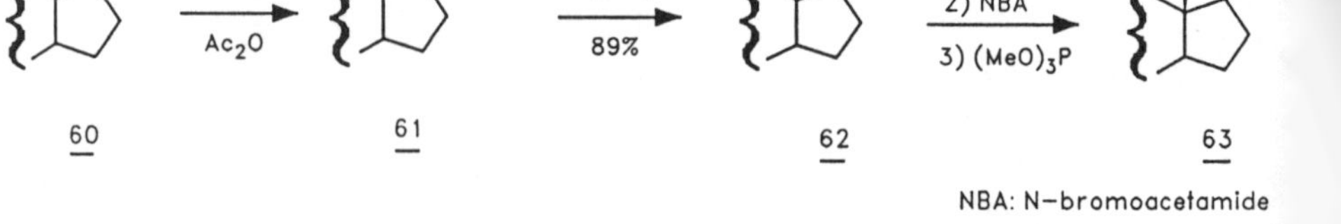

NBA: N-bromoacetamide

**Figure 15.**

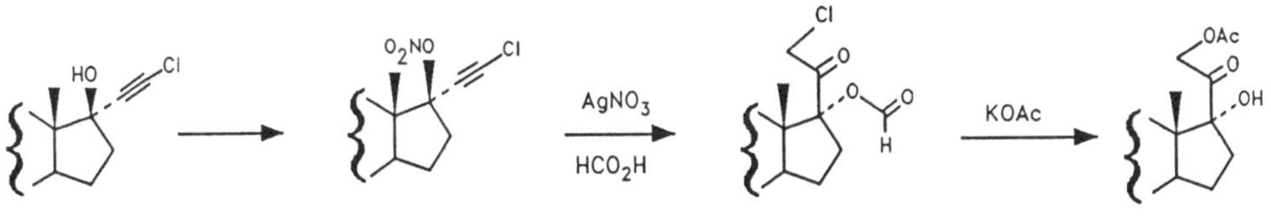

**Figure 16.**

tive, here again by starting with the acetylene in a higher oxidation state (Fig. 16; see also Fig. 13). In this case, the lability of 17-formate (**66**) allows direct conversion to the corticosteroid in good yield.

As far as hydroxyl inversion methods are concerned, probably the most efficient way to establish the correct relative stereochemistry between the 17-hydroxyl and the carbon side chain is by formation of the 17-cyanohydrin. It is a surprising but general property of steroid 17-cyanohydrins that the β-cyanohydrin (e.g., **68**) is more crystalline than the α-cyanohydrin (e.g., **67**, Fig. 17). Although addition of cyanide to the 17-ketone occurs first from the α-face, an equilibrium between the α- and β-cyanohydrins can be set up under conditions where the β-cyanohydrin will selectively crystallize from the reaction mixture in a classic second-order asymmetric transformation.

By this method, the equilibrium can be driven almost completely to the β-isomer, which is isolated by filtration of the crude reaction mixture. Addition to the 17-ketone is usually regioselective with respect to 4-ene-3-ones (such as AD, **24**). Even under equilibrating conditions, very little addition either to the 5-position (Michael fashion) or to the 3-position is found in the product.[41]

The first application of the cyanohydrin equilibration method in pregnane synthesis was due to Gasc and Nedelec (Roussel) in the early 1970s, as part of a commercial synthesis of $\Delta^{9,11}$-17α-hydroxyprogesterone (**49**) from 19-norsteroid (**69**) (Fig. 18).[42]

De Ruggieri and Ferrari (Richter) had earlier applied a 17-cyanohydrin protection/methylation/hydrolysis sequence (Fig. 19) in the conversion of 17-

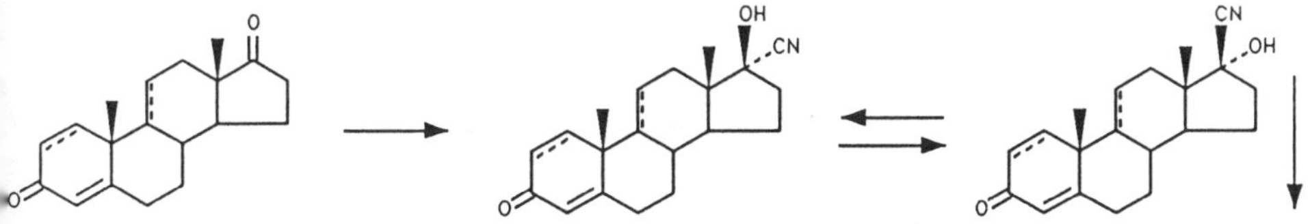

**Figure 17.**

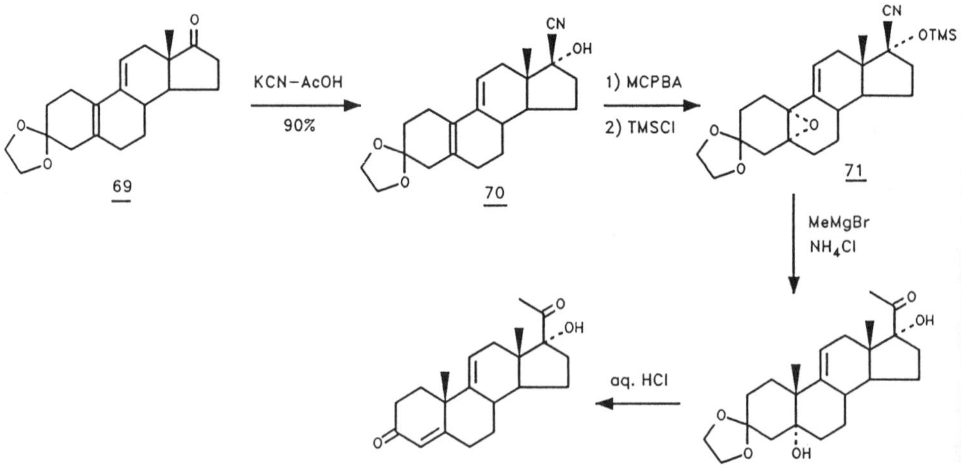

**Figure 18.**

ketosteroids to 17α-hydroxyprogesterones. Their synthesis proceeded from the epimeric mixture of 17-cyanohydrins, however, with separation effected by crystallization at the stage of protected cyanohydrin (**74**).

The above methodology was applied to AD (**24**) in the early 1980s by Teichmueller and co-workers at Jenapharm (DDR),[44] and then by Nitta and co-workers at Mitsubishi[45] for the synthesis of 17α-hydroxyprogesterone (**76**, Fig. 20). The latter workers further expanded the generality of the selective cyanohydrin preparation, and carried the product to the corticosteroid (Reichstein's S acetate, **77**), by Stork iodination[46] of the 17α-hydroxyprogesterone. This is a known hydrocortisone precursor, through microbial 11β-hydroxylation with *Curvularia lunata*.[47]

Van Rheenen and co-workers at Upjohn further demonstrated the generality of the method, also in the early 1980s; a wide variety of cyanohydrins are claimed in the Upjohn patents.[48] They also worked out an alternative synthesis of corticosteroids from the corresponding protected cyanohydrin (Fig. 21). The inter-

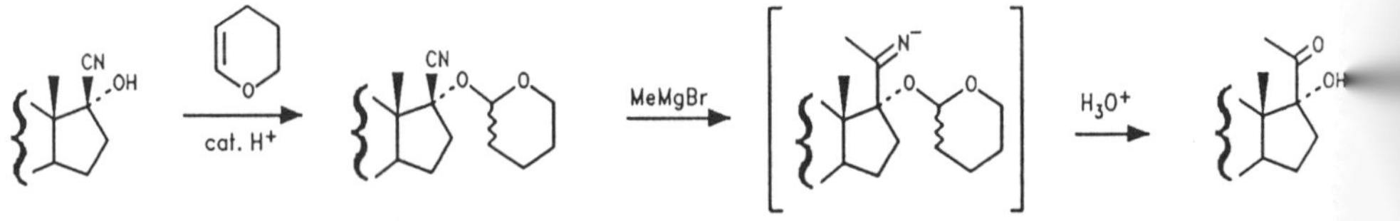

**Figure 19.**

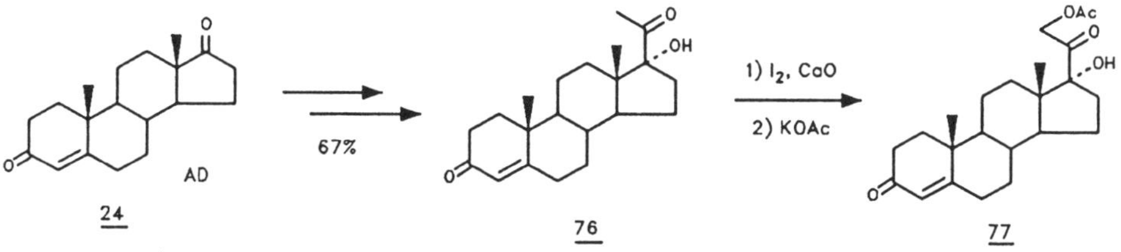

**Figure 20.**

mediate imine anion (**78**) that results from the addition of the methyl anion (see Fig. 19) can be trapped with acetic anhydride, giving the imine acetate (**79**). This can be rearranged with a base such as DBU to the 20,21-enamine acetate (**80**). Oxidation with bromine and hydrolysis yields the 21-bromo ketone. During the course of the hydrolysis, the acetate migrates to the 17-position, giving the 17-acetoxy derivative (**63**), rather than the 17-hydroxy derivative. I have already described the problems inherent in converting intermediate **63** to the corticosteroid.[49]

A limitation to all of the above methodology is the difficulty in adding nucleophiles to the hindered pro-20 position. The nitrile in the substrates is flanked both by the angular methyl at C-18, and by the protecting group covering the 17-oxygen (structure **82**, Fig. 22).

Although addition of methyllithium occurs cleanly to these substrates, the cheaper and more readily available Grignard reagents add with difficulty, and lower

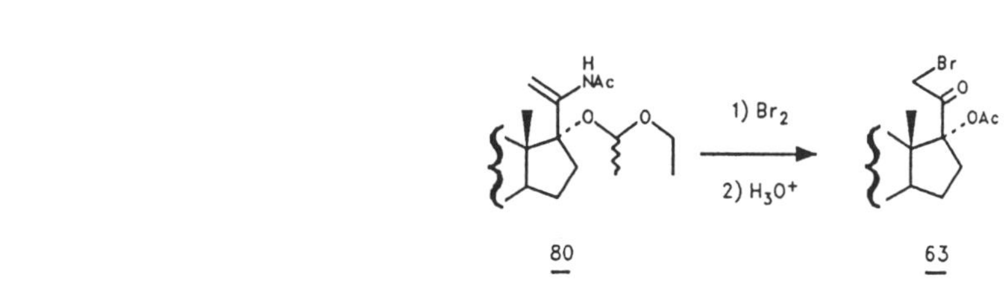

**Figure 21.**

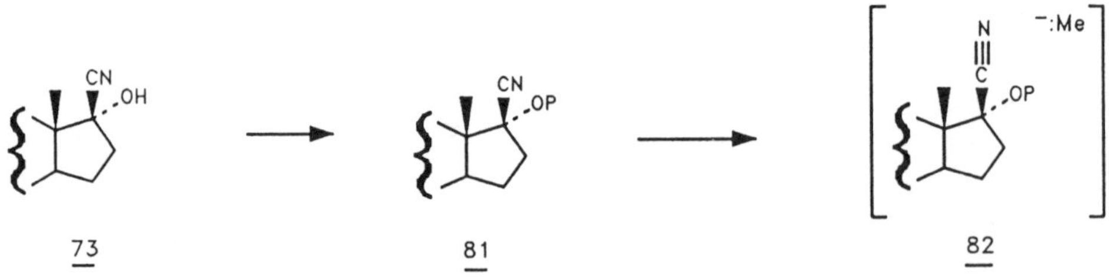

*Figure 22.*

yields can be expected. For example, Gasc and Nedelec (see Fig. 18) were able to selectively open the epoxide in **71** with methylmagnesium bromide at room temperature in 95% yield. Addition to the cyanohydrin, however, required overnight reflux in THF, and the yield of recovered product dropped to 66%.

Even more significant, attempts to prepare corticoids directly from the protected cyanohydrins by addition of "methanol anion" synthons[50] generally have resulted in failure. Those that have been tried to date undergo destruction, presumably by α-elimination to the carbene, before they are capable of adding. Thus, in this methodology, one is currently limited to adjusting the oxidation state at the 21-position as a subsequent step to the elaboration of the nitrile group.

A very recent approach that gets around this problem is due to Reid and Debiak-Krook (Upjohn).[51] This involves prior reduction of the nitrile and

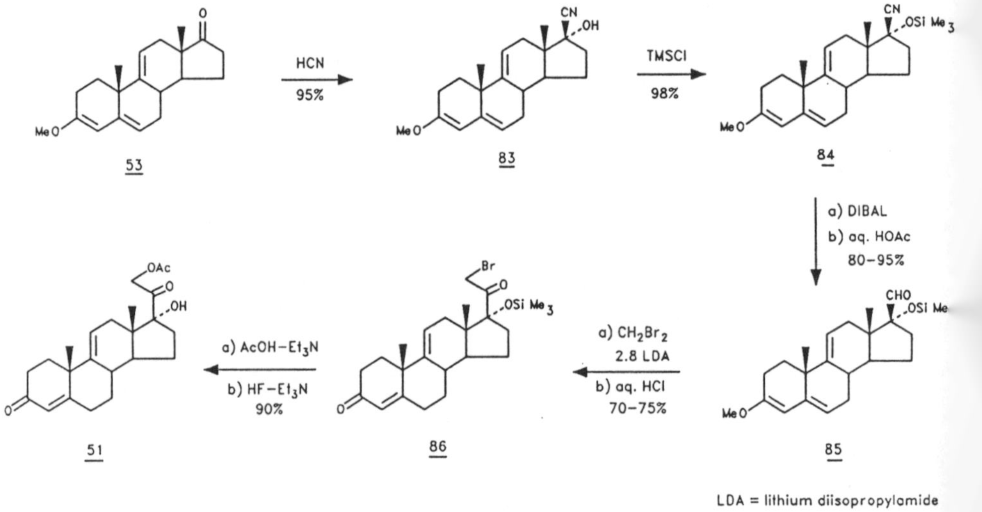

LDA = lithium diisopropylamide

*Figure 23.*

hydrolysis to the more electrophilic aldehyde (structure **85**, Fig. 23). Subsequent addition of lithio dibromomethane, generated in situ, with rearrangement induced by LDA, gives the 21-bromide (**86**). Displacement with acetate and desilylation gives the hydrocortisone precursor (**51**).

The recognition of this difficulty of adding carbon nucleophiles to the nitrile formed the conceptual origin of our "Silicon Nucleophile Annelation Process", the so-called SNAP chemistry. In the remainder of this chapter, I will address the development of this chemistry from various perspectives.

## V. SNAP CHEMISTRY

Our involvement in the cyanohydrin project began late in 1984 with nothing more than a concept: We sought to facilitate nucleophilic addition to the nitrile by linking the entering nucleophile to the 17-hydroxyl (Fig. 24). Theoretically, this could be accomplished without adding steps to the sequence, as the activation of the 17-hydroxyl would substitute for the 17-protection step in the previous chemistry.

One can imagine a number of potential connecting groups to link the 17-oxygen to the nucleophile. We preferred a one-atom tether, which would result in the formation of a 5-membered ring. It was anticipated that this would be faster and more selective in the addition than the formation of a 6- or 7-membered ring.

The requirements for the ideal connecting group are fairly stringent. For instance, one obvious choice would be an acetate group at the 17-position (Fig. 25). This would be both inexpensive and easy to form, would stabilize the carbanion, and could be removed by decarboxylation once the cyclization was complete. However, elimination to cyanide anion and ketene was the predominant pathway in this system, probably because of the very constrained cyclization transition state, due to orbital overlap requirements.[52]

We began experiments along these lines early in 1985, primarily directed toward the preparation of 17α-hydroxyprogesterone from AD (**24**, Fig. 20). It was during that time that we developed an appreciation for the importance of selecting a suitable linker. In April of that year, my colleague Bruce Pearlman, consultant Scott

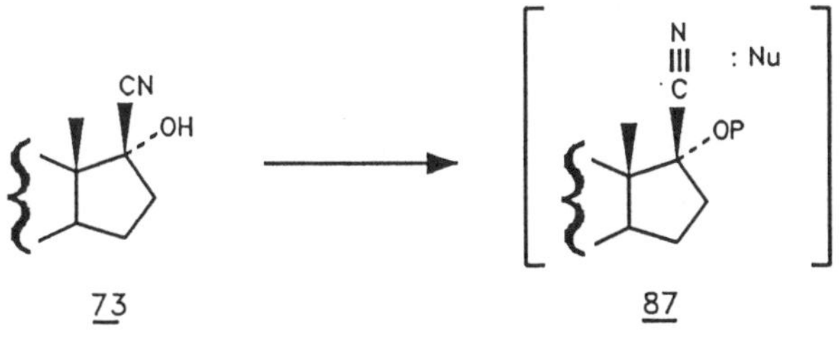

*Figure 24.*

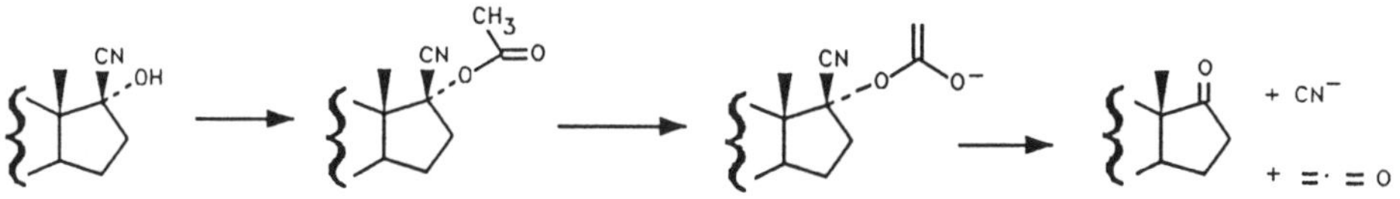

**Figure 25.**

Denmark of the University of Illinois, and I were involved in an informal discussion of the problem, when it became clear that the ideal connecting atom should be silicon.

Indeed, silicon possesses a number of features that make it a perfect candidate for our application:

1.  The precursor silyl ethers could be very easily prepared from the corresponding chlorosilanes. A variety of these chlorosilanes are readily available, and the advent of the silicone polymer industry has made them inexpensive (see below).
2.  The two pendant groups on silicon can, in principle, be varied within the limits of the commercially available silanes, as the linker/protecting group is to be discarded once it has fulfilled its function. This offered the possibility of designing a wide variety of structural variations of the intermediate silyl ethers, in turn offering a handle with which to control the physical and toxicological properties of the intermediates and the waste stream.
3.  Silicon stabilizes an adjacent carbon-metal bond, and it does so without undergoing elimination. Elimination in this case is unfavorable, due to the instability of the silicon–carbon double bond in any product silene.
4.  Silicon stabilizes an adjacent carbon–metal bond without putting difficult steric requirements on the cyclization transition state. Whether the stabilization is due to overlap with the silicon d-orbitals, or with the antibonding orbitals on the silicon substituents, is arguable. However, there is no requirement that the three atoms in the side chain be coplanar following the generation of the carbanion, as there is for acetate. It is this coplanarity requirement that hinders the transition state for the attack of the acetate group.
5.  Removal of the silicon substituent could potentially be accomplished by any of a number of carbon–silicon bond cleaving reactions.

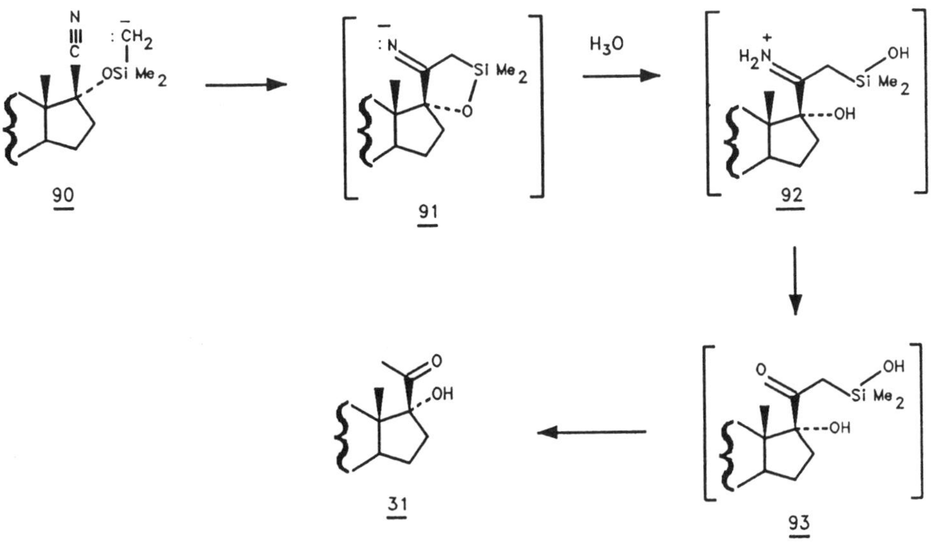

**Figure 26.**

Thus, we envisioned the overall process for the preparation of 17α-hydroxyprogesterone derivatives occurring *via* the pathway presented in Figure 26.

Following the initial cyclization of the carbanion **90** onto the nitrile, the intermediate silacycle **91** could be broken down, at least in principle, by simply quenching with acid. The silicon–oxygen bond would be hydrolyzed (**92**), the imine would be protonated and hydrolyzed (**93**), and finally protiodesilylation would remove the silicon appendage. It was also recognized at the time that an oxidative cleavage of the silicon-carbon bond in one of the intermediates **91–93** could yield a synthesis of corticosteroids from the same intermediate. This was precedented by the then recently published work of Kumada and Tamao (Fig. 30).[53] Ideally, we hoped that addition would prove to be so facile that the methyl group could be added in a more highly oxidized state (but faster than α-elimination to the carbene), leading to a direct synthesis of the corticoid. This was in fact accomplished early in the project, but for the sake of organization, discussion of this alternative has been deferred to the section on "deprotonation methods", below.

## A.  Reductive Methods

We first decided to generate the requisite α-silyl carbanion by the reductive metal cleavage of a carbon–halogen bond. The precursor halomethyl dimethylchlorosilanes (**94, 95**) are readily available from free radical chlorination (or bromination) of trimethylchlorosilane, which is an inexpensive item of com-

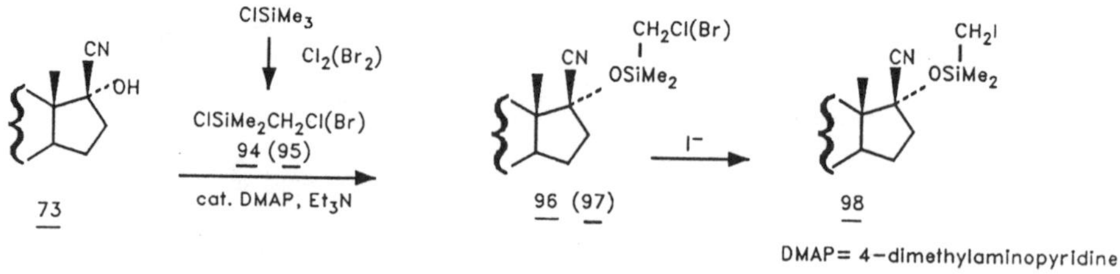

**Figure 27.**

merce. Silylation of the very hindered 17-hydroxyl in **73** was achieved in good yield, with care to avoid excessively basic conditions, which results in reversal of the cyanohydrin to the ketone (Fig. 27). This gives directly the required chloromethyl or bromomethyl dimethylsilyl ethers (**96** and **97**, respectively).

The corresponding iodomethyl compound **98** was prepared most readily by displacement of the bromomethyl compound with sodium iodide in acetone. The neighboring silicon group makes this displacement particularly facile.

Protection of other reducible functionalities in the molecule, such as a 3-ketone, is necessary with many reducing agents. Typically, this is accomplished by formation of the 3-dienol ether or ketal. The best point at which to do this is usually after the cyanohydrin formation, which has already accomplished differentiation of the 3- and 17-ketones.

A number of metals were found to effect cyclization (with varying efficiencies) on each of these halomethyl derivatives.[54] However, the least expensive of the substrates is the chloromethyl derivative. The carbanion from this derivative was best prepared by using one of the available dissolving metal-type reagents, for instance lithium or sodium naphthalenide, lithium biphenylide, lithium di-*tert*-butyl biphenylide, etc.[55]

Our earliest attempts to prepare 17α-hydroxyprogesterones *via* reductive cyclization, by using excess dissolving metal reagent followed by an acid quench, resulted in the appearance of an unexpected steroid as the major product of the reaction. This was characterized as progesterone (**17**), rather than the 17α-hydroxyprogesterone that we had expected. What we believe happens under these conditions is depicted in Figure 28.

Following the initial reduction and cyclization (**91**), the addition of acid effects protonation of the intermediate imine anion at a rate faster than that of destruction of the excess reducing agent by acid (Fig. 28). Presumably, the electroreduction potential of the anion is too high for it to undergo further reduction by the reagent. The neutral imine is capable of further reduction by the excess reducing agent, which occurs with elimination of the 17-ether (**99**). This leads to an enamine anion (**100**) which eventually is hydrolyzed with protonation from the α-face. Further

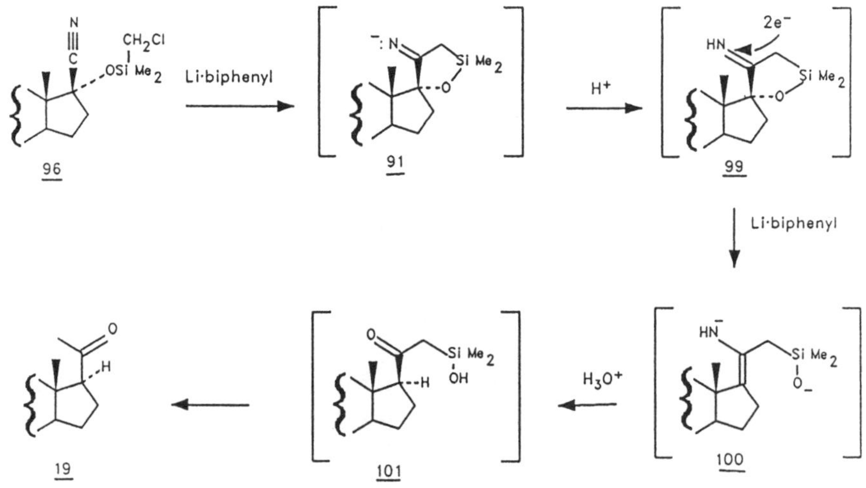

**Figure 28.**

hydrolysis and protiodesilylation yields the progesterone **19**, with the natural configuration at C-17.

This overreduction to progesterone could be overcome by using an exactly stoichiometric amount of the reducing agent, which we found to be operationally inconvenient. Alternatively, the excess reducing agent could be destroyed with an aprotic quenching agent, such as dichloroethane, prior to the acidic hydrolysis (Fig. 29). Good yields of the 17α-hydroxyprogesterones (i.e. **76**) were obtained with this modification.

In order to prepare corticosteroids *via* the above method, an oxidation is required. We decided to investigate taking advantage of the presence of the silicon substituent at carbon-21 in the cyclic intermediate, recognizing that competing protiodesilylation may well pose a problem.

As mentioned previously, Kumada and Tamao had reported the oxidation of a silicon–carbon bond by using alkaline hydrogen peroxide. This was part of a

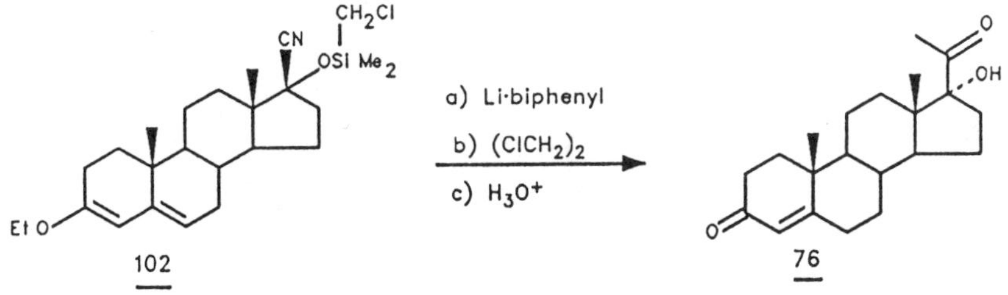

**Figure 29.**

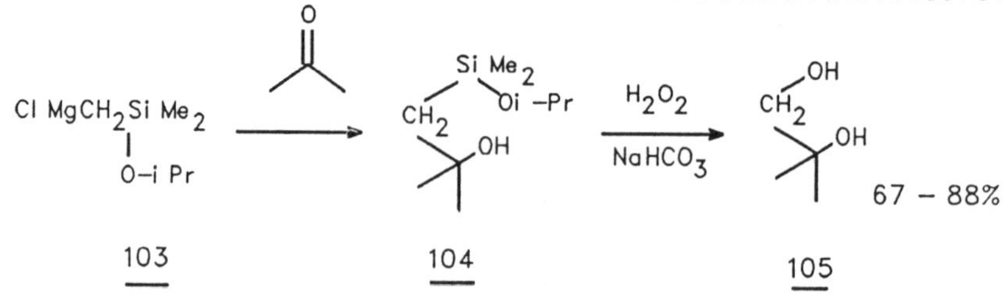

**Figure 30.**

synthesis of glycols from the reaction of a modified Peterson reagent (**103**) with ketones (Fig. 30).

It occurred to us that a similar oxidation of the 21-silyl intermediate (i.e. **93**, Fig. 26) in the SNAP reaction might lead directly to the corticosteroid side chain. Unfortunately, we were never able to isolate the 21-silyl intermediates without concurrent protiodesilylation to the 17α-hydroxyprogesterone. It is not surprising, then, that attempts to effect oxidative cleavage by simply adding hydrogen peroxide to the reaction mixture (after the dichloroethane quench) resulted in protiodesilylation as the major reaction pathway (Fig. 31).

Theoretically, the oxidation could be facilitated by altering the substituents to give a more electropositive silicon. Silanes with substituents other than methyl or chlorine on silicon are more expensive. This originates with the industrial synthesis of silicone polymers from silicon metal, and compels a choice of three inexpensive chloromethylsilyl chlorides (Fig. 32).

In addition to trimethylchlorosilane, dimethyldichlorosilane and methyltrichlorosilane are produced on a very large scale (ca. 500,000 tons/year) as intermediates in the synthesis of silicone oil and other silicone polymers. Free-radical chlorination of the latter methylsilanes produces chloromethylsilanes with one (**108**) or two (**109**) "replaceable" sites on silicon, respectively. These compounds can be monocondensed with the steroid 17-cyanohydrins. Presumably, *bis* addition is avoided because of the steric hindrance around the 17-position on the steroid. Quenching the reaction mixture with alcohols resulted in formation of the mixed chloromethyl silyl ethers (**111** and **113**, Fig. 33).

Initially, we found that the monomethoxy and dimethoxy ethers were too unstable to handle. The more sterically hindered isopropoxy ethers were much

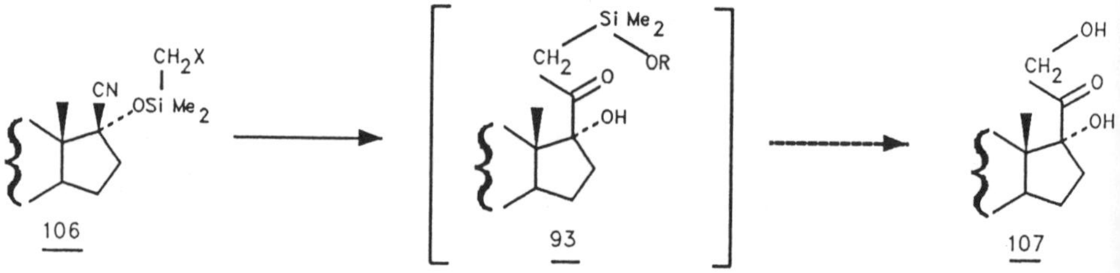

**Figure 31.**

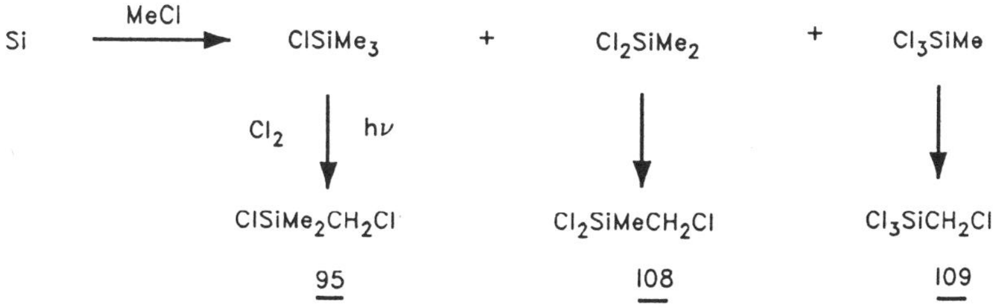

$$Si \xrightarrow{MeCl} ClSiMe_3 \quad + \quad Cl_2SiMe_2 \quad + \quad Cl_3SiMe$$

Figure 32.

more stable, and were isolated as well-behaved crystalline compounds. Note that the monomethyl, monoisopropoxy silyl ether has a stereogenic center at the silicon atom, which gave rise to a mixture of diastereomers in the reaction. Thus, we favored the diisopropoxy compounds for further investigations (**113**, R = *i*-Pr).

Reductive cyclization of these compounds occurred uneventfully under the usual conditions. We were somewhat disappointed (but not surprised) to find that oxidative cleavage of the silicon carbon bond was also not the exclusive pathway for these silyl ethers. Although we did obtain the corticosteroid **107** as the major product, it was always contaminated with an unacceptably large amount of the corresponding 17α-hydroxyprogesterone (**31**, Fig. 34).

One final point should be emphasized regarding these alternative substrates for cyclization. The ability to alter the substituents on silicon allows a certain degree of control over the physical properties of the intermediate, and there is obviously a change in the chemical structure of the silicon polymer that is obtained as a byproduct in the reaction. In some cases this may offer advantages in the handling of the intermediate, in controlling solubility, in isolating the product, and in waste disposal.

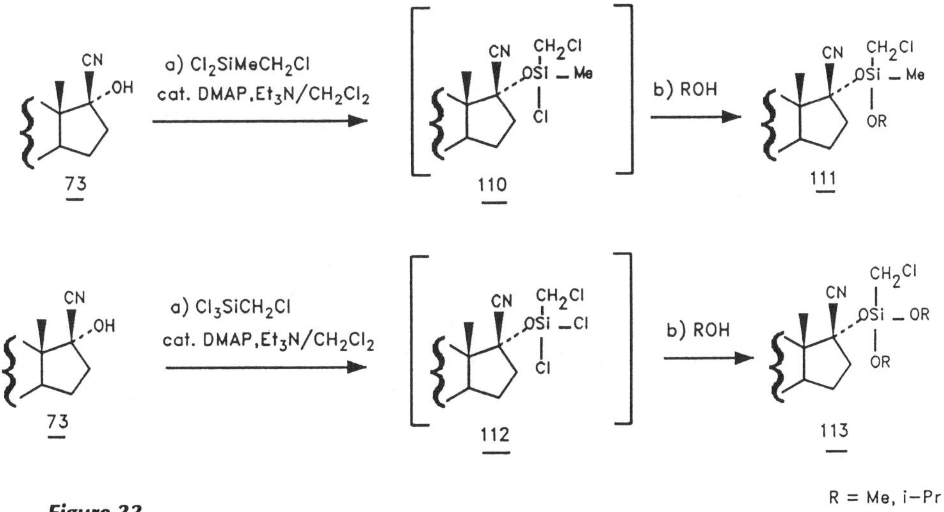

R = Me, i–Pr

Figure 33.

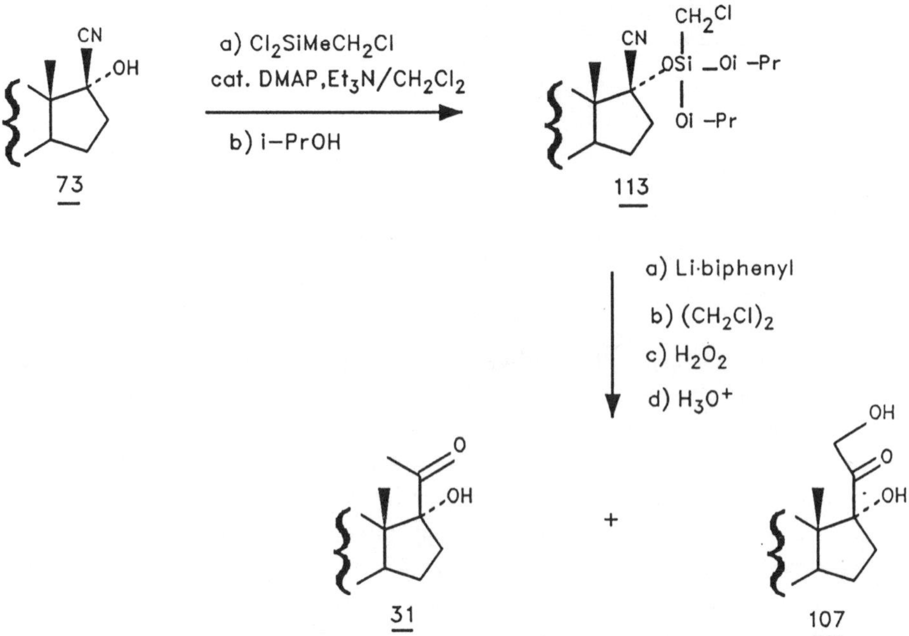

**Figure 34.**

## B.  Deprotonation Methods

The possibility of producing the requisite α-silyl carbanion simply by deprotonation was best precedented by the work of Magnus[56] and Olofson.[57] In the course of developing a method for the preparation of α-silyl epoxides (**116**), Magnus and co-workers studied the deprotonation of chloromethyltrimethylsilane (**114**, Fig. 35). They determined that the best base for deprotonation is sec-butyl-lithium, and that the addition of tetramethylethylenediamine (TMEDA) speeded up this deprotonation. n-Butyllithium tended to attack nucleophilically at the silicon center, and tert-butyllithium tended to undergo halogen-metal exchange.

There are inherent difficulties in the handling of sec-butyllithium such as its pyrophoricity and instability, and we thus wished to avoid its use. Magnus had also looked at deprotonation with hindered amide bases, but found that the yields of α-silyl epoxides were significantly lower than with sec-butyllithium, which gave 95% deprotonation. For instance, lithium diisopropylamide (LDA) gave only about 8% upon trapping with ketones. Olofson and co-workers used lithium 2,2,6,6-tetramethylpiperidide (LiTMP) at above ambient temperatures to effect deprotonation and (irreversible) α-elimination to carbenoids, which were trapped with various alkenes in yields of 28–33%.

Despite the somewhat discouraging precedent, Bruce Pearlman, who has con-

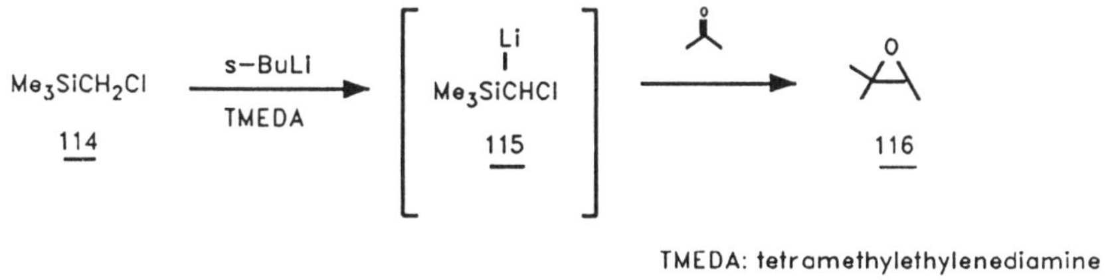

TMEDA: tetramethylethylenediamine

**Figure 35.**

siderable expertise in the carbenoid field,[58] strongly encouraged us to attempt the LDA deprotonation. He argued that the lower yield in the Magnus case was simply due to an unfavorable equilibrium for the deprotonation, and that the ketone was trapping out what little carbanion was present. In our case, as soon as an anion were generated, it would be trapped out in the ensuing cyclization reaction. Thus, any base capable of deprotonating the chloromethylsilyl ether at a reasonable forward rate should be operable, regardless of the position of the equilibrium.

It should be noted that LDA may be prepared very directly *via* the "Reetz procedure",[59] which actually dates back to Ziegler in the 1930s[60] (Fig. 36). This involves the reaction of lithium metal with an unsaturated hydrocarbon, such as styrene or isoprene, in the presence of diisopropylamine as a proton donor, yielding lithium diisopropylamide (**119**), and half an equivalent of the reduced hydrocarbon (i.e. **120**). Importantly, in this procedure each equivalent of lithium generates one equivalent of LDA. In contrast, during the preparation of alkyllithiums [such as *sec*-butyllithium (**122**)], one equivalent of lithium metal goes to forming lithium chloride in the course of the halogen-metal exchange. The LDA synthesis is thus more efficient with respect to the use of metallic lithium.

As it turned out, this deprotonation strategy worked extremely well. Cyclization of the chloromethylsilyl ether upon treatment with LDA occurred almost instan-

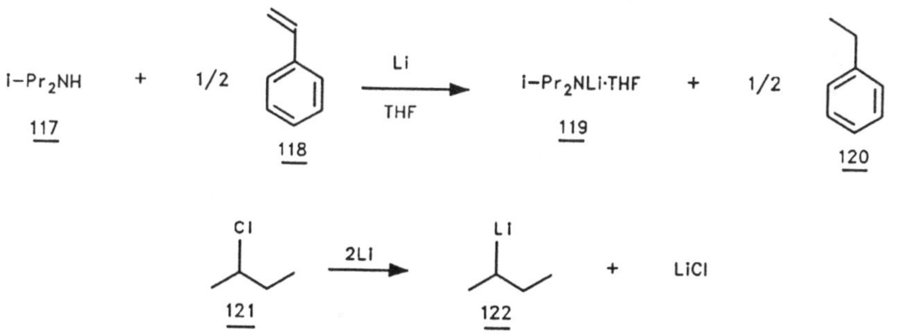

**Figure 36.**

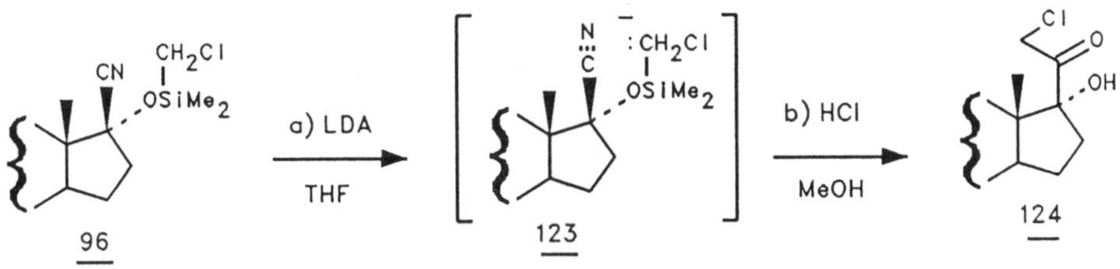

**Figure 37.**

taneously. A simple quench of the reaction mixture with aqueous acid promoted the usual set of hydrolyses, however, throughout this process we have succeeded in retaining the halogen (Fig. 37).

We were delighted to find that after a few trivial operations, we could separate the 21-chloroketone **124** directly from the crude reaction mixture by filtration, in a very pure state and in quite high yield. In the exhilaration of the moment, the acronym SNAP (Silicon Nucleophile Annelation Process) was coined to describe the novel procedure.

We found that the order of addition is not important, i.e., the LDA can be added to the steroid, or the steroid can be added to the LDA. The isopropoxy silyl ethers (**111** and **113**, Fig. 33), and the bromomethyl silyl ether (**97**, Fig. 27), also undergo an analogous cyclization. In the latter case, one must be careful in choosing the acid used for the quench; for example, the use of hydrochloric acid gives the 21-*chloro*ketone as the major product, *via* displacement of the 21-bromide by the chloride ion present in the hydrochloric acid. This ease of displacement of the 21-halides makes conversion to the corticoid very facile, by displacement with acetate anion, as we have seen previously (Fig. 38). Note that the 21-chloride **124** is also a precursor to the corresponding 17α-hydroxyprogesterone (**31**) through zinc reduction, which proceeds in nearly quantitative yield.

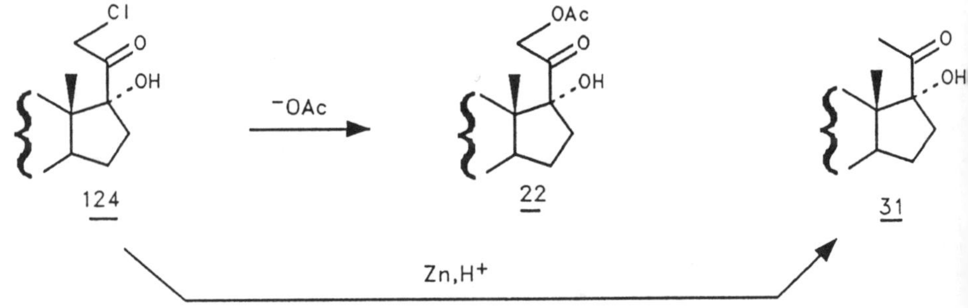

**Figure 38.**

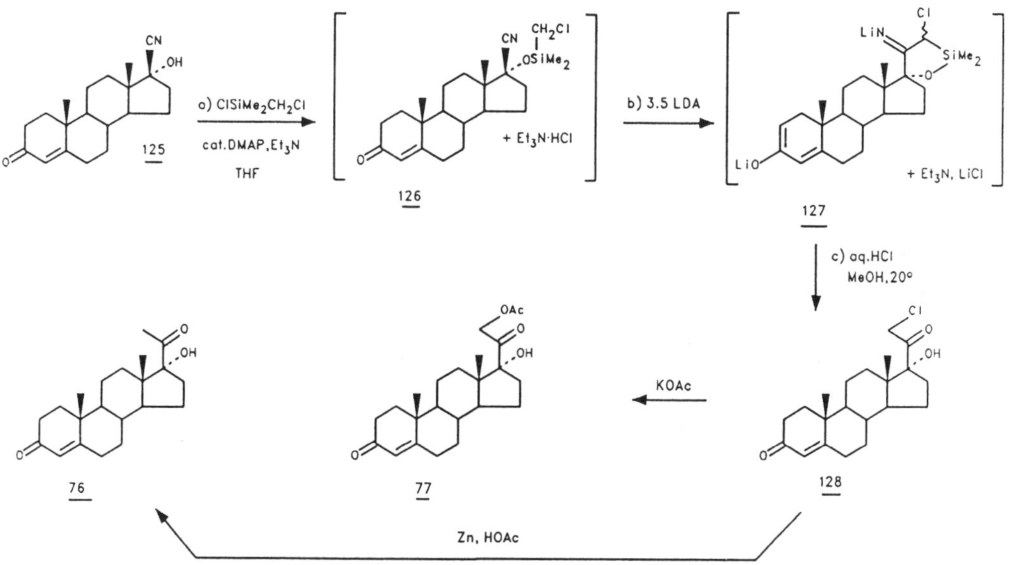

**Figure 39.**

One of the important features of the base-promoted cyclization is its versatility. A wide variety of functionality is capable of surviving treatment with LDA without protection. For instance, ketones are generally enolized, hydroxyls are simply deprotonated, etc. One example of this versatility is a three-pot synthesis of the "parent" corticosteroid, Reichstein's S acetate (**77**), from androst-4-ene-3,17-dione (AD) (Fig. 39).

The cyanohydrin of AD (**125**) is formed under the usual conditions and isolated. This is slurried in tetrahydrofuran (THF) and triethylamine with a catalytic amount of 4-dimethylaminopyridine (DMAP), and then treated with chloromethyldimethylchlorosilane. A mixture of triethylamine hydrochloride and the silylated cyanohydrin (**126**) in THF results. The crude reaction mixture is added directly to an excess of LDA; the triethylamine hydrochloride is deprotonated and the 3-ketone is enolized. The final equivalent of LDA effects the SNAP reaction, and the mixture is then quenched with aqueous hydrochloric acid and warmed to room temperature. The 21-chloride crystallizes from the reaction mixture, and is recovered in 93% yield. This is then displaced with potassium acetate, giving substance S acetate quantitatively; zinc-acetic acid reduction gives 17α-hydroxyprogesterone.

Similar procedures have been used to prepare a wide variety of other important substituted corticosteroids, including hydrocortisone, prednisolone, and methylprednisolone.

This versatility actually presents some problems. For instance, the ability to emplace the corticosteroid side chain at any number of places in the synthesis of a particular product means that considerable research is required in order to determine the optimum sequence of steps. The expenditure of resources needed for such development work is always limited by the degree of payoff expected, and the level of cost savings—if any exists—is often hard to gauge. This is particularly true at the beginning of a project, when the number of available options is greatest, and information on whether or not to continue development is most needed. The answer, of course, is efficient laboratory work, decisive management, and good communication in both directions.

A little of what we have learned in the course of our ongoing evaluation of the scope and limitations of this new chemistry is described below. As with anything new, serendipity has generated some interesting observations.

## C. Scope and Limitations

When using an unprotected 3-ketone, the intermediate dianion (or trianion, in the case of hydroxylated substrates) is often insoluble in the reaction mixture. To facilitate processing, we designed an "in situ" protection scheme (Fig. 40). TMSCl is added to the mixture prior to the addition of the LDA, trapping the 3-enolate as it is formed (cf. **113**).[61]

In the case of 3-keto-$\Delta^{1,4}$ substrates, the protection of the 3-ketone as the enolate was complicated by the tendency of diisopropylamine to add into the A-ring. There was some evidence that this was due to an electron transfer process; the problem was ultimately overcome by performing the in-situ deprotonation/protection with a base known not to undergo electron transfer–lithium hexamethyldisilazide (LiHMDS).

Although lithium hexamethyldisilazide is effective in enolizing the 3-ketone, it is not a strong enough base to effect the SNAP reaction. We had to resort to the subsequent addition of two equivalents of LDA; the first to deprotonate the hexamethyldisilane formed, and the second to effect the SNAP reaction itself. This was not a particularly satisfactory solution to the problem, so we became involved

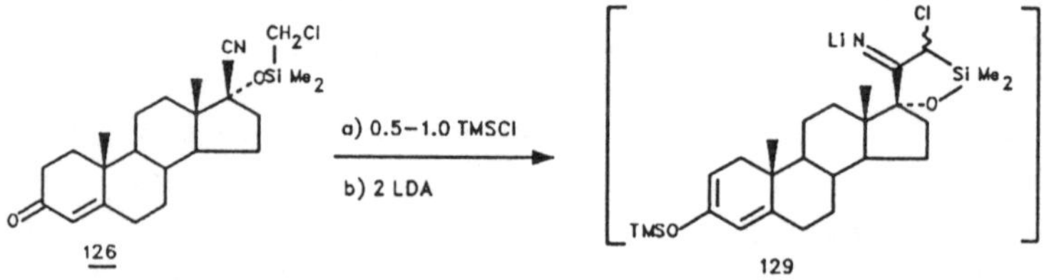

TMSCl: trimethylchlorosilane

**Figure 40.**

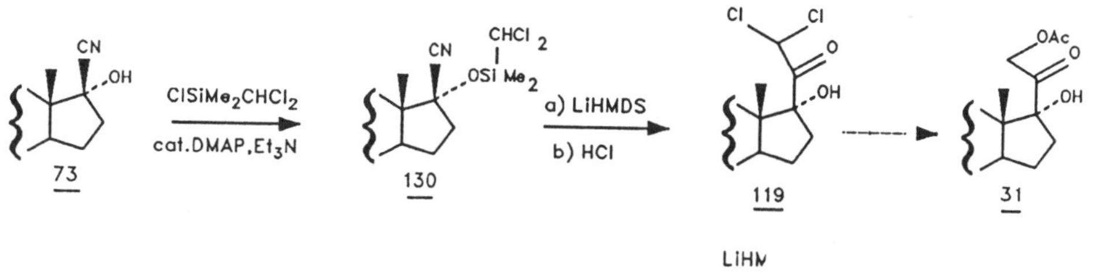

**Figure 41.**

in an ongoing investigation aimed at acidifying the requisite hydrogen α to the silicon.

One method for doing this is the addition of another chlorine atom to the methyl group bearing the hydrogen. Thus, the dichloromethyl silyl ether (**130**) was successfully deprotonated with lithium hexamethyldisilazide, and undergoes cyclization in good yield, leading to the 21-dichloro compound (**131**, Fig. 41).

Efficient methods for direct conversion of the 21-dichloro functionality into a 21-acetate are of current interest in our laboratory. Note that this is precedented by the Stork iodination (Fig. 20), which proceeds largely through the 21-diiodide.

Another functionality that causes problems is the 16-methylene that is found, for instance, in the progestational agent melengestrol acetate (**137**). Attempts to add cyanide reversibly to the 17-ketone in **132** invariably resulted in large amounts

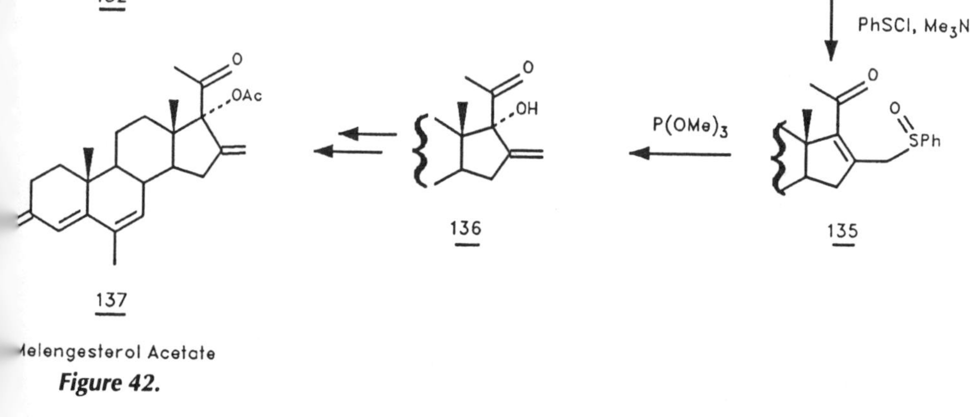

**Melengesterol Acetate**

**Figure 42.**

of Michael addition to the exocyclic methylene. This is an irreversible process, and eventually most of the product is drained off along this pathway. One way around this was worked out by Pearlman and his group. The kinetically formed α-cyanohydrin was silylated under the usual conditions to give **133** (Fig. 42). This neatly avoids the problem of having to set up equilibrium conditions.

This "inverted" substrate cyclized under the usual conditions, resulting in the 17-hydroxyl in the wrong configuration (**134**). Having the 16-methylene, however, now allows inversion via a vinyl sulfoxide rearrangement (**134** to **136**), in a process that was worked by Van Rheenen.[62]

## D.  Serendipity

The study of the side reactions in this chemistry has led to some interesting observations. For example, in evaluating the in situ trapping procedure for the 3-ketone (Fig. 40), it was discovered that if excess TMSCl and LDA were added, and the reaction mixture was allowed to warm up prior to the quench, a finite amount of the pentacyclic steroid **141** would be isolated as a by-product. Under forcing conditions, this interesting compound[63] even became the major product of the reaction.

We can account for this as follows. The intermediate imine anion is trapped as **138** (Fig. 43). Deprotonation and α-elimination gives the carbene **139**, which inserts into the nearest carbon–hydrogen bond, on the 18-methyl group (**140**).[64] This leads, upon hydrolysis, to cyclosteroid (**141**), wherein the 18-methyl and the

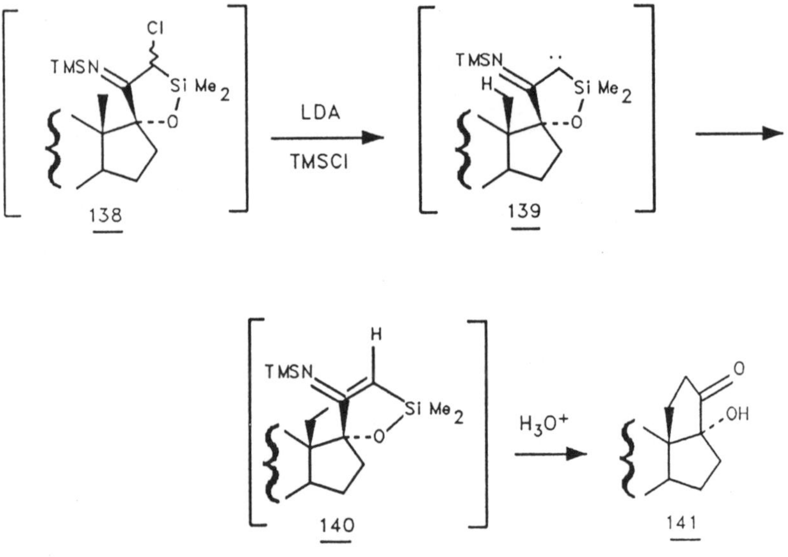

*Figure 43.*

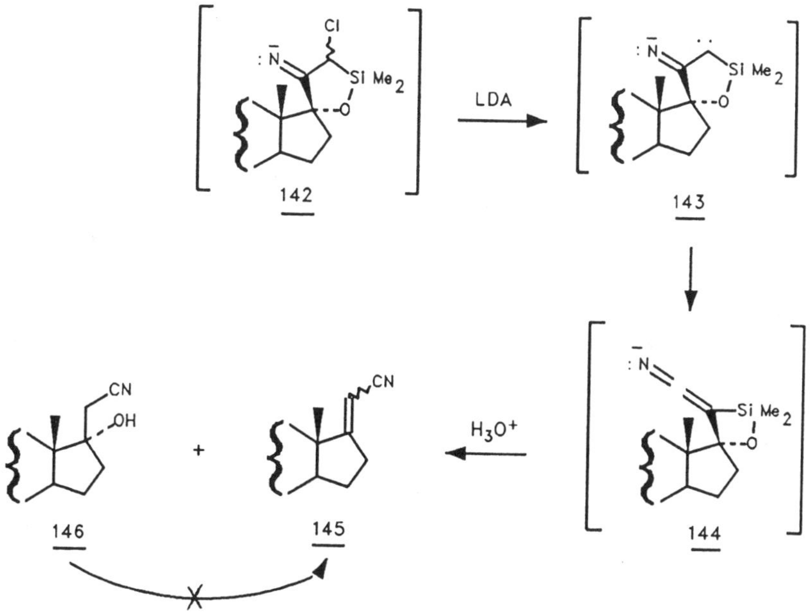

**Figure 44.**

21-methyl have been fused together. Not unexpectedly, the hindered imine in **140** is more difficult to hydrolyze than the imine from the usual reaction.

From a medicinal chemistry point of view, this is interesting because it creates a structure wherein the steroid side chain has been fixed rigidly in space, with a minimum of structural modification to the molecule.

We also observed that three other significant byproducts can form if the reaction mixture is allowed to warm prior to quenching with acid, if excess LDA is used, and when TMSCl or other alkylating agents are absent. These were the two isomeric vinyl nitriles **145**, and the β-hydroxy nitrile (**146**, Fig. 44). We could demonstrate that the 17-hydroxy nitrile was not the precursor of the vinyl nitriles by resubjecting it to the reaction conditions. These could be formed *via* a similar carbene or carbenoid intermediate (**143**), which is produced by the deprotonation of intermediate silylcycle anion (**142**).

In a reaction analogous to a Wolff rearrangement, anion **142** could now generate silaoxetane intermediate **143**, driven along this pathway by α-nitrile anion resonance. The oxetane is capable either of hydrolyzing and protiodesilylating to the 17α-hydroxynitrile, **146**, or eliminating the elements of dimethylsilicone in a Peterson fashion, leading (after protonation) to the isomeric vinyl nitriles (**145**). Prequenching with deuteriomethanol gave deuterium incorporation at the vinyl position (C-20) in both isomers.

This rearrangement cannot be driven to completion even under forcing conditions, i.e. with the addition of large amounts of LDA or on warming for extended periods. The amount of rearrangement that occurs is dependent mostly upon the exact structure of the substrate, and to a certain degree upon the reaction conditions during the initial cyclization.

This may be a reflection of the ratio of diastereomers that is formed in the initial cyclization. One of these diastereomers has a hydrogen that is much more accessible for the second deprotonation. This ratio is, in turn, determined by which of the two diastereotopic protons on the chloromethyl group is initially removed, a process that is probably subject to a very subtle interplay of structural and environmental factors. Serious mechanistic studies of either of the above rearrangements have not been done; thus, we cannot rule out other rearrangements or radical or intermolecular processes.[65] The above mechanistic explanations account for all of our observations to date, and seem to be the most straightforward, however.

# VI. SUMMARY

At Upjohn, we have developed a unique, efficient, and versatile new technology for the synthesis of commercially important corticosteroids (22), 17α-hydroxy progesterones (31), and progesterones (19) from the readily available 17-ketones (32), *via* a common silylated cyanohydrin intermediate (96) as depicted in Figure 45.

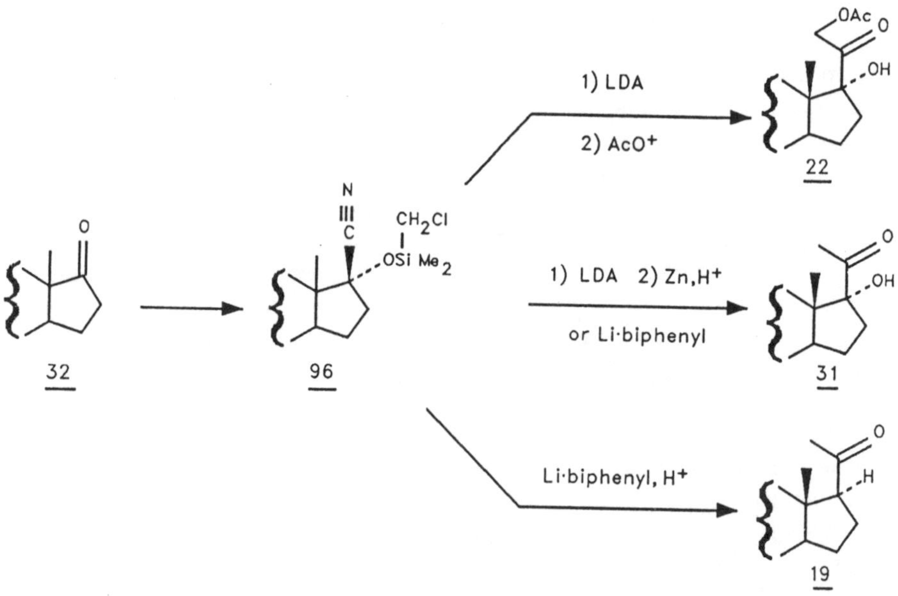

*Figure 45.*

Only time and further evaluation will tell if our SNAP chemistry has met the criterion for elegance, and will contribute substantially to the sitosterol utilization problem. Regardless, attempting to meet the challenge has certainly been an enriching experience for ourselves and many other synthetic chemists.

## ACKNOWLEDGMENTS

Beyond those already mentioned in the text, I would like to thank my colleagues whose contributions were described without specific attribute: Kyung Lee, Jan Petre, Carol Bergh, Bill Perrault, John DeMattei, Bill Hartley, Greg Reid, Therese Debiak-Krook, Brad Hewitt, Randy Putt, Gordon Snow, Curt Gillespie, Ed Morseau, Necdet Kuruoglu, and Steve Trank.

## REFERENCES AND NOTES

1. (a) Livingston, D.A.; Petre, J.E.; Bergh, C.L. *J. Am. Chem. Soc.* **1990**, *112*, 6449–6450. See also: (b) U.S. Patent 4,921,638, May 1, 1990 (Upjohn) and (c) U.S. Patent 4,977,255, Dec 11, 1990 (Upjohn).

2. The pregnane numbering system is depicted as follows; androstanes are truncated at the 17-position, but numbered analogously.

Stereochemistry not indicated is taken to be in the "natural" configuration:

3. There are a number of historical accounts available. For a monumental one dealing with the earlier accomplishments from a chemical perspective, see: (a) Fieser, L.F.; Fieser, M. *Steroids*, Reinhold, New York, 1959. For a more anecdotal description from an Upjohn perspective, see: (b) Engel, L., *Medicine Makers of Kalamazoo*, McGraw-Hill, New York, 1961.

4. The leaders in the Upjohn effort were Marvin H. Kuizenga and George F. Cartland. Tadeusz Reichstein at the University of Basel, Edward C. Kendall at the Mayo Foundation, and Oskar Wintersteiner at Columbia University (later Squibb) are generally regarded as having been the world leaders in this effort. An excellent review of this work from its historical context is found in Reference 3(a).

5. (a) Sarett, L.H., *J. Biol. Chem.* **1946**, *162*, 601. (b) Sarett, L.H., *J. Am. Chem. Soc.* **1948**, *70*, 1454–1458. (c) Sarett, L.H., *J. Am. Chem. Soc.* **1949**, 2443–2444.

6. Hench, P.S.; Kendall, E.C.; Slocumb, C.H.; Polley, H.F. *Proc. Staff Meetings Mayo Clinic* **1949**, *24*, 181.

7. Reference 3(a), p. 650.

8. *Chemical Marketing Reporter*, April 9, 1989, p. 36, lists hydrocortisone acetate at $0.67/gram.

The 350% inflation figure is based on the relative producer purchasing power figure for the years in question.

9.  (a) Heyl, F.W.; Centolella, A.P.; Herr, M.E. *J. Am. Chem. Soc.* **1947**, *69*, 1957–1961. (b) Heyl, F.W.; Herr, M.E. *J. Am. Chem. Soc.* **1950**, *72*, 2617–2619. (c) Herr, M.E.; Heyl, F.W. *J. Am. Chem. Soc.* **1952**, *74*, 3627–3630.

10. (a) Van Rheenen, V. *Chem. Commun.* **1969**, 314–315. (b) Huber, J.E. *Tetrahedron Lett.* **1968**, 3271–3272.

11. Peterson, D.H.; Murray, H.C. *J. Am. Chem. Soc.* **1952**, *74*, 1871–1872.

12. (a) Peterson, D.H.; Murray, H.C.; Eppstein, S.H.; Reineke, L.M.; Weintraub, A.; Meister, P.D.; Leigh, H.M. *J. Am. Chem. Soc.* **1952**, *74*, 5933–5936. (b) Peterson, D.H. *Steroids* **1985**, *45*, 1–17.

13. (a) Hogg, J.A.; Beal, P.F.; Nathan, A.H.; Lincoln, F.H.; Schneider, W.P.; Magerlein, B.J.; Hanze, A.R.; Jackson, R.W. *J. Am. Chem. Soc.* **1955**, *77*, 4436–4438. (b) Hogg, J.A.; Lincoln, F.H.; Nathan, A.H.; Hanze, A.R.; Schneider, W.P.; Beal, P.F.; Korman, J. *J. Am. Chem. Soc.* **1955**, *77*, 4438–4439. (c) Schneider, W.P.; McIntosh, A.V.; Upjohn Co., U.S. Patent 2,769,824.

14. For a very detailed description of this operation, and the "leaching plant" in which it is carried out, see: Poulos, A.; Greiner, J.W.; Fevig, G.A. *Ind. Eng. Chem.* **1969**, *53*, 949–962.

15. Reviews: (a) Kieslich, K. *J. Basic Microbiol.* **1985**, *25*, 461–474. (b) Martin, C.K.A. *Adv. Applied Microbiol.* **1977**, *22*, 29–58. (c) Marsheck, W.J., Jr. *Prog. Ind. Microbiol.* **1971**, *10*, 49–103. See also: (d) Arima, K.; Nagasawa, M.; Bae, M.; Tamura, G. *Agr. Biol. Chem.* **1969**, *33*, 1636–1643. (e) Nagasawa, M.; Bae, M.; Tamura, G.; Arima, K. *Agr. Biol. Chem.* **1969**, *33*, 1644–1650. (f) Nagasawa, M.; Watanabe, N.; Hashiba, H.; Tamura, G.; Arima, K. *Agr. Biol. Chem.* **1970**, *34*, 798–800. For a recent reference in this area, see: (g) Wang, K.C.; Young, L.-H.; Wang, Y.; Lee, S.-S. *Tetrahedron Lett.* **1990**, *31*, 1283–1286.

16. (a) Fujimoto, Y.; Chen, C.-S.; Szeleczky, Z.; DiTullio, D.; Sih, C.J. *J. Am. Chem. Soc.* **1982**, *104*, 4718–4720. (b) Fujimoto, Y.; Chen, C.-S.; Gopalan, A.S.; Sih, C.J. *J. Am. Chem. Soc.* **1982**, *104*, 4720–4722. See also preceding references.

17. Marsheck, W.J.; Kraychy, S.; Muir, R.D. *Appl. Microbiol.* **1972**, *23*, 72–77.

18. Wovcha, M.G.; Antosz, F.J.; Knight, J.C.; Kominek, L.A.; Pyke, T.R. *Biochim. Biophys. Acta* **1978**, *531*, 308–321.

19. (a) Redpath, J.; Zeelen, F.J. *Chem. Soc. Rev.* **1983**, *12*, 75–98. (b) Nitta, I.; Ueno, H. *Org. Synth. Chem.* **1987**, *45*, 445–461.

20. Hessler, E.J.; Van Rheenen, V.H.; Upjohn Co. U.S. Patent 4,216,159, Aug. 5, 1980.

21. Walker, J.A.; Upjohn Co. U.S. Patent 4,284,827, Aug. 18, 1981.

22. (a) Walker, J.A.; Upjohn Co. U.S. Patent 4,568,492, Feb. 4, 1986. (b) Horiguchi, Y.; Nakamura, E.; Kuwajima, I. *J. Org. Chem.* **1986**, *51*, 4323–4325. (c) Horiguchi, Y.; Nakamura, E.: Kuwajima, I. *Tetrahedron Lett.* **1989**, *30*, 3323–3326.

23. (a) Bernstein, S.; Lenhard, R.H.; Allen, W.S.; Heller, M.; Littell, R.; Stolar, S.M.; Feldman, L.I.; Blank, R.H. *J. Am. Chem. Soc.* **1959**, *81*, 1689–1696. (b) Allen, W.S.; Bernstein, S. *J. Am. Chem. Soc.* **1956**, *78*, 1909–1913. (c) Bernstein, S.; Lenhard, R.H.; Allen, W.S.; Heller, M.; Littell, R.; Stolar, S.M.; Feldman, L.I.; Blank, R.H. *J. Am. Chem. Soc.* **1956**, *78*, 5693–5694.

24. (a) Van Rheenen, V.H.; Huber, J.E.; Upjohn Co. U.S. Patent 4,704,455, Nov. 3, 1987. (b) Van Rheenen, V.H.; Huber, J.E.; Upjohn Co. U.S. Patent 4,891,426, Jan. 2, 1990. (c) Van Rheenen, V.H.: Huber, J.E.; Upjohn Co. U.S. Patent 4,929,395 May 29, 1990.

25. (a) Daniewski, A.R.; Wojciechowska *J. Org. Chem.* **1982**, *47*, 2293–2995. (b) Daniewski, A.R.; Wojciechowska *Synthesis* **1984**, 132–134.

26. (a) Neef, G.; Eder, U.; Seeger, A.; Wiechert, R. *Chem. Ber.* **1980**, *113*, 1184–1188. (b) Barton, D.H.R.; Motherwell, W.B.; Zard, S.Z. *J. Chem. Soc., Chem. Comm.* **1982**, 551–552.

27. (a) van Leusen, D.; van Leusen, A.M. *Tetrahedron Lett.* **1984**, *25*, 2581–2584. (b) Nedelec, L.; Torelli, V.; Hardy, M.; Roussel Uclaf, U.S. Patent 4,565,656, Jan. 21, 1986. (c) Nedelec, L.; Torelli, V.; Hardy, M. *J. Chem. Soc., Chem. Comm.* **1981**, 775–777. (d) Barton, D.H.R.; Motherwell, W.B.; Zard, S.Z. *J. Chem. Soc. ,Chem. Comm.* **1981**, 774–775.

28. (a) Solyom, S.; Szilagy, K; Lajos, T. *Liebigs Ann. Chem.* **1987**, 153–160. (b) Solyom, S.; Lajos, T.; Szilagyi-Farago, K.; Richter Gedeon Vegyeszeti Gyar R.T., U.S. Patent 4,668,437, May 26, 1987.

29. (a) Walker, J.A.; Upjohn Co. U.S. Patent 4,600,538, July 15, 1986. (b) Aberhart, D.J.; Hsu, C.T. *J. Org. Chem.* **1978**, *43*, 4374–4376. (c) Freerksen, R.W.; Raggio, M.L.; Thoms, C.A.; Watt, D.S. *J. Org. Chem.* **1979**, *44*, 702–710. (d) Haffer, G.; Eder, U.; Neef, G.; Sauer, G.; Wiechert, R. *Chem. Ber.* **1978**, *111*, 1533–1539.

30. Poos, G.I., Lukes, R.M., Arth, G.E.; Sarett, L.H. *J. Am. Chem. Soc.* **1954**, *76*, 5031–5034.

31. (a) Van Rheenen, V.; Shephard, K.P. *J. Org. Chem.* **1979**, *44*, 1582–1584. (b) Shephard, K.P.; Van Rheenen, V.H.; Upjohn Co., U.S. Patent 4,041,055, Aug. 9, 1977.

32. Barton, D.H.R.; Basu, N.K.; Hesse, R.H.; Morehouse, F.S.; Pechet, M.M. *J. Am. Chem. Soc.* **1966**, *88*, 3016–3021.

33. Beaton, J.M.; Huber, J.E.; Padilla, A.G.; Breuer, M.E.; Upjohn Co., U.S. Patent 4,127,596, Nov. 28, 1978.

34. Shephard, K.P.; Upjohn Co., U.S. Patent 4,102,907, July 25, 1978.

35. Van Rheenen, V.H.; Cha, D.Y. Upjohn Co., U.S. Patent 4,526,720, July 2, 1985.

36. Sacks, C.E.; Upjohn Co. U.S. Patent 4,613,463, Sept. 23, 1986.

37. Walker, J.A.; Upjohn Co. U.S. Patent 4,342,702, Aug. 3, 1982.

38. (a) Walker, J.A.; Upjohn Co. U.S. Patent 4,357,279, Nov. 2, 1982. (b) Walker, J.A.; Hessler, E.J.; Upjohn Co., U.S. Patent 4,401,599, Aug. 30, 1983. (c) Walker, J.A.; Hessler, E.J.; Upjohn Co. U.S. Patent 4,404,142, Sept. 13, 1983. (d) Walker, J.A.; Hessler, E.J.; Upjohn Co. U.S. Patent 4,411,835, Oct. 25, 1983. (e) Walker, J.A.; Hessler, E.J.; Upjohn Co. U.S. Patent 4,412, 955, Nov. 1, 1983.

39. (a) Nitta, I.; Fujimori, S.; Haruyama, T.; Inoue, S.; Ueno, H. *Bull. Chem. Soc. Japan* **1985**, *58*, 981–986. See also: (b) Hofmeister, H.; Annen, K.; Laurent, H.; Wiechert, R. *Chem. Ber.* **1978**, 3086–3093. Direct inversion is possible, in analogy with cyanohydrin equilibrations, but the yields are low: (c) Runge, J.R.; Upjohn Co. U.S. Patent 4,582,644, April 15, 1986.

40. Hofmeister, H.; Annen, K.; Laurent, H.; Wiechert, R. *Liebigs Ann. Chem.* **1987**, 423–426.

41. Ercoli, A.; de Ruggieri, P. *J. Am. Chem. Soc.* **1953**, *75*, 650–653.

42. Gasc, J.C.; Nedelec, L. *Tetrahedron Lett.* **1971**, 2005–2008.

43. De Ruggieri, P.; Ferrari, C. *J. Am. Chem. Soc.* **1959**, *81*, 5725–5727.

44. Teichmueller, G.; Haessler, G.; Barnikol-Oettler, K.; Grinenko, G.S.; Dolginova, E.M.; V.E.B. Jenapharm, E. Ger. Pat. DD 147,669 Apr. 15,1981 *Chem. Abstr.* **96**:20370f.

45. Nitta, I.; Fujimori, S.; Ueno, H. *Bull. Chem. Soc. Japan* **1985**, *58*, 978–980.

46. (a) Ringold, H.; Stork, G. *J. Am. Chem. Soc.* **1958**, *80*, 250. For other references, see: (b) Fried, J.; Edwards, J.A. "Organic Reactions in Steroid Chemistry", vol. 2, Van Nostrand Reinhold Company, **1972**, pp. 206–209.

47. Zuidweg, M.H.J. *Biochim. Biophys. Acta* **1968**, *152*, 144–158.

48. Van Rheenen, V.H. Upjohn Co., U.S. Patents 4,500,461, Feb. 19, 1985; 4,548,748 Oct. 22, 1985; 4,585,590 Apr. 29. 1986 (*Chem. Abstr.* 103:22844a).

49. See ref. 36.

50. Reviews: (a) Krief, A. *Tetrahedron* **1980**, *36*, 2531–2640. (b) Beak, P.; Reitz, D.B. *Chem. Rev.* **1978**, *78*, 275–316. Other references: (c) Sawyer, J.S.; Kucerovy, A.; Macdonald, T.L.; McGarvey, G.J. *J. Am. Chem. Soc.* **1989**, *110*, 842–853. (d) Katritzky, A.R.; Sengupta, S. *Tetrahedron Lett.* **1987**, *28*, 1847–1850. (e) Corey, E.J.; Eckrich, T.M. *Tetrahedron Lett.* **1983**, *24*, 3163–3164. (f) *ibid.* 3165–3168.

51. Reid, J.G.; Debiak–Krook, T. *Tetrahedron Lett.* **1990**, *31*, 3669–3672.

52. Such cyclizations do have precedent, however. See: Hiyama, T.; Oishi, H.; Saimoto, H. *Tetrahedron Lett.* **1985**, *26*, 2459–2462.

53. (a) Tamao, K.; Ishida, N.; Kumada, M. *J. Org. Chem.* **1983**, *48*, 2120–2122. (b) Tamao, K.; Ishida, N.; Tanaka, T.; Kumada, M. *Organometallics* **1983**, *2*, 1694–1696. (c) Tamao, K.; Ishida, N.

*Tetrahedron Lett.* **1984**, *25*, 4245–4248. (d) Tamao, K.; Ishida, N. *J. Organometallic Chem.* **1984**, *269*, C37–C39.

54. Larcheveque, M.; Debal, A.; Cuvigny, Th. *J. Organometallic Chem.* **1975**, *87*, 25–31.
55. Freeman, P.K.; Hutchinson, L.L. *J. Org. Chem.* **1980**, *45*, 1924–1930.
56. (a) Burford, C.; Cooke, F.; Roy, G.; Magnus, P. *Tetrahedron* **1983**, *39*, 867–876. (b) Magnus, P.; Roy, G. *J. Chem. Soc. Chem. Comm.* **1978**, 297–298. (c) Magnus, P.; Cooke, F. *J. Chem. Soc. Chem. Comm.* **1977**, 513. (d) Burford, C.; Cooke, F.: Ehlinger, E.; Magnus, P. *J. Am. Chem. Soc.* **1977**, *99*, 4536–4537.
57. Olofson, R.A.; Hoskin, D.A.; Lotts, K.D. *Tetrahedron Lett.* **1978**, 1677–1680.
58. (a) Pearlman, B.A.; Putt, S.R.; Fleming, J.A. *J. Org. Chem.* **1985**, *50*, 3622–3624. (b) Pearlman, B.A.; Putt, S.R.; Fleming, J.A. *J. Org. Chem.* **1985**, *50*, 3625–3626.
59. Reetz, M.T.; Maier, W.F. *Liebigs Ann. Chem.* **1980**, 1471–1473.
60. Ziegler, K.; Jacob, L.; Wollthan, H.; Wentz, A. *Liebigs Ann. Chem.* **1934**, *511*, 64.
61. Corey, E.J.; Gross, A.W. *Tetrahedron Lett.* **1984**, *25*, 495.
62. Van Rheenen, V.H.; Upjohn Co. U.S. Patent 4,567,001, Jan. 28, 1986.
63. Similar pentacyclic steroids are known, but apparently not with the 17α-hydroxyl in place. (a) Milliet, P.; Picot, A.; Lusinchi, X. *Tetrahedron* **1981**, *37*, 4201–4208. (b) Miyano, M. *J. Org. Chem.* **1981**, *46*, 1854–1857. (c) Marti, F.; Wehrli, H.; Jeger, O. *Helv. Chim. Acta* **1973**, *56*, 276–277. (d) Jeanniot, J. P.; Lusinchi, X.; Millet, P.; Parello, J. *Tetrahedron* **1971**, *27*, 401–410.
64. The known stability of lithiated α-chloroimines suggests that there are subtleties in this process that are yet understood. DeKimpe, N.; Sulman, P. Schamp, N. *Angew. Chem. Int. Ed.* **1985**, *24*, 881–882.
65. (a) DeKimpe, N.; Yao, Z.; Schamp, N. *Tetrahedron Lett.* **1986**, *27*, 1707–1710. (b) Newcomb, M.; Varick, T.R.; Goh, S.H. *J. Am. Chem. Soc.* **1990**, *112*, 5186–5193.

# HUPERZINE A - A POSSIBLE LEAD STRUCTURE IN THE TREATMENT OF ALZHEIMER'S DISEASE

Alan P. Kozikowski, Edda Thiels,

X.C. Tang, and Israel Hanin

Advances in Medicinal Chemistry,
Volume 1, pages 175–205.
Copyright © 1992 by JAI Press Inc.
All rights of reproduction in any form reserved.
ISBN: 1-55938-170-1

175

# ABSTRACT

This review chapter details the current state of our knowledge of the alkaloid natural product, huperzine A, a molecule which has attracted widespread attention because of its possible use in the treatment of Alzheimer's dementia. A summary of the chemical synthesis of huperzine A in both its racemic and enantiomerically pure forms is provided together with a description of the preparation and acetylcholinesterase inhibitory properties of a variety of analogues. Details of huperzine A's in vitro as well as in vivo pharmacological action then are discussed, and a critical evaluation of the compound's effects on memory function in both animals and humans is presented. The chapter finally deals with the possibility of designing additional huperzine A analogues that may serve as more efficacious candidates for use in the clinic.

# I. INTRODUCTION

The phones started ringing continuously in Alan P. Kozikowski's office one day, and continued to do so weeks later. Even after a year since the popular press first reported our work on the alkaloid huperzine A, calls keep coming in from concerned individuals looking for a drug to help alleviate the memory disorders and confusion of friends or relatives afflicted by Alzheimer's disease (AD).[1] Why all this fuss about a relatively simple alkaloid? Is this simply much ado about nothing, or is in fact huperzine A the nootropic agent it is claimed to be?[2] In this chapter, we summarize much of the chemistry, pharmacology, and behavioral experiments that have been carried out on huperzine A and its analogues. We hope that in the context of this article we will begin to answer some of the questions posed above.

The herbal medicine known as Qian Ceng Ta apparently was prescribed over the centuries for a variety of maladies.[3] Researchers at the Shanghai Institute of Materia Medica, being aware of the history of Qian Ceng Ta, examined the effects of extracts of the club moss *Huperzia serrata* (Thunb.) Trev. = *Lycopodium serratum* Thunb. in crude tests of memory in rodents.[4] In this manner the Chinese workers identified two alkaloids, huperzine A and huperzine B, as being the agents exhibiting facilitatory effects on memory. The structures of these alkaloids were

Huperzine A                                          Huperzine B

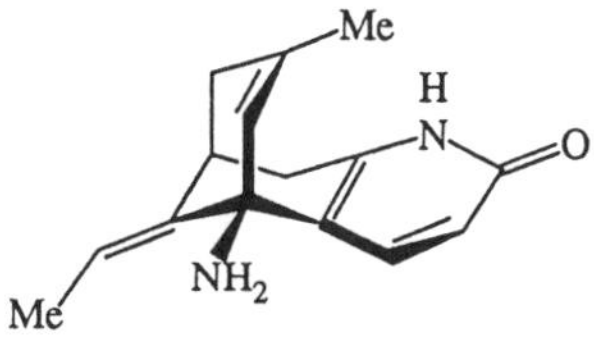

Selagine 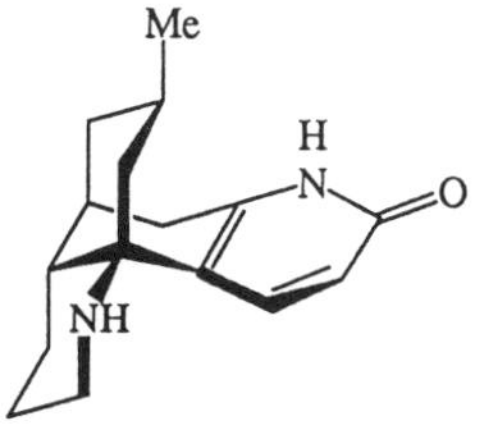                    β-Obscurine

assigned on the basis of a variety of spectroscopic and chemical studies. Huperzine B was found to contain an α-pyridone ring as well as a double bond between carbon atoms 8 and 15. However, unlike huperzine A, the $^1H$ and $^{13}C$ NMR spectra of huperzine B were devoid of signals that correspond to the exocyclic double bond at $C_{11}$–$C_{12}$. These data, together with the knowledge that dehydrogenation of huperzine B over Pd/C at 300 °C affords 7-methylquinoline and 6-methyl-2(1*H*)-pyridone, supported the structure shown above for huperzine B.[4] Although huperzine A initially was reported to be closely related to another pyridone containing alkaloid, selagine, a compound whose structure was elucidated in 1960 by Wiesner and co-workers, recent studies have revealed the earlier structural assignment to be incorrect.[5] The alkaloid isolated from *Lycopodium selago* L. and named selagine is, in fact, identical to huperzine A. Within the family of Lycopodium alkaloids, it is further worth noting that huperzine B is structurally similar to β-obscurine, from which it differs by only two hydrogen atoms.[6]

While our own synthetic studies together with an X-ray analysis of (±)-huperzine A provide complete verification of huperzine A's structure,[7] the details of such matters will be reserved for further discussion later in this chapter.

As can be seen from the NMR spectrum provided in Figure 1, huperzine A appears to be the major constituent of the alkaloid fraction extracted from *Huperzia serrata*, which we obtained from Taiwan. An NMR spectrum is also provided of the alkaloid extracts obtained from *Lycopodium lucidulum*, a club moss common to the United States. As is apparent, the United States club moss contains no huperzine A.[8] While other species of club moss are now being investigated for their huperzine content, we note that *Lycopodium selago* collected near Tarna in Sweden, in the Tonquin valley near Jasper, Alberta, and near Summit Lake in northern British Columbia has been shown by Valenta et al. to contain huperzine A (1.43 kg of dried *Lycopodium selago* afforded 262 mg of huperzine A).[5]

Huperzine A has captured the interest of many researchers because of its so-called "nootropic" effects. A nootropic agent is a compound that acts on cognitive functions and which facilitates learning and memory and/or prevents impairment of cognitive function.[2] How does huperzine A work its magic? In pharmacological studies conducted by Tang and co-workers as well as by us, huperzine A appears to function as a potent inhibitor of the degradative enzyme

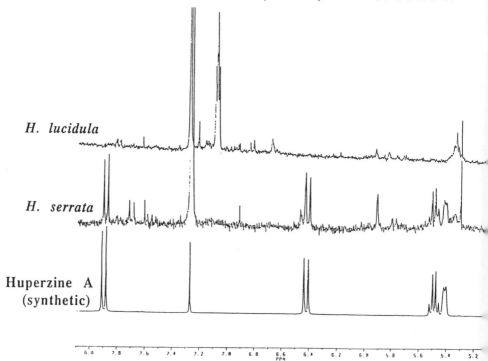

**Figure 1.** Downfield region of $^1$H NMR spectra of huperzine A and crude extracts from club moss.

acetylcholinesterase (AChE).[9] This enzyme breaks apart the neurotransmitter acetylcholine into choline plus acetate ion. Because one of the major deficits in the brain of the Alzheimer's-afflicted individual appears to involve the cholinergic system, and specifically the degeneration of those cells committed to the synthesis of acetylcholine, huperzine A is capable of facilitating cholinergic neurotransmission through its ability to increase local concentrations of this important neurotransmitter within the synaptic cleft. The possibility that huperzine A might interact with the N-methyl-D-aspartate (NMDA) receptor, a brain receptor known to be intimately linked to memory and learning,[10] was also investigated, for in a structural sense this alkaloid bears some similarity to phencyclidine, a drug of abuse which binds within the NMDA receptor-associated ion channel.[11] However, in the MK-801 binding assay, which provides both a measure of the affinity of a molecule for PCP/MK-801 recognition sites, and which also detects any possible modulatory action at other sites on the receptor, huperzine A was found to have no effect at concentrations <100 μM.[12] The only other pharmacological action which has been described for huperzine A is its ability to displace [$^3$H]-(−)nicotine binding at feeble concentrations around $6 \times 10^{-4}$M.[13]

The following sections of this review have been organized into chemistry, pharmacology, and behavior. In the chemistry section the methods used to assemble huperzine A in the laboratory from inexpensive and simple starting materials are outlined together with modifications in the scheme which allow the generation of analogous structures. The effects of these structural modifications on AChE inhibitory potency are also discussed. In the pharmacology section details of huperzine A's in vitro pharmacological profile are provided together with the results of certain in vivo pharmacological studies. Lastly, the action of huperzine A on animal performance in several different memory paradigms, together with its therapeutic action in man, are summarized in the behavior section.

## II.  CHEMISTRY

### A.  Total Synthesis

Because it was unlikely that huperzine A could ever be isolated from natural sources in an ecologically sound manner to meet the needs required for clinical studies, a total synthesis approach to this molecule appeared to be warranted. Due to the rather special conditions required for the growth of the club moss, it is furthermore unlikely that *Huperzia serrata* could be cultivated. Although plant cell culture methods might possibly provide a source of huperzine A, such methods remain untested. The total synthesis approach does, of course, embody a further distinct advantage over product isolation from natural sources, in that one can readily modify the "target structure" in ways which might lead to pharmacologically improved versions of the molecule.

Additionally, because huperzine A represents a relatively rigid structure, modifications can be engineered in a manner that will help to define more precisely the steric and electronic topography of the acetylcholine binding site of AChE.

Our retrosynthetic analysis of huperzine A is shown in Scheme 1. The main feature of the synthesis plan is the projected use of the ring-fused pyridone structure which provides an appropriate platform upon which to construct the unsaturated carbon bridge. The ester group of this substrate not only serves as an activating group for incorporation of the carbon bridge by a combined Michael/aldol process, but also provides a handle for introduction of the amino group through execution of a Curtius rearrangement near the end of the synthesis. Subsequent to generation of the endocyclic double bond by dehydration of the intermediate aldol product **1**, it was anticipated that the exocyclic olefin at $C_{11}$–$C_{12}$ could be introduced through a Wittig reaction. Although predictions regarding the stereoselectivity of this process can not be made with certainty,[13] we realized that a thermodynamically controlled isomerization process could be called upon if required. Lastly, the ring-fused pyridone **2** needed to initiate this project could come from the mono-ethylene ketal of cyclohexane-1,4-dione (**3**) by way of an enamine intermediate.

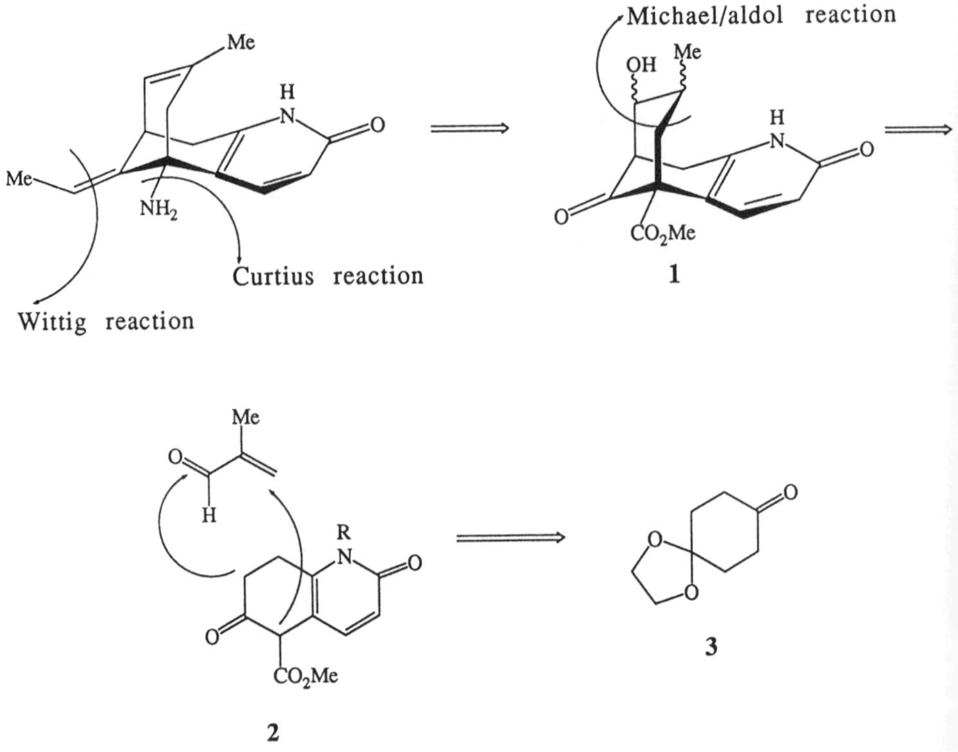

**Scheme 1.** A retrosynthetic analysis of huperzine A.

Before describing the successful synthetic route to huperzine A, we wish to note here that some effort was taken to examine the use of the readily available ring-fused pyridone **6** in the preparation of huperzine A (Scheme 2). The keto-pyridone **6** can be generated from cyclohexane-1,3-dione (**4**) by reaction with ammonia followed by heating with methyl propiolate.[14] After protection of its pyridone ring nitrogen, **6** was subjected to an α-hydroxylation reaction to provide hydroxy ketone **7**.[15] Next, the oxime **8** was generated (Z/E ratio of oximes = 9:1), and this intermediate was reacted with 2,2-dimethoxypropane and p-toluenesulfonic acid (PTSA) to yield acetonide **9**. We had hoped that methallylmagnesium bromide would add to **9** to yield **10**, a possible precursor to the keto-aldehyde **11** which should undergo facile intramolecular aldol reaction to provide huperzine A's tricyclic skeleton. Although in model studies methallylmagnesium bromide readily added to the oxime ether **12** (Scheme 3), this reagent unfortunately failed to react with **9** to provide the 1,2-addition product **10**. When Lewis acid catalysts (e.g., BF₃ • OEt₂) were employed, **10** was isolated only as a minor product. The major product

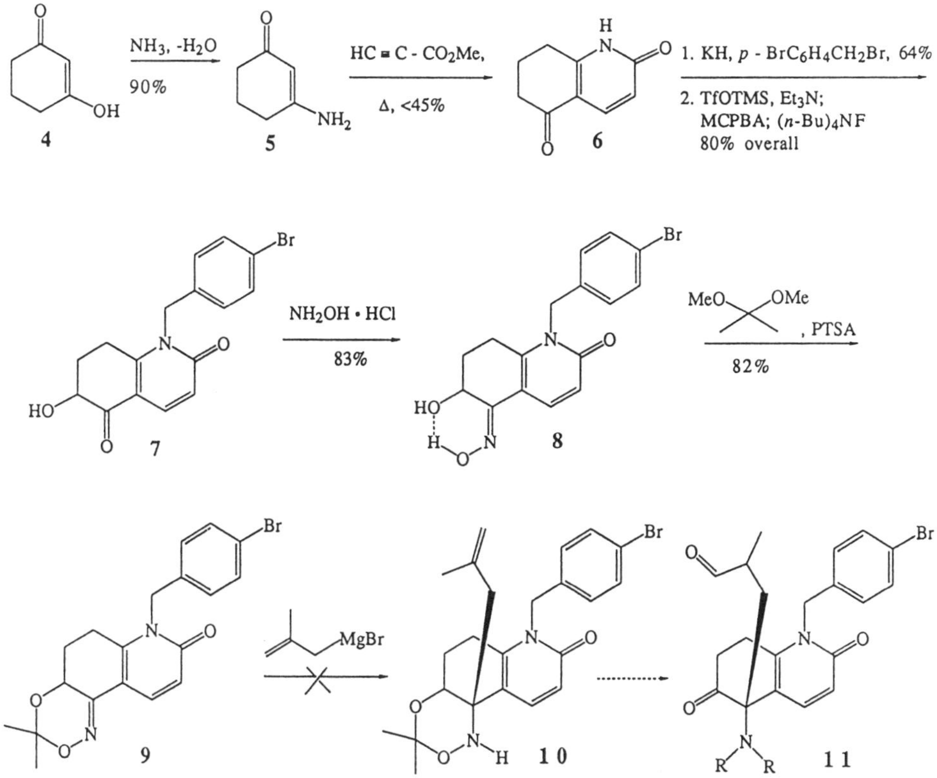

**Scheme 2.** Synthesis of the ring-fused pyridone **6** and its synthetic elaboration.

to be isolated was **14**, a compound formed by 1,4-addition of the Grignard reagent to the pyridone ring.

Other strategies were explored to utilize pyridone **6** in the synthesis of huperzine A. However, these also proved fruitless, and we eventually turned our attention to the synthesis and employment of the carbonyl-transposed pyridone **2** of Scheme 1.

In preparing a pyridone structure related to **2**, we initially applied Stork's aza-annelation method,[16] to the commercially available mono-protected diketone **3**. The pyrrolidine enamine of **3** was heated with acrylamide followed by hydrolysis to provide an 85:15 mixture of the lactams **16** and **17** (Scheme 4). After protection of the lactam nitrogen by benzylation, a dehydrogenation sequence was brought about to afford pyridone **20a**. The dehydrogenation reaction can be accomplished in a single step by the use of DDQ (2,3-dichloro-5,6-dicyano-1,4-benzoquinone) in benzene; however, the yield is only 40% and side products arising from aromatization of the cyclohexene ring are also isolated. Thus, a higher yielding, multi-step process that consists of selenylation, oxidative elimination, and

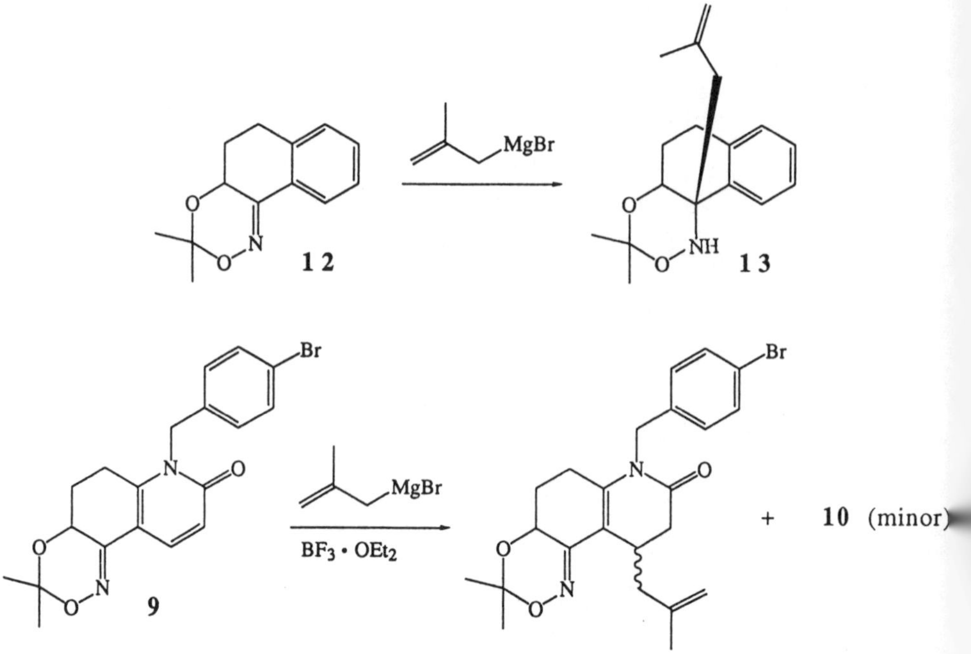

**Scheme 3.** Grignard addition reactions to oxime acetonides.

isomerization (for **18**) was adopted.[17] Next, intermediate **20a** was hydrolyzed and the ketone **21** reacted with potassium hydride and dimethyl carbonate to produce the β-keto ester **2**.[18] Unfortunately, the yield of this reaction was quite poor, with substantial amounts of the benzenoid product **22** being isolated. Because in model reactions employing β-tetralone the carbomethoxylation was found to proceed cleanly and in yields of >95%, we suspected that the pyridone ring was responsible for the wayward chemistry.[19] Proton abstraction is, of course, possible at C-8, an event which can initiate a subsequent dehydrogenation step.[20] To prevent this deprotonation reaction from taking place, we decided to protect the pyridone ring on oxygen rather than nitrogen, thus creating an alkoxypyridine derivative. Protection of the pyridone ring as a pyridine derivative also leads to the further advantage in that pyridines often are less polar than their pyridone counterparts, thus making them easier to isolate and purify.

The α-methoxypyridine **23** was synthesized from **20a** by hydrogenolysis of its N-benzyl group, followed by O-methylation with methyl iodide and silver carbonate (Scheme 5). This time the α-carbomethoxylation proceeded readily, and an

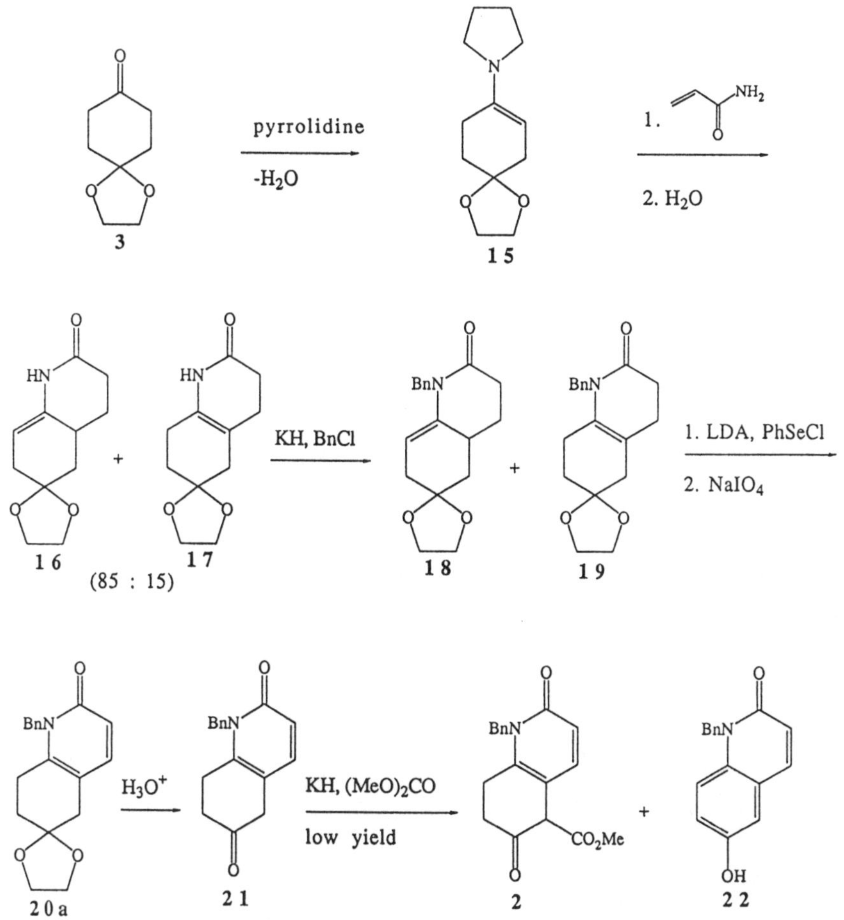

**Scheme 4.** Synthesis of ring-fused pyridone **5**.

87% yield of the desired β-keto ester **24** plus a very small amount of the hydroxy-quinoline **25** were isolated.

At this stage, the Michael/aldol reaction sequence could be studied. A variety of Michael reaction catalysts (e.g., NaOMe, *n*-Bu₄NF, Et₃N, ZnCl₂, and HOAc) were examined in order to bring about the reaction of **24** with methacrolein.[21] Unfortunately, no reaction was observed. Although several alternative strategies for introducing the carbon bridge were examined, including the palladium-

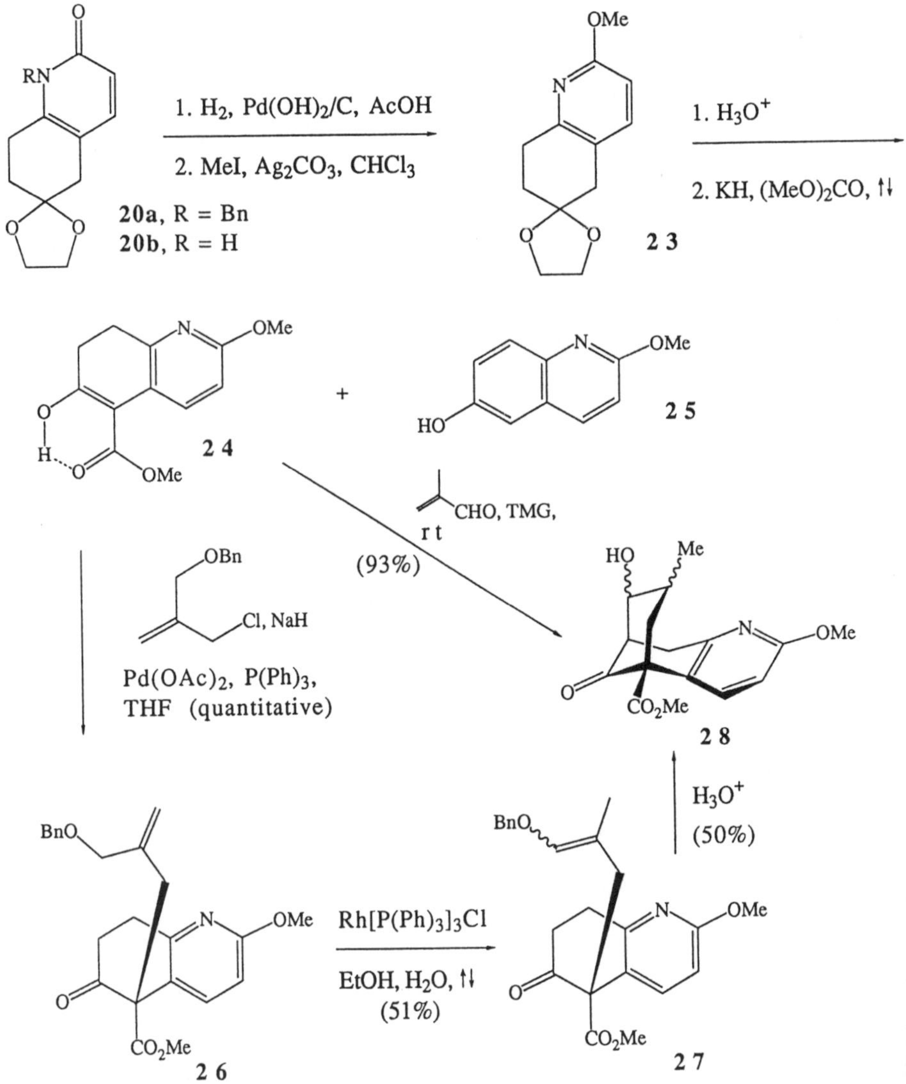

**Scheme 5.** Synthesis of β-keto ester **24** and two possible routes to bridged intermediate **28**.

mediated process pictured in Scheme 5,[22] we eventually returned to the original Michael/aldol process because of its brevity. Tetramethylguanidine (TMG) is known to catalyze the Michael reaction of nitroalkenes with electron-deficient olefins. For example, the reaction of nitromethane with methyl β,β-dimethylacrylate proceeds in 45% yield in the presence of TMG, but fails when carried out using the usual Michael catalysts.[23] Because β-keto esters have a pKa (~ 10) comparable to that of nitroalkanes, we examined the reaction of **24** with methacrolein using TMG as the base catalyst. Under these conditions cyclic ketol **28** was formed as a mixture of stereoisomers in a single step in greater than 90% yield. To the best of our knowledge this reaction represents the first example of a TMG-catalyzed Michael addition of a β-keto ester to an electron-deficient olefin. The reaction is easy to execute and only requires that the reactants and catalyst be stirred in methylene chloride at room temperature for several hours. Concentration of the reaction mixture, followed by silica gel chromatography, cleanly affords the desired ketols. The TMG-catalyzed Michael/aldol sequence probably owes its success not only to the higher basicity of TMG relative to triethylamine, but also to the ability of the resonance delocalized TMG-H$^+$ cation to pair with and thus to stabilize the resonance delocalized enolate anion intermediate generated during the course of the "bridging" process.[24]

The ketol mixture **28** was now dehydrated to the alkene **30** by reaction of its derived mesylates with sodium acetate and acetic acid at 120°C (Scheme 6). The mesylate elimination reaction does, for the moment, represent the lowest yielding step of the entire synthetic sequence. Although a few other methods have been examined to bring about the conversion of **28** → **30**, no significant improvements as yet have been made.

With the endocyclic double bond in position, introduction of huperzine's exocyclic double bond was carried out next. In spite of the fact that the ketone carbonyl group of **30** is sterically encumbered, its Wittig reaction with ethylidenetriphenylphosphorane took place smoothly in THF at 0°C to room temperature to provide a 90:10 mixture of the *Z*- and *E*-alkenes **31** and **32** in 73% yield. This olefin mixture consisted largely of the incorrect isomer needed to assemble huperzine A, and an isomerization reaction was therefore brought about using AIBN and thiophenol.[26] The extent of isomerization was dependent on the reaction temperature: at 130°C this process led to an 80:20 mixture of the *E*- and *Z*-isomers **31** and **32**, and at 170°C it led to a 95:5 mixture.

The last operations that remained to complete the synthesis of (±)-huperzine A were the conversion of the ester group of **32** to an amino group and cleavage of the *O*-methyl ether to reveal the pyridone ring.

As shown in Scheme 6, the hindered ester group of **32** was first saponified to acid by heating with 20% aqueous sodium hydroxide in THF/MeOH. Under these conditions, only the isomer of *E*-olefin stereochemistry was converted to acid, whereas the more hindered *Z*-olefinic ester failed to react. The acid **33** was converted in a one-pot operation to the carbamate **34** by acid chloride synthesis, acyl azide formation with concomitant Curtius rearrangement, and methanolysis of the resulting isocyanate.[27] Lastly, cleavage of both the methyl ether and the methyl carbamate of **34** was accomplished by the use of iodotrimethylsilane. A small amount (~ 8%) of the partially deprotected carbamate **35** was also isolated. The synthetic huperzine A was identical to a sample of the natural material by high field $^1$H NMR, IR, mass spectral, and TLC comparisons. An X-ray analysis (Fig.

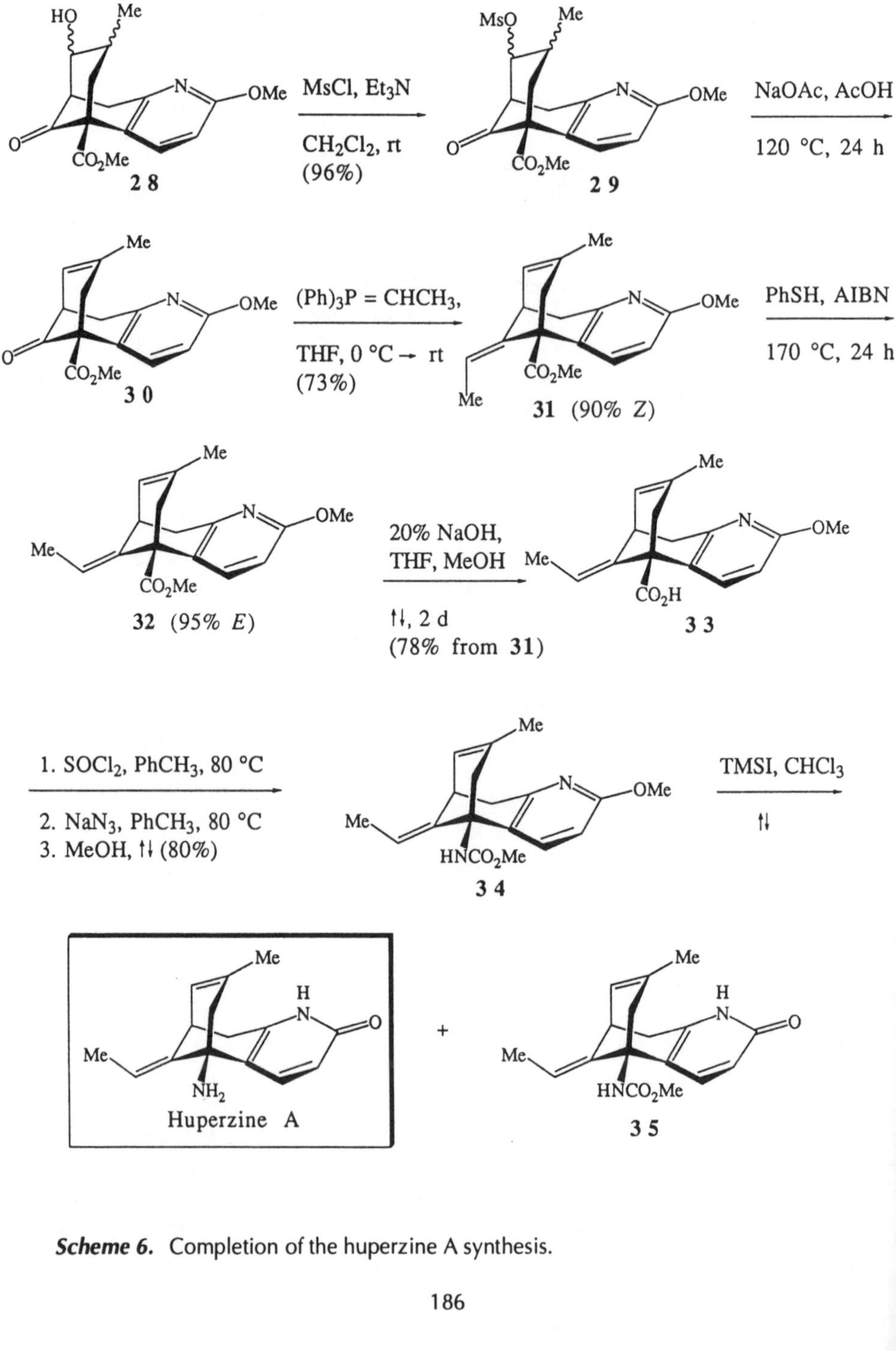

**Scheme 6.** Completion of the huperzine A synthesis.

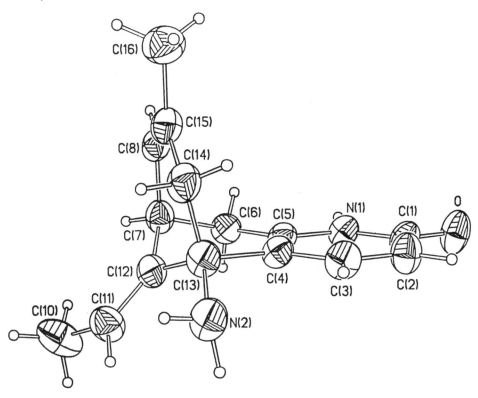

**Figure 2.** Perspective drawing of (±)-huperzine A as determined by single crystal X-ray analysis.

2) carried out recently at the University of Pittsburgh on (±)-huperzine A provides further verification of structure.[28]

Although the overall scheme presented above offers a reasonably efficient route to (±)-huperzine A, we felt that further improvements were necessary in order to facilitate scale-up procedures as well as analogue synthesis. Specifically, the chemical route to pyridone **20b** (Scheme 5) was unduly laborious and costly. After considerable experimentation, we eventually discovered that **20b** could be obtained in a single-step reaction in 40–70% yield by heating ketone **3**, methyl propiolate, and ammonia in a Parr reactor at 100°C for 10 h (200 psi internal pressure). Mechanistically, this chemistry resembles the pyridone synthesis reported by Speckamp,[14] but unlike the earlier procedure does not require prior formation and isolation of the enamine intermediate. The use of this new pyridone-forming route thus eliminates three of the synthetic steps required to produce huperzine A in the laboratory, and allows one to obtain multi-gram amounts of the bridged ketol **28** within several working days.

The foregoing route to huperzine A provides access to the racemate. By simply carrying out an acid-catalyzed ester exchange to convert **24** to its (–)-8-phenylmenthol ester **24a** and conducting a nearly identical sequence of operations, we were able to produce (–)-huperzine A. The only change in the synthetic strategy is that at the stage of **32a**, the 8-phenylmenthol ester is reduced to an alcohol and then reoxidized to an acid by Jones reagent prior to the Curtius rearrangement. From ${}^1$H NMR analysis of the chromatographically separable mixture of isomers **30a** formed by dehydration of **28a**, a mixture that was generated by carrying out the TMG-catalyzed Michael/aldol reaction at –20°C, a 9:1 ratio of diastereomers in favor of the natural stereochemistry was detected. By conducting the Michael/aldol reaction at room temperature, a lower level of diastereoselectivity was found (ratio of isomers = 4:1).[29]

Although this route to enantiomerically pure huperzine A is fairly efficient, it still is our aim to examine the use of chiral base catalysts for bringing about the Michael/aldol reaction. The development of a catalytic, enantioselective synthesis of (–)-huperzine A certainly would be of considerable practical interest from a manufacturing standpoint.[30]

## B. Analogue Synthesis

Having completed the synthesis of huperzine A, we were able to show that the racemic compound possessed potent AChE inhibitory activity (*vide infra*). Because of the promise of huperzine A as a palliative agent in the treatment of Alzheimer's dementia, we chose to define certain structure activity relationships (SAR) for this molecule. Acetylcholine is believed to adopt the {150°, 180°} conformation during its enzymatic hydrolysis to choline and acetate ion.[31] From an examination of a graphical overlay of huperzine A with this conformation of acetylcholine, one can observe a good spatial coincidence of their respective heteroatom groups. Because the amino group of huperzine A will be protonated at physiological pH, it may mimic the $Me_3N^+$ group of acetylcholine. From such a simple structural comparison, it is reasonable to conclude that an aminomethyl-substituted pyridone, a "bare-bones version" of huperzine A, might exhibit AChE inhibitory activity. Compound **36a** thus was synthesized in addition to its *N,N*-dimethylamine counterpart **36b**, for this later compound can be viewed as an even closer mimic of acetylcholine. These two compounds were prepared conveniently from 2-hydroxy-5-pyridinecarboxylic acid through a fairly standard sequence of operations.[32] Unfortunately, neither of these compounds was found to be a very good inhibitor of AChE, as is discussed below.

**36a**, R = H
**36b**, R = Me

1. $NH_2OH$

2. $H_2$, Raney Ni

**6**

**3 7**

One additional simplified version of huperzine A was constructed from ring-fused pyridone **6** by transforming the ketone to oxime, and then reducing the oxime to amine **37** over Raney nickel.[33] Again, only poor AChE inhibitory activity was found for compound **37**.

From these findings one is forced to conclude that the structural rigidity conferred upon huperzine A by its tricyclic skeleton, and the contribution of these additional carbon atoms to hydrophobic binding forces and van der Waals interactions must play a significant role in huperzine A's interaction with the active site of the enzyme.

With regard to modifications to huperzine A itself, we have investigated alterations at $C_{12}$, $C_{13}$, and $C_{15}$, as well as the pyridone ring. In the context of developing our total synthesis route to huperzine A, we employed β-tetralone as a model compound and converted this compound to the benzenoid analogue **38** of huperzine A.[19] Not unexpectedly, this compound was a very poor inhibitor of AChE.

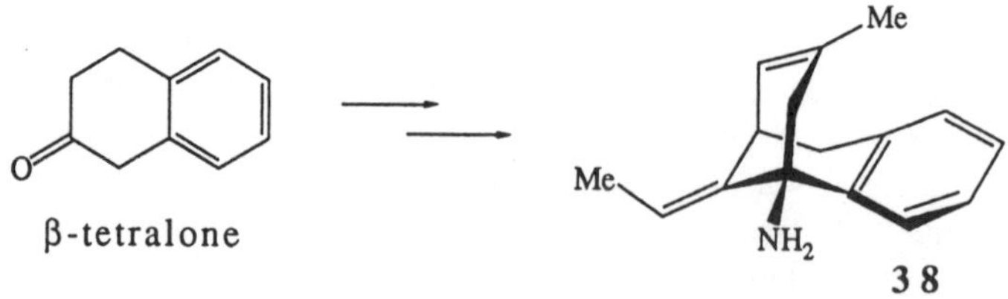

β-tetralone

**38**

To examine the contribution of the ethylidene group at $C_{12}$ to huperzine A's anti-AChE activity, this group was replaced by both a methylene and a propylidene group. Rather curiously, both compounds were found to be about a hundred-fold less active than huperzine A in inhibiting AChE. Apparently, the molecular volume of the propylidene analogue **39** may be too large for optimal binding to the enzyme, whereas removal of the methyl group leads to a compound, **40**, of diminished surface area with the subsequent loss in hydrophobic binding and van der Waals interactions.[34a] By contrast, (±)-Z-huperzine A (viz. **31**) showed activity ($6 \times 10^{-6}$ M) more comparable to the standard (±)-E isomer (Table 1).[34b]

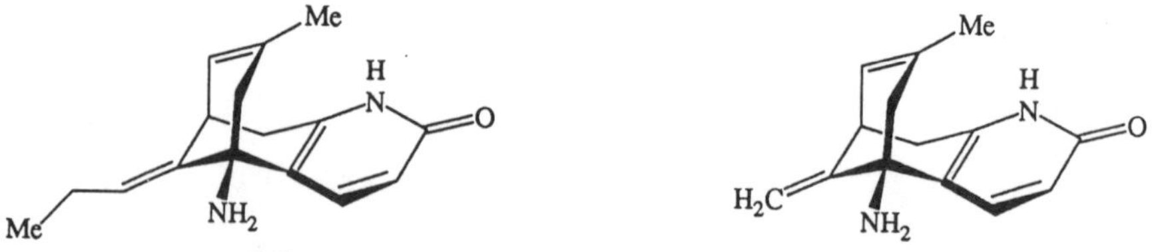

**39**

**40**

Likewise, when the methyl group at $C_{15}$ was replaced by the sterically larger phenyl group, a compound that was prepared conveniently by substituting 2-phenylacrolein[35] for methacrolein in the TMG-catalyzed Michael/aldol process, AChE inhibitory activity largely was lost. As apparent from the summary data of Table 1, Section III, the $IC_{50}$ of the phenyl bearing analogue is ~2000 times larger

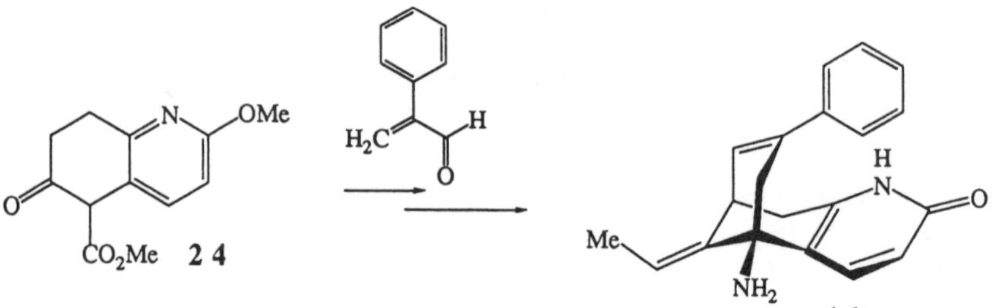

**24**

**41**

than the IC$_{50}$ of (±)-huperzine A. Apparently, substantial steric bulk in this area of the carbon bridge of huperzine A cannot be tolerated by the AChE binding site.

By reducing the ester function of intermediate 30 to an alcohol, then activating the alcohol as its mesylate, displacing with azide, reducing azide 42 to amine 43, generating the urethane, and carrying out the Me$_3$SiI-promoted deprotection step, the homologue 44 of huperzine A was obtained. This compound contains a one-carbon spacer between huperzine A's amino group and the bridgehead carbon. If the computer overlay of huperzine A with acetylcholine is at all correct, then this homologue would be anticipated to be poorly active as an AChE inhibitor. Indeed, this compound was at least 1000-fold less potent than huperzine A as an AChE inhibitor. Compounds 42 and 43, the immediate precursors to 44, were tested also and found to exhibit no activity.

Lastly, we investigated the effect of introducing two methyl groups on the -NH$_2$ group of huperzine A.[36] Our logic here was that the *N,N*-dimethylamino derivative would more closely resemble acetylcholine, and that the extra possible van der Waal's interactions might lead to improved anti-AChE activity. Unfortunately, this dimethyl analogue 45 was only poorly active (IC$_{50}$ = 4.5 × 10$^{-4}$M).

In summary, although a number of obvious analogues of huperzine A have been prepared to date, none of the compounds have shown any AChE inhibitory activity rivaling that of the parent structure. Because the analogues synthesized represent rather trivial and obvious modifications to the parent structure, we remain hopeful that more "creative" alterations of huperzine A will lead to more active compounds.

The current analogue work does, however, support conclusions drawn from overlay comparisons of huperzine A, acetylcholine, physostigmine, and neostigmine that the structural features essential to the design of good AChE inhibitors are the presence of a charged or protonatable nitrogen atom located at ~5Å distance from a carbonyl carbon embodied with a hydrophobic (bicyclic) framework. These overlay comparisons in combination with comparative molecular field analyses (CoMFA) have led to the design of additional analogue structures. Indeed, synthetic efforts are currently underway to prepare electronically altered versions of this nootropic agent that are expected to exhibit improved in vitro activity.

Of course, the concentration of a drug needed to inhibit the enzyme in vitro is but one criterion of activity, for to be useful clinically the drug candidate must also pass certain hurdles relating to toxicity, metabolism, pharmacokinetics, and bioavailability.[37]

Below, we review some of these matters as we turn to the effects of huperzine A both in vitro and in vivo.

## III. PHARMACOLOGY

Literature on the neurochemical effects of huperzine A administration is sparse, primarily due to the fact that this compound is so scarce, and hence not too many studies have been conducted on its neurochemical effects to date. Nevertheless, a number of reports have appeared, in which at least an effect of huperzine A on the cholinergic system, both in vivo and in vitro, has been well established.

Huperzine A has been shown, in vitro, to be a mixed and reversible inhibitor of AChE, as well as of butyrylcholinesterase (BuChE).[38] The negative logarithm of the molar concentration causing 50% inhibition of cholinesterase, the $pI_{50}$, indicated that huperzine A has an anti-AChE effect approximately 3-fold that of physostigmine (7.2 vs 6.7), when measured on erythrocyte membrane AChE. Huperzine A also exhibited a weak anti-BuChE activity; this effect was only evident at a thousand-fold increase in concentration of the compound ($pI_{50} = 4.1$) above that required for inhibition of AChE.

The first comprehensive investigation of the effect of huperzine A in vivo was reported in 1989, by Tang and his coinvestigators.[13] Following intramuscular (im) administration of 2 mg/kg huperzine A, whole brain AChE was shown to be inhibited by 42% at 60 min post injection. This inhibition lasted for up to 4 hr and was still significant (32%) at 6 hr post-treatment. Erythrocyte AChE and plasma

BuChE inhibition peaked at 30 min (39% and 47%, respectively), but dropped rapidly, and both returned to normal levels within 4 hr.

The same treatment with huperzine A also resulted in a concurrent increase in whole brain acetylcholine levels (reaching a maximal increase of 40% by 60 min), which was still significantly elevated at 4 hr, but which returned to normal by 6 hr. This elevation in acetylcholine levels most probably was due to the inhibition of AChE, which reached a nadir of 42% within 30 min, and remained at approximately that level of inhibition up to 6 hr post huperzine administration.

Interestingly, the extent of AChE inhibition, and of acetylcholine elevation showed considerable regional variability in the brain. Measures were conducted in different brain regions, including frontal cortex, parietal cortex, hippocampus, striatum, medulla oblongata, medial septum and cerebellum, at 5 different dose levels of huperzine A, spanning a concentration range of 0.1–2.0 mg/kg, intraperitoneally (ip). Each brain area exhibited a different magnitude and time course of effects in response to huperzine A administration. In all brain regions tested, acetylcholine levels returned to normal within 2 hr.

Electrically evoked fractional release of radiolabelled acetylcholine from cortical slices, also conducted in the same study by Tang et al.[13] was not affected by up to $10^{-4}$ M huperzine A in the medium. Neither was there any effect of huperzine A on muscarinic or nicotinic receptor binding in rat brain cortex in the physiological range of the compound. In both cases the $IC_{50}$ of the compound exceeded $5 \times 10^{-4}$ M.

Based on the above described comprehensive study, Tang and his coinvestigators concluded that huperzine A produces significant AChE inhibition at optimal doses, with less peripheral side effects than other AChE inhibitors. They suggested, on the basis of their findings, that huperzine A may be a more suitable drug for the treatment of AD than, for example, physostigmine.

Laganiere et al. have also studied the in vivo effects of huperzine A in rats.[39] Their results confirm and amplify those of Tang and co-investigators, which have been described above. Specifically, they have also studied the effects of huperzine A on the high affinity uptake of choline (HAChT) in synaptosomal homogenates of rat brain regions, following both its acute and chronic administration. In both cases, a dose of 0.5 mg/kg huperzine A significantly decreased hippocampal HAChT by 20% at 45 min, and this inhibition was reversed completely back to normal by 90 min. This effect was not evident at the lower dose of 0.1 mg/kg of huperzine A, nor was it seen in striatal synaptosomes at either dose, or treatment strategy (acute; chronic) used. In vitro, on the other hand, huperzine A had no direct effect on hippocampal synaptosomal HAChT, following incubation with doses of $10^{-7}$ to $10^{-4}$ M. These combined findings indicate that the effect of huperzine A on synaptosomal HAChT activity in vivo is indirect, and probably reflects a compensatory response, via a presynaptic feedback system, to the inhibition of AChE seen in vivo.

As summarized in Section I, we have successfully worked out a synthetic

**Table 1.** Extent of AChE Inhibition by the Compounds Tested*

| Compound | | $IC_{50}$ (M) |
|---|---|---|
| Natural huperzine A | | $10^{-7}$ |
| (±)-Huperzine A | | $3 \times 10^{-7}$ |
| | 35 | $>10^{-4}$ |
| | 36a | $9.5 \times 10^{-4}$ |
| | 36b | $>3 \times 10^{-3}$ |
| | 37 | $1–1.5 \times 10^{-4}$ |
| | 38 | $>10^{-4}$ |
| | 39 | $2 \times 10^{-5}$ |
| | 40 | $3 \times 10^{-5}$ |

**Table 1.** (*continued*) Extent of AChE Inhibition by the Compounds Tested*

| Compound | | $IC_{50}$ (M) |
|---|---|---|
| | **41** | $8 \times 10^{-4}$ |
| | **42** | no activity |
| | **43** | no activity |
| | **44** | $9.5 \times 10^{-4}$ |
| | **45** | $4.5 \times 10^{-4}$ |

*All compounds were tested using rat hippocampal crude homogenates over a concentration range of $10^{-11}$ M to $10^{-3}$ M. These substances were all dissolved in 10% DMSO made up in the incubation buffer medium. In addition, those substances which did not dissolve readily were treated with mild hydrochloric acid and sonicated briefly until they dissolved in the solution. In each case the control medium always consisted of the same ingredients as those which were used to dissolve the compound under investigation. AChE was measured as described in Mantione *et al.*, *J. Neurochem.*, **41**, 251 (1983).

procedure for huperzine A,[7] and also have prepared a series of 12 analogues of huperzine A, based upon specific structural designs. These compounds have been tested for their effect on crude hippocampal synaptosomal homogenate AChE activity in vitro. To date, none of these analogues has been found to have even a fraction of the AChE inhibitory activity of huperzine A. The synthesized racemic huperzine A, however, has an $IC_{50}$ ($3 \times 10^{-7}$ M) comparable to that of the natural product ($10^{-7}$ M; Table 1).[40]

In the interest of establishing whether the synthesized product has effects in vivo comparable to those of natural huperzine A, we compared the effect of these two compounds, administered intraperitoneally to rats (0.1, 0.25, and 0.5 mg/kg), on brain levels of acetylcholine, choline, norepinephrine, serotonin, 5-hydroxy-indoleacetic acid, and AChE in vivo.[41] A dose of 0.5 mg/kg of either racemic or natural huperzine A induced a significant increase in hippocampal acetylcholine levels, that lasted for at least 60 min, and returned to normal within 4 hr. In the same tissue none of the other neurotransmitter substances measured was altered by such treatment. On the other hand, by 60 min AChE activity was significantly reduced in hippocampus, frontal cortex, and striatum of rats treated with either racemic or natural huperzine A, at doses as low as 0.25 mg/kg.

Thus, racemic huperzine A appears to have a similar pharmacological profile to the natural alkaloid. It could therefore probably be used as a substitute for natural huperzine A in future biological applications. Further studies with this compound are required, however, to establish its comparable effect to that of the natural product, on behavior of experimental animals, according to some of the paradigms which are described in the following section of this chapter.

# IV. BEHAVIOR

## A. Introduction

Around the same time that huperzine A was isolated chemically and its pharmacological properties studied in vitro, other researchers in China investigated its pharmacological effects in vivo. Because of evidence that huperzine A could alleviate memory deficits in experimental animals, the behavioral studies immediately focussed on the compound's effect on various aspects of memory function in a variety of species, including human subjects. Thus, Zhang reported that daily intramuscular administration of 30-60 µg/day of huperzine A caused notable improvements in recognition memory of elderly humans (46 to 82 years of age), some of whom were diagnosed as afflicted with AD.[3c] These initial results were encouraging because not only did they reveal the apparent nootropic properties of huperzine A, but they also suggested that this drug could improve memory at low doses and over a prolonged period of time (8 hr). Other AChE inhibitors, such as physostigmine or tetrahydroaminoacridine, which also have been shown to improve cognition in humans, do so only at high doses (20–200 mg/day; oral), and their beneficial effects tend to dissipate within 1 hr.[42–45]

A series of investigations followed, using rodents and non-human primates as test subjects. The animal studies differ with respect to the species, behavior paradigm, and route of drug administration employed, and the drug doses tested. Despite the differences in protocol, the findings generally show that huperzine A treatment improves performance in memory paradigms of both cognitively impaired and normal animals. In the following pages, we discuss briefly the most prominent paradigms and findings obtained with them. We will conclude this section by suggesting directions for future research aimed at further characterizing the behavioral effects of huperzine A, notably its behavioral specificity and time course of action.

## B.  Experimental Findings

The studies discussed below generally include a comparison of the effects of huperzine A with those of vehicle solution (0.15 M NaCl) and of the classical AChE inhibitor, physostigmine. Comparison of the behavioral effects of the two AChE inhibitors is useful for estimating the value of huperzine A as a therapeutic agent; we accordingly will summarize the results for both compounds. The studies also share in common the feature that the dose-effect relation of both compounds is expressed as an inverted U-shaped function. The failure of high doses of huperzine A to enhance performance in memory tests appears to be, in part, attributable to non-specific effects of AChE inhibition, including peripheral effects. We will return to the important, although easily overlooked issue of side effects below. Finally, in the vast majority of the studies the test compounds were administered intraperitoneally (ip). For purposes of comparison between studies, our review focusses on this route of drug administration; however, it is noteworthy that in those cases in which oral or intragastric administration of huperzine A was included, the pattern of results was comparable to that for ip drug administration, except that oral administration required higher doses for the same behavioral outcome.

### Escape and/or Avoidance Paradigms

The class of learning and memory paradigm used most frequently in the assessment of huperzine A's effects on cognition capitalizes on the fact that animals quickly learn and memorize behaviors that enable them to escape and/or avoid an aversive event. Because escape/avoidance responses can be acquired within one training session, avoidance paradigms allow the investigator to determine whether a drug exerts effects on memory acquisition (i.e., learning), retention, or retrieval. Specifically, when a drug given *before training* accelerates the rate of acquisition of criterion ("perfect") performance, then it is considered to enhance memory acquisition. When a drug administered immediately *after training* improves performance during subsequent testing, it is considered to enhance memory storage, i.e., retention. When a drug administered immediately *before testing* improves performance during testing, it is considered to enhance memory retrieval. This logic has been used extensively in the analysis of huperzine A's effect on memory in escape/avoidance paradigms.

*Acquisition.* The influence on memory acquisition by pretreatment with huperzine A has been examined in both intact animals and animals with learning deficits induced by exposure to $CO_2$. Mice trained to choose one of two spatially distinct alley-ways of a Y-shaped maze in order to escape electric shock, were found to attain correct choice performance significantly faster when injected 15 min before training with either huperzine A (75 µg/kg) or physostigmine (140 µg/kg) than with the vehicle solution, 0.15 M NaCl.[46] Similarly, mice trained to locate a spatially distinct, hidden platform in a water pool learned to swim to the safe platform in significantly fewer trials when injected 30 min before training with huperzine A (1–30 µg/kg) than with either physostigmine or vehicle solution.[3b]

Using the spatial Y-maze task, Zhu and Tang investigated the ability of huperzine A and physostigmine to block the effect of $CO_2$ exposure on response acquisition.[47] Administration of either AChE inhibitor to mice 10 min before $CO_2$ exposure and 15 min before training prevented the learning deficit associated with hypercapnia, the most effective doses being 75–125 µg/kg and 200 µg/kg for huperzine A and physostigmine, respectively. The two agents also were found to alleviate a $CO_2$-induced learning deficit in rats trained to choose one of two visually distinct alley-ways of a Y-maze in order to escape electric shock. Rats injected with either huperzine A (100–200 µg/kg) or physostigmine (200 µg/kg) prior to $CO_2$ exposure required significantly fewer trials to criterion than similarly treated animals injected with vehicle solution.[48] Moreover, the learning rates of the $CO_2$-exposed drug-treated animals were found to be comparable to those of animals never exposed to $CO_2$.[48]

Taken together, these studies suggest that pretreatment with huperzine A accelerates memory acquisition, and that this effect applies to both cognitively intact and cognitively impaired animals. However, it would be inappropriate to conclude that the beneficial effect of huperzine A results from the compound's action on the neuronal substrates that underlie memory acquisition without ruling out alternative explanations for the results. For instance, it is conceivable that huperzine A (and physostigmine) enhance sensitivity to shock, so that the perceived intensity of the aversive event was considerably higher for drug-treated than for control animals. Consequently, drug-treated animals would acquire the desired response faster than control animals. Alternatively, the AChE inhibitors may have increased the animals' level of attention and/or arousal which, in turn, could have promoted more rapid acquisition.[49,50] Thus, the notion that huperzine A improves memory acquisition by direct effect on the corresponding neuronal substrates has yet to be substantiated with the appropriate control experiments.

*Retention.* Demonstration of improved memory function when the drug is administered during the retention interval is less vulnerable to alternative explanations. For instance, drug effects on stimulus intensity, attention, or arousal during exposure to the training material do not apply. The observation of a positive drug effect in retention paradigms therefore can be attributed with greater confidence to direct drug action on the neuronal processes involved in memory storage.

Using the spatial Y-maze task, Zhu and Tang found that mice displayed significantly fewer incorrect choices in a retention test 24 hr after training when injected at the beginning of the retention interval with either huperzine A (150 µg/kg) or physostigmine (120 µg/kg) than with vehicle solution.[47] Similarly, rats trained on the visual Y-maze task committed significantly fewer errors during retention testing 48 hr after training when treated at the beginning of the retention interval with either huperzine A or physostigmine than with vehicle solution.[48] The beneficial effects of the AChE inhibitors were observed in both young adult and aged rats; however, young adult rats tended to show improved retention at lower doses (100 µg/kg and 100–200 µg/kg for huperzine and physostigmine, respectively) than did aged rats (100–200 µg/kg and 300 µg/kg, respectively).

Drug-dependent improvement of retention also was shown in aged mice that were trained on a passive avoidance task.[46] In this paradigm, the animal has to learn and remember to not engage in a high-probability behavior, such as stepping down from a platform, in order to avoid electric shock. Zhu and Tang found that aged mice injected with either huperzine A (125–150 µg/kg) or physostigmine (150 µg/kg) at the beginning of a 24-hr retention interval displayed significantly longer step-down latencies during the retention test than did aged mice injected with vehicle solution.[46]

A number of studies examined the effect of huperzine A on retention in animals with a retention deficit caused by treatment with the selective muscarinic blocker, scopolamine. Administration of scopolamine immediately after training is associated with severe performance deficits during subsequent retention testing, presumably because of scopolamine-induced interference with the neuronal processes underlying information storage. Rats trained on the visual Y-maze task and treated with scopolamine (500 µg/kg; ip) 10 min after completion of training exhibited significantly poorer discrimination 1 hr post training than did control animals.[48] The amnestic effect of scopolamine was completely eliminated by oral administration of either huperzine A (400 µg/kg) or physostigmine (700 µg/kg).[48]

Administration of scopolamine during the retention interval also produces inferior test performance in the step-down passive avoidance task. Zhu and Tang noted that the performance deficit (i.e., rapid stepping down from the platform) displayed by mice injected subcutaneously (sc) with 2 mg/kg of scopolamine at the beginning of a 4 hr retention interval could be alleviated by prior treatment with either huperzine A (75–100 µg/kg) or physostigmine (125 µg/kg).[46] Similarly, Vincent et al. demonstrated that a scopolamine-induced (1 mg/kg; sc) retention deficit of a learned passive avoidance response in rats tested 2 hr after training could be overcome by concurrent administration of either huperzine A (0.3–1 µg/kg) or physostigmine (3-10 µg/kg).[3b]

The findings summarized in this section indicate that memorization of newly-learned material is enhanced under the influence of huperzine A, as well as of physostigmine. Furthermore, the beneficial effect of the AChE inhibitors appears to apply to both cognitively intact and cognitively impaired animals, including aged animals. The parallel findings for the intact and the aged animals are remarkable, as they suggest that huperzine A can aid retention in populations of differing levels of cognitive function. The observations with the scopolamine-treated animals, however, are open to alternative explanations, because in these studies retention testing occurred within 1 to 4 hr of drug administration. It is feasible that drug effects on neuronal transmission persisted into the test phase, which then renders unclear whether the test performance deficits (and their blockade by huperzine A or physostigmine) arose from drug action on retention, retrieval, or memory-unrelated performance factors. As we will show below, the effects of cholinergic manipulation immediately before testing do not rule out such alternative explanations. The studies in which both the AChE inhibitors and scopolamine were administered, therefore, would be more informative if they included markedly longer retention intervals.

*Retrieval.* Similar to drug administration before training, administration of drugs shortly before testing may enhance performance for a variety of reasons, only one of which could be a drug-dependent alteration of the neuronal processes underlying memory retrieval. The following findings therefore should be interpreted cautiously. Tang and associates reported that rats retrained in the visual Y-maze task after a 48 hr retention interval required significantly fewer trials to reattain perfect performance when they were injected with either huperzine A (36 to 167 µg/kg) or physostigmine (150 to 180 µg/kg) 20 min before retraining than when injected with vehicle solution.[3a] However, in the same study, huperzine A (100 µg/kg) was not effective in alleviating a scopolamine-induced (200 µg/kg) retrieval deficit.[3a] In contrast, using electrical brain stimulation to disrupt retrieval of an active shock avoidance response acquired 24 hr earlier, Vincent et al. noted that administration of either huperzine A (1 to 10 µg/kg) or physostigmine (100 µg/kg) 20 min before testing reinstated correct avoidance behavior to levels that were comparable to those of mice that never received disruptive brain stimulation.[3b]

It remains to be demonstrated that these outcomes do not reflect merely drug-induced increases in attention and/or arousal which then provided for superior performance during testing. For instance, it would be useful if it could be shown that similar doses of the AChE inhibitors enhance retrieval of a *passive* response task. Similarly, it would be useful if it could be demonstrated that increases in attention and/or arousal level alone are not sufficient to produce improvements in test performance comparable to those observed after huperzine A or physostigmine treatment.

## Approach Paradigms

Although useful because of the rapid rate at which the test response is learned, avoidance paradigms have the disadvantage of presenting conditions that give rise to a high level of anxiety/arousal which, in turn, may interact with the drug effect in some unknown way.[51,52] Conclusions drawn from avoidance paradigms consequently may not generalize to situations that do not involve a high level of anxiety or arousal. It therefore is important that putative nootropic agents be evaluated in other, non-anxiety-provoking memory paradigms. A variety of memory paradigms exist in which the desired response is maintained by a positive event; however, the effect of huperzine A on cognition has been examined in only two of those.

*Spatial Working Memory.* A popular appetite-motivated memory paradigm is the eight-arm radial maze, which consists of a center platform and 8 arms radiating in 8 equidistant directions. The animal's task is to remove food pellets placed at the distal ends of the arms, by entering each arm only once. Correct performance commonly is defined as the removal of all food with a maximum of 10 arm entries (i.e., two errors). To accomplish this task, the animal thus has to remember which arms it already has visited within a trial.

We have investigated the effect of huperzine A on acquisition of correct performance on the radial maze, using rats as subjects. Huperzine A (5 µg/kg; ip) administered 35 min before training slightly accelerated the rate of acquisition in comparison to animals injected with vehicle solution; however, the effect was only marginally significant.[53] The same dose of huperzine A failed to alleviate a scopolamine-induced (250 µg/kg) learning deficit which appears to have been caused, at least in part, by a general response suppression after concurrent administration of the two compounds.

*Delayed-Matching-To-Sample.* Another approach paradigm used to evaluate putative nootropic agents is the "Delayed-Matching-To-Sample" task. In this task, the animal is shown a sample stimulus and, after a variable delay (i.e., retention interval), a set of choice stimuli one of which matches the sample stimulus. The animal is positively rewarded for selecting the matching stimulus. Thus, similar to the eight-arm radial maze, this task probes retention of information acquired within a trial. Vincent et al. found that squirrel monkeys treated with huperzine A (3–30 µg/kg; im) 20 min before testing performed 5–13% more accurately than did vehicle-injected control animals.

In summary, huperzine A has been shown to aid short-term (i.e., within-trial) memory retention in two types of appetite-motivated tasks, and in both situations the beneficial effect was observed in cognitively-intact animals. However, at least with respect to the radial-maze task it is necessary to evaluate a broader range of drug doses before definite conclusions can be drawn.

## IV. CONCLUSIONS AND SUGGESTIONS FOR FUTURE RESEARCH

It is clear from the preceding discussion that huperzine A can improve memory retention in both aversive and appetitive paradigms in intact as well as cognitively impaired animals, including aged animals. With respect to the influence of huperzine A on memory acquisition and retrieval, the appropriate control experiments have yet to be conducted to delineate drug effects on memory processes vs. memory-unrelated performance factors, such as perceived stimulus intensity, attention, arousal, and motivation. As discussed above, the properties of huperzine A observed in in vitro assays and in the intact brain suggest that this compound is 3-fold more potent in the ability to inhibit AChE than any previously tested AChE inhibitor, and its inhibitory action on AChE persists for a significantly longer time period.[13,38,54,55] At the behavioral level, the effective dose range of huperzine A generally was found to be only comparable to or 2-fold lower than that of physostigmine (but see ref. 3b). We are not aware of behavioral studies that compare the temporal profiles of action of the two agents. Based on the neuropharmacological analyses, one might expect a more profound superiority for huperzine A pretreatment over pretreatment with other AChE inhibitors some time after peak action of the compounds when, due to a differential rate of decay, differences in bioavailability are magnified. Thus, a greater dose advantage of huperzine A may not have been noted in some studies due to the time, relative to drug administration, at which the observations were conducted.

Another important aspect of drug action that may render huperzine A therapeutically more favorable than other AChE inhibitors concerns the issue of side effects. AChE inhibition in brain can cause restlessness, anxiety, delirium, and at higher doses, drowsiness, disorientation, ataxia, and convulsions; peripherally, AChE inhibition is associated with increased muscle tone and various alterations of autonomic nervous system output, including increased salivation and sweating, and decreased blood pressure and heart rate. Yan and associates reported that a comparable dose of huperzine A (35 µg/kg; intravenously) and physostigmine (47 µg/kg; iv) produced a 50% increase in contractions of the sciatictibialis muscle in anesthetized rats.[56] However, the $LD_{50}/ED_{50}$-index was found to be $\geq$ 10-fold higher for huperzine A compared to physostigmine in rats, and $\geq$ 5-fold higher for huperzine A compared to physostigmine in mice, in all cases using iv administration of the compounds.[55] We noted no effect of huperzine A pretreatment (5 µg/kg; ip) on either motivation to feed or various indices of sensorimotoric function in adult rats.[53] Clearly, additional systematic examinations of the compound's side effects would be useful, and these examinations should include both young and aged animals, as drug effects on the aforementioned organismic variables may well vary with the overall biological state of the subject population.

In conclusion, a substantial number of behavioral studies demonstrate beneficial effects of huperzine A pretreatment on performance in memory paradigms. However, additional investigations geared towards isolating the specificity on memory processes of huperzine A treatment, delineating its time course of action at the cognitive-behavioral level, and characterizing its potential side effects would greatly enhance our ability to evaluate huperzine A treatment as a means to relieve the cognitive symptoms of AD and related disorders. Moreover, the synthesis of carefully designed analogues may uncover new compounds that possess a longer duration of action and a wider margin of safety, yielding even more efficacious candidates for use in the clinic than the parent compound, huperzine A.

## REFERENCES AND NOTES

1. See, *inter alia*: Washington Post, September 19, 1988 and Chicago Tribune, September 16, 1988.
2. Schindler, U.; Rush, D.K.; Fielding, S. *Drug Dev. Res.* **1984**, *4*, 567.
3. (a) Tang, X.C.; Han, Y.F.; Chen, X.P.; Zhu, X.D. *Acta Pharm. Sin.* **1986**, *7*(6), 507; (b) Vincent, G.P.; Rumennik, L.; Cumin, R.; Martin, J.; Sepinwall, J. *Soc. Neurosci. Abstr.* **1987**, *13*(2), 844; (c) Zhang, S.L. *New Drugs and Clinical Remedies* **1986**, *5*(5), 260; (d) Cheng, Z.S. et al., *ibid.* **1986**, *5*(4), 197.
4. Liu, J.S.; Zhu, Y.L.; Yu, C.M.; Zhou, Y.Z.; Han, Y.Y.; Wu, F.W.; Qi, B.F. *Can. J. Chem.* **1986**, *64*, 837 and references cited therein.
5. Ayer, W.A.; Browne, L.M.; Orszanska, H.; Valenta, Z.; Liu, J.S. *Can. J. Chem.* **1989**, *67*, 1538. Yoshimura, H.; Valenta, Z.; Wiesner, K. *Tetrahedron Lett.* **1960**, *12*, 14. For synthetic approaches to selagine, see: Gravel, D.; Bordeleau, L.; Landouceur, G.; Rancourt, J.; Thoraval, D. *Can. J. Chem.* **1984**, *62*, 2945; Kende, A.S.; Ebetino, F.H.; Battista, R.; Boatman, R.J.; Lorah, D.P.; Lodge, E. *Heterocycles* **1984**, *21*, 91.
6. Ayer, W.A.; Berezowsky, J.A.; Iverach, G.G. *Tetrahedron* **1962**, *18*, 567.
7. For a preliminary account of the huperzine A synthesis, see: Xia, Y.; Kozikowski, A.P. *J. Am. Chem. Soc.* **1989**, *111*, 4116. For another synthesis of huperzine A, see: Qian, L.; Ji, R. *Tetrahedron Lett.* **1989**, *30*, 2089. Noted added in proof: For a full paper on the synthesis of huperzine A, see: Kozikowski, A.P.; Xia, Y.; Reddy, R.; Tückmantel, W.; Hanin, I.; Tang, X.C. *J. Org. Chem.* **1991**, *56*, 4636–4645.
8. Unpublished work of W. Tueckmantel, MindLabs, Inc.
9. For a review on acetylcholinesterase, see: Quinn, D.M. *Chem. Rev.* **1987**, *87*, 955.
10. Collingridge, G.L.; Singer, W. *Trends Pharmacol. Sci.* **1990**, *41*, 290 and references cited therein.
11. Kozikowski, A.P.; Pang, Y.P. *Mol. Pharm.* **1990**, *37*, 352.
12. Unpublished experiments of I.J. Reynolds, Department of Pharmacology, University of Pittsburgh.
13. Tang, X.C.; DeSarno, P.; Sugaya, K.; Giacobini, E. *J. Neuroscience Research* **1989**, *24*, 276.
14. Dubas-Sluyter, M.A.T.; Speckamp, W.N.; Huisman, H.O. *Recl. Trav. Chim. Pays-Bas.* **1972**, *91*, 157.
15. Rubottom, G.M.; Vasquez, M.A.; Pelegrina, D.R. *Tetrahedron Lett.* **1974**, 4319.
16. Stork, G. *Pure Appl. Chem.* **1968**, *17*, 383; also see: Ninomiya, I.; Naito, T.; Higuchi, S.; Mori, T. *J. Chem. Soc., Chem. Comm.* **1971**, 457.
17. Reich, H.J.; Renga, J.M.; Reich, I.L. *J. Am. Chem. Soc.* **1975**, *97*, 543.
18. Colvin, E.W.; Martin, J.; Shroot, B. *Chem. Ind. (London)* **1966**, 2130.

19. Xia, Y.; Reddy, E.R.; Kozikowski, A.P. *Tetrahedron Lett.* **1989**, *30*, 3291.

20. Komatsu, M.; Yamamoto, S.; Ohshiro, Y.; Agawa, T. *Tetrahedron Lett.* **1981**, *22*, 3769; Hazai, L.; Deak, G.; Toth, G.; Volford, J.; Tamas, I. *J. Heterocyclic Chem.* **1982**, *19*, 49; Kasturi, T.; Krishnan, L.; Prasad, R.S. *J. Chem. Soc., Perkin Trans. I* **1982**, *63*.

21. For reviews on the Michael reaction, see: House, H.O. *Modern Synthetic Reactions*, 2nd Ed., W.A. Benjamin, Inc., 1972, pp. 595–623; Gawley, R.E. *Synthesis* **1976**, 777; Jung, M.E. *Tetrahedron* **1976**, *32*, 3. For the fluoride ion-catalyzed Michael reaction, see: Clark, J.H.; Miller, J.M. *J. Chem. Soc., Perkin Trans. I* **1977**, 1743.

22. Backvall, J.-E. *Pure Appl. Chem.* **1983**, *55*, 1669; Nystrom, J.-E.; Backvall, J.-E. *J. Org. Chem.* **1983**, *48*, 3947; Kende, A.S.; Battista, R.A.; Sandoval, S.B. *Tetrahedron Lett.* **1984**, *25*, 1341; Kende, A.S.; Roth, B.; Sanfilippo, P. J. *J. Am. Chem. Soc.* **1982**, *104*, 1784.

23. Nysted, L.N.; Burtner, R.R. *J. Org. Chem.* **1962**, *27*, 3175; Pollinio, G.P.; Barco, A.; De Giuli, G. *Synthesis* **1972**, *44*; Ono, N.; Kamimura, A.; Miyake, H.; Hamamoto, I.; Kaji, A. *J. Org. Chem.*, **1985**, *50*, 3692; Nakagawa, Y.; Stevens, R.V. *J. Org. Chem.* **1988**, *53*, 1871.

24. The bicyclic amidine base DBU was also found to catalyze the conversion of **24** to **28**. For the use of DBU as a catalyst for the aldol reaction, see: Corey, E.J.; Anderson, N.H.; Carlson, R.M.; Paust, J.; Vedejs, E.; Vlattas, I.; Winter, R.E.K. *J. Am. Chem. Soc.* **1968**, *90*, 3245.

25. Colvin, E.W.; Martin, J.; Parker, W.; Raphael, R.A.; Shroot, B.; (in part) Doyle, M. *J. Chem. Soc., Perkin Trans. I* **1972**, 860.

26. Bhalerao, U.T.; Rapoport, H. *J. Am. Chem. Soc.* **1971**, *93*, 4835.

27. Sunagawa, M.; Katsube, J.; Yamamoto, H. *Tetrahedron Lett.* **1978**, 1281.

28. Geib, S.; Tueckmantel, W.; Kozikowski, A. P. *Acta Cryst.* **1991**, *C47*, 824.

29. Kozikowski, A.P.; Yamada, F.; Reddy, E.R.; Pang, Y.P.; Miller, J.H.; McKinney, M. *J. Am. Chem. Soc.* **1991**, *113*, 4695.

30. For examples of chiral catalysis of the Michael reaction, see: Cram, D.J.; Sogah, G.D.Y. *J. Chem. Soc., Chem. Commun.* **1981**, 625.

31. Beveridge, D.L.; Radna, R.J. *J. Am. Chem. Soc.* **1971**, *93*, 3759; Chothia, C.; Pauling, P. *Proc. Natl. Acad. Sci.. USA* **1973**, *70*, 3103; Chothia, C.; Pauling P. *Nature (London)* **1969**, *223*, 919. For a caveat on the use of molecular modeling strategies in structure-activity studies, see: Behling, R.W.; Yamane, T.; Navon, G.; Jelinski, L.W. *Proc. Natl. Acad. Sci. USA* **1988**, *85*, 6721.

32. Forrest, H.S.; Walker, J. *J. Chem. Soc.* **1948**, 1939.

33. Martin, Y.C.; Jarboe, C.H.; Krause, R.A.; Lynn, K.R.; Dunnigan, D.; Holland, J.B. *J. Med. Chem.* **1973**, *16*, 147.

34. (a) *Modern Drug Research—Paths to Better and Safer Drugs*, Martin, Y.C.; Kutter, E.; Austel, V. Eds., Marcel Dekker: New York, 1989. (b) Kozikowski, A.P.; Yamada, F.; Tang, X.-C.; Hanin, I. *Tetrahedron Lett.* **1990**, *31*, 6159.

35. Crossland, I. *Org. Synth.* **1981**, *60*, 6.

36. Liu, J.S.; Yu, C.M.; Zhou, Y.Z.; Han, Y.Y.; Wu, F.W.; Qi, B.F.; Zhu, Y.L. *Acta Chim. Sin.* **1986**, *44*, 1035; Pine, S.H.; Sanchez, B.L. *J. Org. Chem.* **1971**, *36*, 829.

37. Notari, R.E. "Biopharmaceutics and Clinical Pharmacokinetics," 4th ed., Marcel Dekker: New York, 1987.

38. Wang Y.E.; Yue, D.X.; Tang, X.C. *Acta Pharm. Sin.* **1986**, *7*, 110.

39. Laganiere, S.; Corey, J.; Tang, X.C.; Wulfert, E.; Hanin, I. *Neurobiology of Aging* **1990**, *11*, 348.

40. Hanin, I.; Tang, X.C.; Corey, J.; Xia, Y.; Reddy, E.R.; Kozikowski, A.P. *FASEB J.* **1990**, *4*, A471.

41. Hanin, I.; Tang, X.C.; Kindel, G.; Xia, Y.; Kozikowski, A.P. *The Pharmacologist* **1990**, *32*.

42. Gauthier, S.; Bouchard, R.; Bacher, Y. et al. *Can. J. Neurol. Sci.* **1989**, *16*, 543.

43. Summers, W.K.; Majovski, L.V.; Marsh, G.M. et al. *New Eng. J. Med.* **1986**, *315*, 1241.

44. Thal, L.J.; Altman-Fuld, P. *New Eng. J. Med.* **1983**, *308*, 720.

45. Whelpton, R.; Hurst, P. *New Eng. J. Med.* **1985**, *313*, 1293.

46. Zhu, X.D.; Tang, X.C. *Acta Physiol. Sin.* **1988**, *9*, 492.

47. Zhu, X.D.; Tang, X.C. *Acta Physiol. Sin.* **1987**, *22*, 812.

48. Lu, W.H.; Shou, J.; Tang, X.C. *Acta Physiol. Sin.* **1988**, *9*, 11.
49. Beller, S.A.; Overall, J.E.; Swann, A.C. *Psychopharm.* **1985**, *87*, 147.
50. Smith, C.M.; Coogan, J.S.; Hart, S. *Psychopharm.* **1986**, *90*, 364.
51. Gamzu, E. *Ann. New York Acad. Sci.* **1985**, *444*, 370.
52. Heise, G.A. *Trends Pharmacol. Sci.* **1987**, *8*, 65.
53. Thiels, E.; Weisz, D.J.; Kozikowski, A.P., in preparation.
54. Hallak, M.; Giacobini, E. *Neurochem. Res.* **1986**, *11*, 1037.
55. Hallak, M.; Giacobini, E. *Neuropharm.* **1989**, *28*, 199.
56. Yan, X.F.; Lu, W.H.; Lou, W.J.; Tang, X.C. *Acta Pharm. Sin.* **1987**, *8*, 117.

# MECHANISM-BASED DUAL-ACTION CEPHALOSPORINS

Harry A. Albrecht and James G. Christenson

**Advances in Medicinal Chemistry,**
**Volume 1, pages 207–234.**
Copyright © 1992 by JAI Press Inc.
All rights of reproduction in any form reserved.
ISBN: 1-55938-170-1

## ABSTRACT

When cephalosporins exert their biological activity by reacting with bacterial enzymes, opening of the β-lactam ring is accompanied by liberation of the 3'-substituent, if that substituent can function as a leaving group. When the eliminated substance possesses antibacterial activity of its own, the cephalosporin should exhibit a dual mode of action. As a rationale for drug design, this mechanism provides an opportunity to expand the antibacterial spectrum, and to create bifunctional cephalosporins with unique properties. Accordingly, dual-action cephalosporins have been synthesized by incorporating antibacterial quinolones as potential leaving groups at the 3'-position either by an ester bond or by a bond through a quaternary nitrogen. These compounds demonstrate a broad spectrum of antibacterial activity derived from both cephalosporin-like and quinolone-like components. The mechanism of action and pharmacokinetic properties of representative compounds have been investigated.

## I. INTRODUCTION

### A. The Dual-Action Concept

According to the generally accepted mechanism by which bacterial enzymes react with β-lactams, cephalosporins exert their biological activity by acylating active-site serine residues of the transpeptidases responsible for cross-linking peptidoglycan.[1,2] Similarly, acylation of active-site serine residues occurs as a key step in the mechanism by which most β-lactamases inactivate these antibiotics.[3,4] In either case, when a cephalosporin contains a potential leaving group (X) at the 3'-position, that group is eliminated (Scheme 1).

Published evidence suggests that opening of the β-lactam ring correlates with elimination of the nucleofugal group, although the reaction is probably not concerted.[5-11] Quantitative release of such 3'-substituents as acetate, azide, and pyridine can be brought about by treatment with β-lactamases,[6,12] and this reaction appears to be quite general. If the leaving group possesses intrinsic antibacterial activity, then the cephalosporin should exhibit a dual mode of action.[13,14] In addition to providing its own β-lactam activity, the cephalosporin should also act

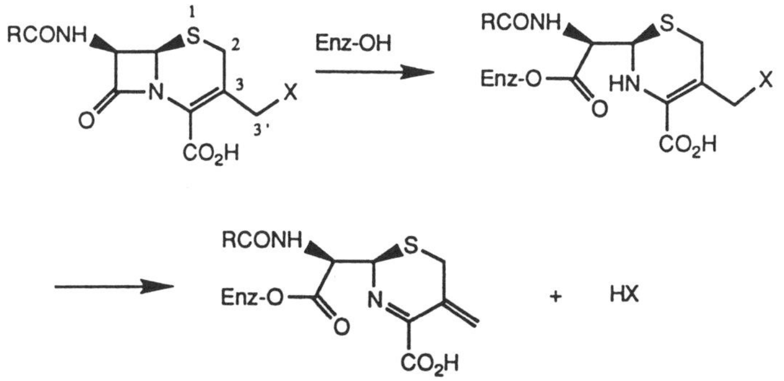

**Scheme 1.** Mechanism of dual action. (Reprinted with permission, copyright 1990 American Chemical Society)

as a targeted prodrug (see Note 1) for the second antibacterial agent, delivering it close to its site of action. The term "dual action" has been used to describe such cephalosporins.[14] This mechanism presents an opportunity to expand the antibacterial spectrum to include organisms that are resistant to the third-generation cephalosporins. Thus, it may be possible to design cephalosporins with significant advantages over those currently in use.

## B. Historical Perspectives

The first mention of the release of a biologically active substituent from a cephalosporin seems to be the 1971 report from workers at the University of Wales describing the unusual mechanism of action of cephalosporin 7/30.[8] Like dimethyl-dithiocarbamate (DMDT), but unlike the reference cephalosporins cephalexin, cephaloglycin, and cephaloridine, compound 7/30 inhibited DNA, RNA, and protein synthesis in *Escherichia coli*. These properties were attributed to DMDT released in solution.

It was in a 1976 publication by O'Callaghan, Sykes and Staniforth that the dual-action hypothesis was first advanced.[13] These workers at Glaxo recognized that the single property which most limited the clinical efficacy of cephalosporins available at that time was susceptibility to β-lactamases. They described the

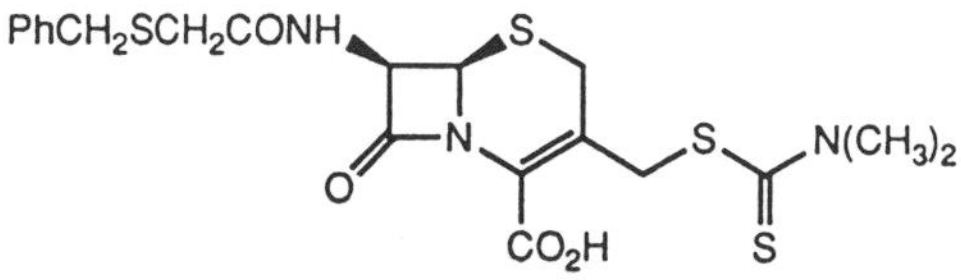

Cephalosporin 7/30

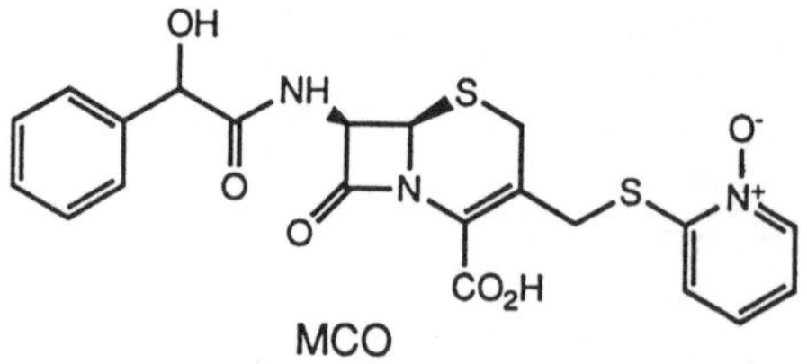

MCO

properties of MCO, a cephalosporin substituted at the 3'-position by pyrithione (2-mecaptopyridine *N*-oxide), a substance with antibacterial and antifungal activities. In the form of its zinc salt, pyrithione is still used today as an antiseptic in shampoos. The Glaxo group demonstrated that the action of β-lactamases on MCO liberated pyrithione. Compared to cephalothin, MCO demonstrated a broadened antibacterial spectrum which seemed to confirm the then-novel principle of dual action. Apparently the intact MCO exerted its activity against susceptible strains, while β-lactamase producers avoided attack by the β-lactam only to succumb to the liberated pyrithione. Because of concerns about the systemic toxicity of pyrithione, MCO was not further developed.

Except for papers relating to MCO and its analogues,[16] no publications on dual-action cephalosporins appeared for a decade. In view of the thought-provoking nature of the dual-action hypothesis, this curious hiatus was due perhaps more to circumstances dictated by the state of the art than by a lack of interest. Rapid advances in chemistry and biology soon provided medicinal chemists with the tools to design and synthesize more and more exotic targets. At the same time, the evolution of third-generation cephalosporins, with improved resistance to β-lactamases and an expanded spectrum of activity, provided opportunities to further test the limits of structure-activity relationships in this field, without regard for the dual-action hypothesis.

In 1986, some ten years after the O'Callaghan paper, a group from the University of Chicago described a cephalosporin (**1**) in which the 3'-position was derived from an antibacterial dipeptide.[17-19] For comparison, **2** was prepared with L-Ala-L-Ala, a dipeptide devoid of antibacterial activity, at the 3'-position. Both **1** and **2** are excellent substrates for TEM β-lactamases, with the dipeptide com-

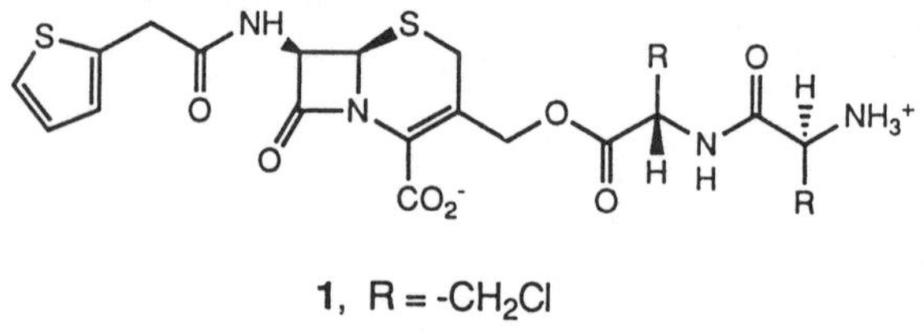

**1**, R = -CH$_2$Cl

**2**, R = -CH$_3$

ponent being released in each case by the action of the enzymes. Against selected gram-positive and gram-negative β-lactamase-producing microorganisms sensitive to the dichlorodipeptide, **1** showed moderate activity in vitro. Against a strain of *Escherichia coli* selected for resistance to the dichlorodipeptide, **1** proved inactive. The authors concluded that the pattern of activity shown by **1** was accounted for by the lactamase-dependent release of dichlorodipeptide. Compound **2** was not active against any of the test strains, implying that the cephem nucleus does not contribute significantly to the observed antibacterial activity of **1**. No data were published on the behavior of **1** and **2** toward cephalothin-susceptible organisms, so it is unclear whether these compounds manifested β-lactam type activity against non-β-lactamase producers.

## II. CEPHALOSPORIN 3'-QUINOLONE ESTERS

In 1980, a limited dual-action program was initiated at Hoffmann-La Roche in Nutley, New Jersey. Due to limited resources, the effort was pursued only sporadically for a number of years, but by the mid-1980s it began to gather momentum. At the start, a synthetic program was designed to incorporate antibacterial quinolones into cephalosporins through an ester linkage to the 3'-position. Since many cephalosporins contain a 3'-acetate function, the ester link was considered to be compatible with antibacterial activity, and seemed a reasonable choice. The possibilities (a) of using acylamino side chains designed to promote susceptibility to β-lactamases — and thus optimize the delivery of the second agent — or (b) of using third-generation cephalosporin-type side chains to confer greater resistance to β-lactamases and optimize the β-lactam component of activity, would both be investigated. The effect that the rather large quinolone substituent would have upon the ability of the cephalosporin to penetrate the bacterial outer membrane and to interact with the target enzymes was uncertain. It was just at that time that research on quinolones at such centers as Abbott, Bayer, Dainippon, Kyorin and Parke-Davis, led to a proliferation of new, highly active products which were far superior to the earlier quinolones.

The quinolones provided a broad class of antibacterials that seemed well suited to the role of second agent for a number of reasons: (1) The antibacterial spectra of the two classes are complementary, with quinolones being active against β-lactam-resistant strains, while cephalosporins are more active against streptococci. (2) Most quinolones are not very soluble under physiological conditions, and might benefit by incorporation into a more soluble codrug (see Note 1) form. Cephalosporins often have excellent solubility and pharmacokinetic properties. (3) The mechanism of action of the quinolones, which inhibit DNA gyrase, seems compatible with that of cephalosporins; i.e., there should be no antagonism.[20,21] (4) Toxic effects of quinolones might be diminished by incorporation into a codrug form. Central nervous system (CNS) toxicity as well as the arthropathic effects that prevent use of quinolones in pediatrics might be overcome.

## A. Synthesis of Cephalosporin 3'-Quinolone Esters

In their initial publications, the Roche workers described the synthesis and properties of nineteen bifunctional cephalosporins, including compounds **3–10** (Tables 1 and 2).[15,22–30] These compounds, in which quinolones **11–15** (Table 3) are incorporated as ester-linked second agents, were prepared according to synthetic strategy which made use of a nucleophilic displacement reaction for the key step in establishing the 3'-ester bond of intermediates **16–23** (Table 4). Although dual-action cephalosporins derived from the more potent second- and third-generation quinolones would prove more interesting, the first target compound **3** contained nalidixic acid as its quinolone component; **3** was initially prepared by a more elaborate route than the later compounds. Bromo ester **24** was prepared (Scheme 2) by the method developed earlier at Eli Lilly, involving an elegant ring expansion of a penicillin to a suitably functionalized cephalosporin.[31,32] Displacement with the sodium salt of nalidixic acid (**11**) provided the intermediate ester **16** in 37% yield as a mixture of $\Delta^2$- and $\Delta^3$-isomers (Scheme 3). The desired $\Delta^3$-isomer was the minor component. Base-catalyzed isomerization of the double bond in esterified cephalosporins to the usually thermodynamically preferred 2-position is a trouble spot frequently encountered in the chemistry of cephalosporins. Nalidixic acid, like the other quinolones, is a rather weak carboxylic acid; consequently, its sodium salt is quite basic. After hydrolysis of the 4-nitrobenzyl ester,[33] the double bond was restored to the desired 3-position through the intermediacy of the $\Delta^2$-sulfoxide, according to the standard oxidation-reduction sequence which is familiar to workers in the field.[34]

Most of the new cephalosporins were prepared by the more efficient processes depicted in Scheme 4, starting from 7-aminocephalosporanic acid (7-ACA). This

*Table 1.* Cephalosporin 3'-Quinolone Esters

| No. | $Q^a$ | Precursor |
|-----|-------|-----------|
| 3 | Q$_1$ | 16 |
| 4 | Q$_2$ | 17 |
| 5 | Q$_3$ | 18 |
| 6 | Q$_4$ | 19 |

[a]For structure of Q see Table 3.

**Table 2.** Cephalosporin 3'-Quinolone Esters

| No. | $Q^a$ | Precursor |
|-----|-------|-----------|
| 7   | $Q_2$ | 20        |
| 8   | $Q_3$ | 21        |
| 9   | $Q_4$ | 22        |
| 10  | $Q_5$ | 23        |

$^a$For structure of Q see Table 3.


**Table 3.** Quinolone Structures
QCO$_2$H

| No. | Q | Structure |
|-----|---|-----------|
| 11  | $Q_1$ | |
| 12  | $Q_2$ | |
| 13  | $Q_3$ | |
| 14  | $Q_4$ | |
| 15  | $Q_5$ | |

213

**Table 4.** Intermediate Esters

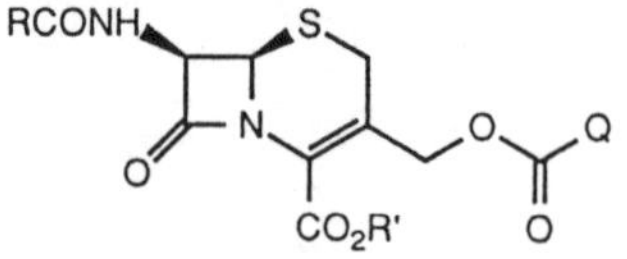

| No. | R | $Q^a$ | R' | % Yield |
|-----|-----|-----|-----|-----|
| 16 | PhOCH$_2$– | Q$_1$ | 4-nitrobenzyl | 37.0 |
| 17 | PhOCH$_2$– | Q$_2$ | $t$-Bu | 59.7 |
| 18 | PhOCH$_2$– | Q$_3$ | $t$-Bu | 39.2 |
| 19 | PhOCH$_2$– | Q$_4$ | $t$-Bu | 45.8 |
| 20 | TATM$^b$ | Q$_2$ | $t$-Bu | 56.0 |
| 21 | TATM | Q$_3$ | $t$-Bu | 32.8 |
| 22 | TATM | Q$_4$ | $t$-Bu | 50.4 |
| 23 | TATM | Q$_5$ | $t$-Bu | 41.0 |

$^a$For Structures of Q see Table 3.

$^b$TATM =

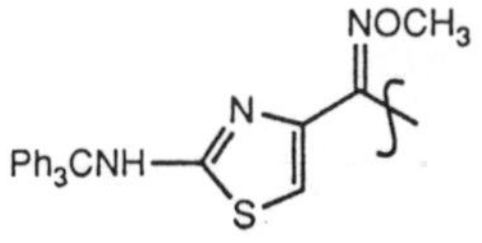

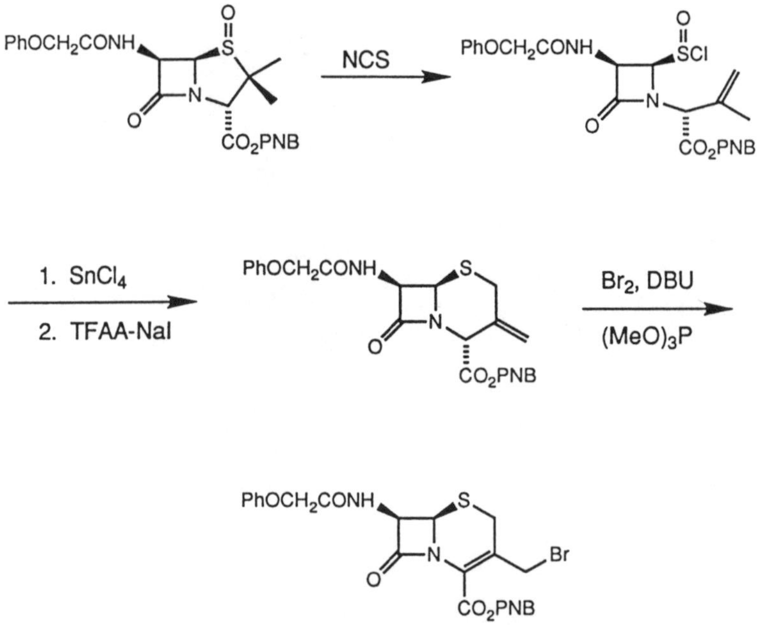

**24,** PNB = 4-nitrobenzyl

**Scheme 2.** NCS = N-chlorosuccinimide; TFAA = trifluoroacetic anhydride.

**Scheme 3.** R = PhOCH$_2$–; PNB = 4-nitrobenzyl; TFAA = trifluoroacetic anhydride; for Q$_1$ see Table 3.

moderately priced and readily available item of commerce was esterified and acylated by known methodology, and then converted to the 3'-iodo derivatives by reaction with iodotrimethylsilane. Cleavage of the 3'-allylic acetate of cephalosporin esters with iodotrimethylsilane, sometimes referred to as the "Bonjouklian reaction," often provides ready access to these key intermediates, although some structural features are not compatible with use of this reagent.[35,36] Certain esters, such as diphenylmethyl and 4-methoxybenzyl, are too labile for these conditions. Other readily removable esters can be used, including benzyl, 4-nitrobenzyl, allyl, and *t*-butyl. When **25** was prepared by this method and compared with **24** in displacement reactions with the sodium salt of nalidixic acid, the $\Delta^2/\Delta^3$-isomeric ratio was reversed with the $\Delta^3$-isomer becoming the major product (Scheme 5). Experiments with other esters, notably *t*-butyl and allyl, soon confirmed that double-bond migration could be almost entirely avoided. Thus, displacement reactions with the sodium or potassium salts of quinolones gave intermediate esters that were essentially free of $\Delta^2$-isomer.

The best ester for maintaining the integrity of the 3-double bond proved to be

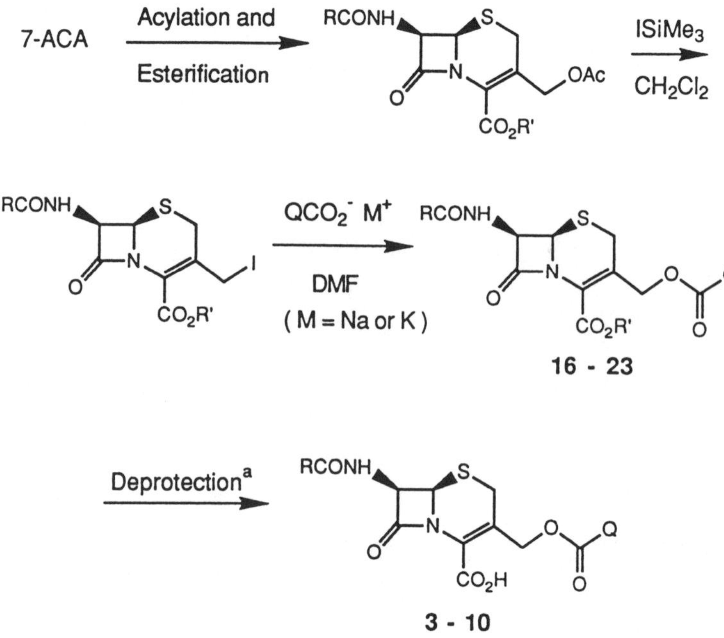

Scheme 4. Deprotection procedures: R' = 4-nitrobenzyl, Na$_2$S in H$_2$O - DMF; R' = t-Bu, CF$_3$CO$_2$H - anisole; R' = allyl, Pd[0] - sodium 2-ethylhexanoate.

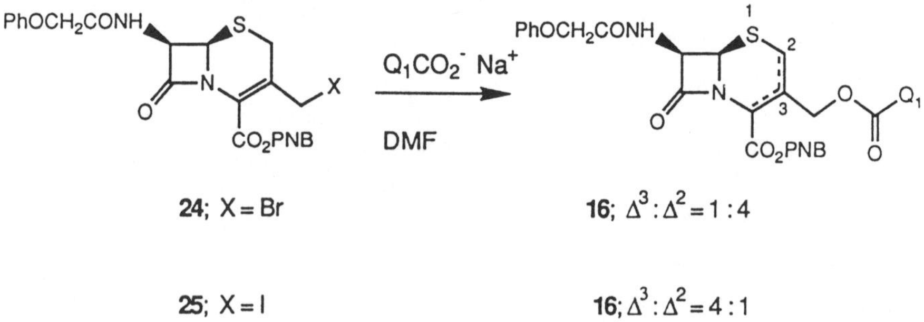

Scheme 5. PNB = 4-nitrophenyl; for structure of Q$_1$ see Table 3.

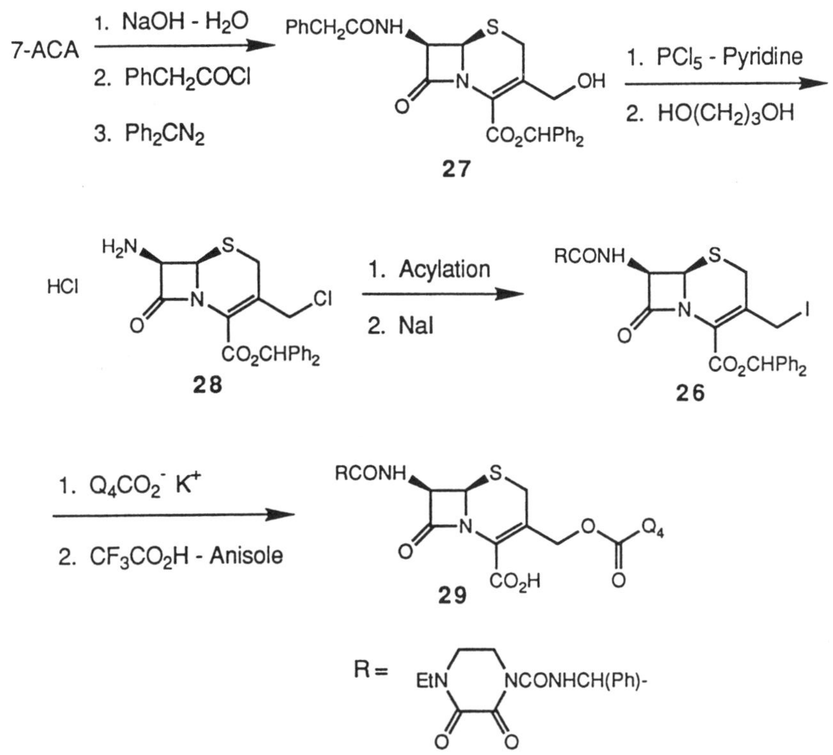

**Scheme 6.** For Structure of Q₄ see Table 3.

*t*-butyl. Even in displacements with the highly alkaline potassium salt of fleroxacin (**15**), double-bond migration was usually not a problem. The intermediate esters were chromatographically purified by using either open column, flash, or HPLC techniques, before being carried along to the final products. Standard conditions were utilized for the deprotection step. Treatment with trifluoroacetic acid - anisole at 0 °C for three to eighteen hours removed *t*-butyl esters, as well as *N*-trityl groups, when present. The final products were generally isolated as sodium salts and purified by $C_{18}$ reverse-phase chromatography.

One iodoester, **26**, was prepared by an alternate procedure (Scheme 6) under conditions adapted from those reported by workers at Fujisawa.[37] This approach is useful when structural features are not compatible with the use of iodotrimethyl-silane. Controlled alkaline hydrolysis of 7-ACA followed by in situ acylation and esterification gave the 3'-alcohol **27**. Reaction with phosphorus pentachloride and pyridine followed by solvolysis with 1,3-propanediol converted **27** to **28**. This versatile intermediate was acylated and then converted to the iodide **26**. The displacement reaction and subsequent deprotection with trifluoroacetic acid-anisole to give **29** were accomplished under conditions similar to those used for the *t*-butyl esters. In this case, as with allyl esters, the displacement reaction generated a few percent of $\Delta^2$-isomer as a contaminant.

## B.   Properties of Cephalosporin 3'-Quinolone Esters

The new cephalosporins **3–10**, reference cephalosporins **30** and **31** (cefotaxime), and quinolones **11–15** were screened for in vitro antibacterial activity.[22] Results are summarized in Tables 5–7. In Table 5, results are shown for bifunctional cephalosporins **3 – 6** and reference compound **30**, all of which have a phenoxyacetyl substituent. In Table 6, results appear for compounds **7–10** and **31**, all of which incorporate a (2-amino-4-thiazolyl)(methoxyimino)acetyl group (ATM). Compared to the reference cephalosporins **30** and **31**, the bifunctional

**30**                                          **31**; Cefotaxime

cephalosporins demonstrated a broadened spectrum of antibacterial activity. As illustrated in the tables, the activity added to the spectrum of the reference cephalosporins generally parallels the activity of the quinolone component (Table

***Table 5.***  In Vitro Activity: MIC (µg/ml)

| Organisms | 3 | 4 | 5 | 6 | 30 |
|---|---|---|---|---|---|
| *Escherichia coli* ATCC 25922 | | 2 | 8 | 1 | 128 |
| *Escherichia coli* TEM-1 | | 1 | 8 | 2 | > 128 |
| *Citrobacter freundii* BS-16 | | 2 | 16 | 8 | > 128 |
| *Enterobacter cloacae* P99 | 16 | 0.5 | 2 | 0.5 | > 128 |
| *Serratia marcescens* 1071 | | 1 | 16 | 1 | > 128 |
| *Proteus vulgaris* 1028βC | | 0.25 | 4 | 1 | > 128 |
| *Proteus mirabilis* 90 | 32 | 4 | 32 | 16 | 64 |
| *Pseudomonas aeruginosa* ATCC 27853 | | >256 | 128 | 128 | >128 |
| *Pseudomonas aeruginosa* 18 S/H | | 16 | 8 | 64 | >128 |
| *Staphylococcus aureus* ATCC 29213 | | 0.25 | 0.5 | 0.125 | 0.5 |
| *Staphylococcus aureus* 1059B | 0.5 | 0.5 | 1 | 0.25 | 0.25 |
| *Staphylococcus aureus* 95 | | 4 | 2 | 0.5 | >128 |
| *Streptococcus pneumoniae* 6301 | ≤ 0.008 | — | — | 2 | 0.25 |
| *Streptococcus pyogenes* 4 | 0.063 | — | — | 4 | 0.063 |
| *Enterococcus faecalis* ATCC 29212 | | 64 | 64 | 64 | 64 |

**Table 6.** In Vitro Activity: MIC (μg/ml)

| Organism | 7 | 8 | 9 | 10 | 31 |
|---|---|---|---|---|---|
| *Escherichia coli* ATCC 25922 | 0.5 | 1 | 0.5 | 0.125 | 0.063 |
| *Escherichia coli* TEM-1 | 0.25 | 0.5 | 0.5 | 0.125 | 0.031 |
| *Citrobacter freundii* BS-16 | 4 | 32 | 8 | 2 | 128 |
| *Enterobacter cloacae* P99 | 0.5 | 4 | 0.5 | 0.125 | 32 |
| *Serratia marcescens* 1071 | 1 | 8 | 2 | 0.5 | 8 |
| *Proteus vulgaris* 1028βC | 0.25 | 8 | 2 | 0.25 | 16 |
| *Proteus mirabilis* 90 | 0.125 | 0.5 | 0.25 | 0.25 | 0.016 |
| *Pseudomonas aeruginosa* ATCC 27853 | 128 | >128 | 128 | 8 | 64 |
| *Pseudomonas aeruginosa* 18 S/H | 16 | 32 | 8 | 4 | 128 |
| *Staphylococcus aureus* ATCC 29213 | 2 | 4 | 0.5 | 1 | 2 |
| *Staphylococcus aureus* 1059B | 2 | 4 | 16 | 1 | 4 |
| *Staphylococcus aureus* 95 | 4 | 8 | 0.5 | 2 | 32 |
| *Streptococcus pneumoniae* 6301 | — | — | 0.016 | ≤0.008 | 0.016 |
| *Streptococcus pyogenes* 4 | — | — | ≤0.008 | 0.031 | ≤0.008 |
| *Enterococcus faecalis* ATCC 29212 | 32 | 32 | 4 | 8 | >128 |

7) in each case. For example, in Table 5, **30** shows significant activity only against gram-positive organisms, while the bifunctional cephalosporins are active against both gram-positive and gram-negative bacteria. Compound **4** displays broad-spectrum activity which reflects a contribution from the spectrum of **12** (oxolinic acid) as well as from that of the parent cephalosporin (**30**). Third-generation cephalosporins with the ATM side chain, such as cefotaxime (**31**), show both an expanded spectrum of activity and certain characteristic weaknesses. As shown in Table 6, the bifunctional compounds clearly demonstrate improved activity relative

**Table 7.** In Vitro Activity of Quinolones: MIC (μg/ml)

| Organisms | 11 | 12 | 13 | 14 | 15 |
|---|---|---|---|---|---|
| *Escherichia coli* ATCC 25922 | 2 | 0.25 | 0.5 | 0.25 | 0.063 |
| *Escherichia coli* TEM-1 | 4 | 0.25 | 0.5 | 0.5 | 0.125 |
| *Citrobacter freundii* BS-16 | 8 | 0.5 | 2 | 16 | 1 |
| *Enterobacter cloacae* P99 | 1 | 0.5 | 0.5 | 0.125 | 0.125 |
| *Serratia marcescens* 1071 | 2 | 1 | 1 | 0.25 | 0.25 |
| *Proteus vulgaris* 1028βC | 2 | 0.063 | 0.5 | 0.5 | 0.25 |
| *Proteus mirabilis* 90 | 8 | 1 | 2 | 16 | 0.5 |
| *Pseudomonas aeruginosa* ATCC 27853 | 128 | 64 | 16 | — | 4 |
| *Pseudomonas aeruginosa* 18 S/H | 32 | 4 | 2 | 1 | 2 |
| *Staphylococcus aureus* ATCC 29213 | 32 | 2 | 1 | 0.063 | 0.5 |
| *Staphylococcus aureus* 1059B | 32 | 2 | 1 | 0.125 | 0.5 |
| *Staphylococcus aureus* 95 | 8 | 2 | 1 | 0.063 | 0.5 |
| *Streptococcus pneumoniae* 6301 | 128 | >128 | 8 | 16 | 4 |
| *Streptococcus pyogenes* 4 | 128 | >128 | 8 | 8 | 1 |
| *Enterococcus faecalis* ATCC 29212 | >128 | >128 | 8 | 1 | 4 |

**Table 8.** Binding of Dual-Action Cephalosporins to Essential PBPs of *E. coli* DC-O Concentration (µg/ml) Required for 90% Inhibition of [$^{14}$C] Pen G Binding

| Compound | PBP 1a 90 kDa | PBP 1b 90 kDa | PBP 2 66 kDa | PBP 3 60 kDa | Morphology | MIC (µg/ml) |
|---|---|---|---|---|---|---|
| 4 | 0.1 | 30 | 100 | 10 | F[a] | 32 |
| 5 | 0.5 | 10 | 100 | 2 | F | 128 |
| 9 | 10 | 10 | >100 | 0.1 | F | 1 |
| 10 | 0.1 | 2 | 100 | 0.1 | F | 0.12 |
| 30 | 0.5 | 100 | >100 | 10 | F | 64 |
| 31 | 0.1 | 0.5 | 100 | 0.1 | F/L[b] | 0.06 |
| 33 | >100 | >100 | 100 | 0.1 | F | 0.2 |
| 34 | 10 | 100 | >100 | 2 | F | 8 |
| 35 | 2 | 30 | 100 | 0.1 | F | 0.2 |

[a]F, Filaments.
[b]L, Lysis

to **31** against such cefotaxime-resistant but quinolone-susceptible strains as *Enterobacter cloacae* P99 and *Staphylococcus aureus* 95, which is a methicillin-resistant β-lactamase producer. Where data are available, increased potency relative to the quinolone component against the streptococci is also evident. This pattern of activity suggests that the quinolone carboxylic acid is being released in situ, since it is generally accepted that quinolone esters show activity only as a consequence of hydrolysis to the acid.[38,39] Thus, the intact quinolone ester itself would not be expected to contribute significantly in terms of quinolone activity. These results lend themselves to the interpretation that a dual mode of action is operative.

Compound **4** proved active in vivo against infections in mice due to *Escherichia coli* and *Streptococcus pneumoniae* (ED$_{50}$ values of 15 and 38 mg/kg s.c., respectively), and showed an elimination half-life in rats of 13 minutes after an i.v. dose of 20 mg/kg, which was comparable to that of cefotaxime. Compound **4** was less toxic than oxolinic acid (LD$_{50}$ values of >500 and 155 mg/kg i.v., respectively, in mice) and showed less potential for CNS liability than oxolinic acid, which is a potent CNS stimulant. In experiments designed to measure CNS stimulation, the increase in spontaneous locomotor activity in mice treated with oxolinic acid (30 mg/kg i.v.) was nearly four times that of the controls. At the same dose level, spontaneous locomotor activity of mice treated with **4** increased only about 50%.[22]

The bifunctional compounds act like typical cephalosporins in their behavior toward penicillin-binding proteins (PBPs). The major determinant of binding appears to be the acylamino function. Compounds with the ATM acyl group (**9**, **10**, and **31**) all show the same affinity for PBP 3 despite the variety of structures at the 3'-positions (Table 8). Compounds with a phenoxyacetyl substituent (**4** and **5**) more closely resemble **30** in PBP binding.

The most active of the cephalosporin 3'-quinolone esters, Ro 23-9424 (**10**),

**Table 9.** In Vivo Activity: ED$_{50}$ (mg/kg,sc) Mouse Protection Test

| Infection | 10[a] | 15[b] | 31[c] |
|---|---|---|---|
| *Escherichia coli* 257 | 1.8 | 0.4 | 1.7 |
| *Klebsiella pneumoniae* A | 3.6 | 0.9 | 28 |
| *Enterobacter cloacae* 5699 | 1.4 | 0.3 | 2.6 |
| *Citrobacter freundii* BS#16 | 7.4 | 2.6 | 43 |
| *Serratia marcescens* SM | 6.4 | 2.7 | >100 |
| *Pseudomonas aeruginosa* 5712 | 60 | 8.9 | 69 |
| *Pseudomonas aeruginosa* 8780 | 29 | 6.4 | >250 |
| *Staphylococcus aureus* Smith (MS) | 5.6 | 1.6 | 5.4 |
| *Staphylococcus aureus* 753 (MR) | 11 | 2.9 | >100 |
| *Streptococcus pneumoniae* 6301 | <1 | 160 | <10 |
| *Streptococcus pyogenes* 4 | 8.8 | 183 | 5.2 |

[a]Ro 23-9424; [b] Fleroxacin; [c]Cefotaxime

incorporates into its structure both a third-generation cephalosporin acylamino function and a third-generation quinolone, fleroxacin (**15**). Compound **10** demonstrated excellent activity in vitro (Table 6[15,24,40]) and in vivo (Table 9[23]), and has been selected for development as a clinical candidate.

Although in vitro structure-activity relationships clearly demonstrate quinolone-like as well as cephalosporin-like contributions, and thus support the dual-action hypothesis, they cannot unequivocally establish that the observed dual mode of action is due to the mechanism of Scheme 1 in which the second agent is enzyme-activated. These cephalosporins are reactive molecules with a certain degree of hydrolytic instability (Table 10). Because of their labile nature, in situ hydrolysis of the allylic ester to yield as major products a 3'-hydroxycephalosporin and free quinolone complicates mechanistic studies. Nevertheless, some data (Table 11) suggest that biologically active hydrolysis products alone cannot explain the observed unique properties of **10**. On a molar basis, against selected strains, **10** is more active than either fleroxacin (**15**) or desacetylcefotaxime (**32**). Against some organisms, **10** is significantly more active than a 1:1 molar mixture of **15** and **32**. Further investigations into the mode of action of Ro 23-9424 are described in a later section.

**Table 10.** Stability of **4** and **10** in Various Media at 37°C
As Determined by HPLC Analysis
t$_{1/2}$ (h)

| Medium | 4 | 10 |
|---|---|---|
| pH 7.4 Phosphate Buffer | 12.5 | 3 |
| Mueller-Hinton Broth | 11.2 | 3.4 |
| Human Serum | 6.0 | 6.3 |

**Table 11.**  In Vitro Activity: MIC (nmol/ml)[a]

| Organisms | 10[b] | 15[c] | 1:1 15 + 32 | 32[d] |
|---|---|---|---|---|
| Micrococcus luteus PCI | 0.0746 | 38.2 | 0.597 | 0.597 |
| Streptococcus pneumoniae 6301 | 0.0187 | 9.53 | 0.0746 | 0.0746 |
| Escherichia coli DC-O | 0.299 | 1.19 | 0.299 | 0.597 |

[a]Reprinted with permission (Albrecht et al., 1990). Copyright 1990 American Chemical Society.
[b]Ro 23-9424
[c]Fleroxacin
[d]Desacetyl cefotaxime:

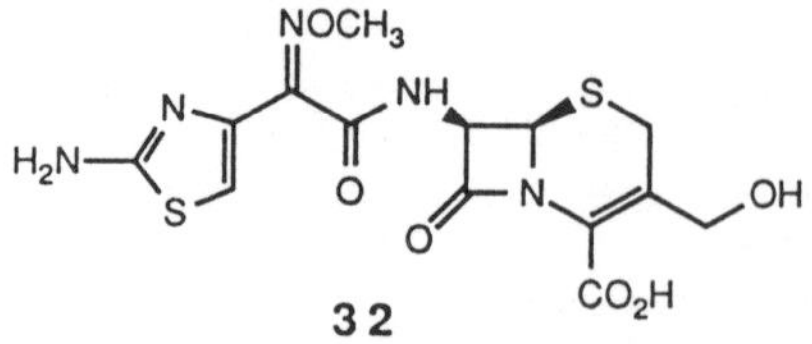

**3 2**

# III. CEPHALOSPORIN 3'-QUATERNARY QUINOLONES

Once the interesting biological activity of the cephalosporin 3'-quinolone esters became evident, work began on another class of dual-action cephalosporins. The dual-action hypothesis (Scheme 1) requires that the 3'-substituent should function as a leaving group. However, it would be advantageous to have a drug with chemical stability superior to that of **10** and which would still function as a dual-action cephalosporin under enzyme-mediated conditions. Some quinolones, such as fleroxacin, have a tertiary amine functionality which can be readily quaternized by reaction with alkylating agents. Since quaternary ammonium substituents can function as leaving groups, and the often beneficial effect of a quaternary nitrogen substituent on the activity of β-lactam antibiotics has been recognized,[41,42] it seemed appropriate to link quinolones to the 3'-position through a quaternary nitrogen.

## A.  Synthesis of Cephalosporin 3'-Quaternary Quinolones

Compounds **33–37** (Table 12) were prepared by the methodologies outlined in Schemes 7 and 8.[43] The same 3'-iodocephalosporin t-butyl esters used to prepare the 3'-quinolone esters[22] can also be used in the quaternization reaction (Scheme 7). Deprotection of the intermediate esters led to the desired products. However, it is usually more convenient to utilize the multistep, one-pot synthesis shown in Scheme 8, which was adapted from methods used by others to introduce substituted pyridinium and N-methylpyrrolidinium groups at the 3'-position.[44–46] In this procedure,[43] initial treatment with a silylating agent provides the trimethylsilyl (TMS) ester in which any sensitive functionality on the acyl side chain is also TMS protected. Reaction with iodotrimethylsilane then gives the 3'-iodo derivative which is used in situ. Addition of the TMS ester of a quinolone such as fleroxacin,

**Table 12.** Cephalosporin 3'-Quaternary Quinolones

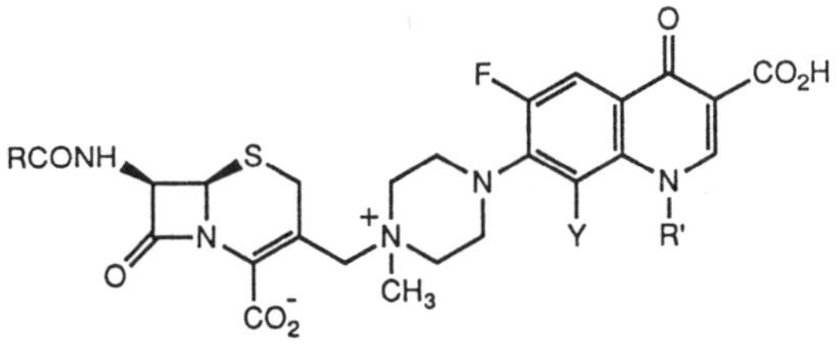

| No. | R | R' | Y |
|-----|-----|-----|-----|
| 33 | ATM | –CH₂CH₂F | F |
| 34 | PhOCH₂– | –CH₂CH₂F | F |
| 35 | ATIBA | –CH₂CH₂F | F |
| 36 | ATM | Et | H |
| 37 | H | –CH₂CH₂F | F |

Structures:

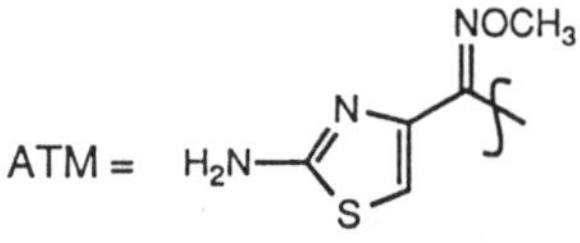

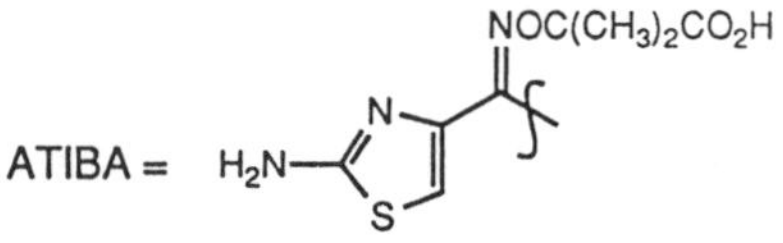

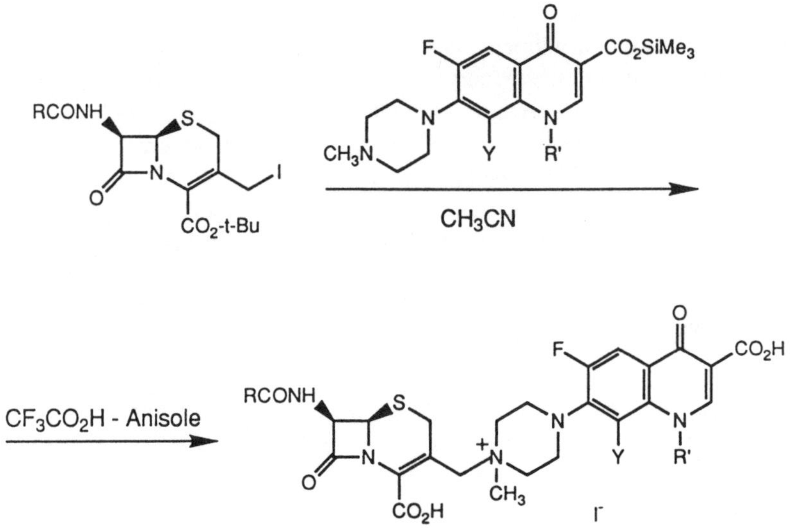

$$ATM = $$

$$ATIBA = $$

CF₃CO₂H - Anisole

**Scheme 7.**

223

**Scheme 8.** MSTFA = CF$_3$CON(Me)SiMe$_3$

which contains a tertiary amine, leads to quaternization. Solvolysis by addition of methanol then precipitates the product as a quaternary iodide. The quaternary iodides can be converted to zwitterions or sodium salts by appropriate treatment with sodium phosphate buffer, sodium bicarbonate, or sodium hydroxide.

The solvent used in the preparation of the TMS-protected iodo intermediates was either methylene chloride or acetonitrile. Acetonitrile was preferred for the displacement step, and was sometimes used for the entire sequence (e.g., for the synthesis of **35**) to avoid the inconvenience of changing solvents. In this reaction sequence, some isomerization of the cephem double bond was occasionally noted. The basic conditions for the displacement reaction were probably primarily responsible, since reducing the amount of quinolone TMS ester seemed to help. Conditions were optimized for each reaction at molar ratios from 0.3 to 0.75; under these conditions products were relatively free of both $\Delta^2$-isomer and unchanged quinolone. The latter impurity could be removed chromatographically, but no practical way of separating the $\Delta^2$- and $\Delta^3$-isomers was found. Other more subtle factors may influence the double-bond shift. In one case, the choice of solvent seemed critical. The TMS ester of cefotaxime showed an alarming solvent-dependent tendency to isomerize. When cefotaxime was treated with N-methyl-N-(trimethylsilyl)trifluoroacetamide in chloroform or methylene chloride, and reprecipitated after five minutes by addition of methanol, no significant isomerization occurred. However, when acetonitrile was used as the solvent in a similar

***Table 13.*** In Vitro Activity: MIC (μg/ml)

| Organisms | 10[a] | 15[b] | 31[c] | 33 | 34 | 35 |
|---|---|---|---|---|---|---|
| *Escherichia coli* 257 | 0.063 | 0.031 | 0.031 | 0.125 | 0.5 | 0.25 |
| *Klebsiella pneumoniae* A | 0.125 | 0.031 | 0.031 | 0.5 | 0.5 | 0.25 |
| *Enterobacter cloacae* 5699 | 0.25 | 0.063 | 0.125 | 0.5 | 0.5 | 0.25 |
| *Enterobacter cloacae* P99 | 0.125 | 0.031 | 64 | 0.5 | 0.5 | 1 |
| *Citrobacter freundii* BS #16 | 0.25 | 0.25 | 128 | 2 | 2 | 2 |
| *Serratia marcescens* SM | 0.125 | 0.25 | 0.25 | 1 | 2 | 1 |
| *Pseudomonas aeruginosa* 5712 | 16 | 4 | 64 | 32 | 32 | 16 |
| *Staphylococcus aureus* Smith (MS) | 1 | 0.25 | 1 | 4 | 0.5 | 16 |
| *Staphylococcus aureus* 95 (MR) | 1 | 0.25 | 128 | 4 | 1 | 8 |
| *Streptococcus pneumoniae* 6301 | ≤0.008 | 8 | 0.016 | 0.063 | 0.25 | 0.5 |
| *Streptococcus pyogenes* 4 | ≤0.008 | 4 | ≤0.008 | 0.125 | 0.063 | 0.5 |

[a]Ro 23-9424; [b]Fleroxacin; [c]Cefotaxime

experiment, up to 30% $\Delta^2$-isomer appeared in the recovered cefotaxime. Fortunately, this behavior was not general; the precursor to **35** showed no tendency toward double-bond migration in similar experiments.

## B. Properties of Cephalosporin 3′-Quaternary Quinolones

The in vitro activities of **33–35** are shown in Table 13 along with those of Ro 23-9424 (**10**), fleroxacin (**15**) and cefotaxime (**31**) for comparison.[43] As in the case of the bifunctional esters, the bifunctional quaternaries showed a pattern of activity which suggested a dual mode of action. Cephalosporins with a 7-(phenoxyacetyl)-amino side chain usually exhibit significant activity only against gram-positive organisms. Such was the case with the model compound **38**, which has a 3′-substituent with no activity of its own. On the other hand, the bifunctional compound **34** showed excellent activity not only against gram-positive bacteria, but also against gram-negative strains, except for *Pseudomonas aeruginosa*. Compared to cefotaxime, **33-35** exhibited increased potency against *Enterobacter cloacae* P99, *Citrobacter freundii* BS#16, and *Staphylococcus aureus* 95, a methicillin-resistant β-lactamase producer. Compared to fleroxacin, they were significantly more potent against the streptococci. Compound **33** differs from **10** only in the type of bonding between the cephalosporin 3′-position and the quinolone component. With bonding through a quaternary nitrogen, **33** is significantly less active in vitro than the

**38**

**Table 14.** In Vivo Activity: $ED_{50}$ (mg/kg,sc) Mouse Protection Test

| Infection | $15^a$ | 35 | $CTZ^b$ |
|---|---|---|---|
| Escherichia coli 257 | 0.26 | 0.28 | <0.5 |
| Klebsiella pneumoniae A | 0.3 | 3 | <2 |
| Enterobacter cloacae 5699 | <2 | <5 | 25 |
| Citrobacter freundii BS#16 | 2 | 9 | 22 |
| Serratia marcescens SM | <2 | 8 | 67 |
| Pseudomonas aeruginosa 5712 | 31 | >250 | 217 |
| Staphylococcus aureus Smith (MS) | 0.65 | 2.6 | <2 |
| Staphylococcus aureus 753 (MR) | 2.8 | 44 | 62 |
| Streptococcus pneumoniae 6301 | >250 | 8 | 3 |
| Streptococcus pyogenes 4 | >250 | 54 | 12 |

[a]Fleroxacin; [b]Ceftazidime

ester-linked compound **10**. Because of limited solubility under physiological conditions and its toxicity toward mice ($LD_{50} = 55$ mg/kg, i.v.), **33** was given only limited in vivo testing. In terms of chemical and biological properties, **35** proved to be a better prospect. On the basis of favorable in vitro activity and good aqueous solubility ($\geq 10\%$) under neutral conditions, **35** was selected for in vivo evaluation. This compound is comparatively nontoxic in mice, with an $LD_{50}$ of 770 mg/kg, i.v., compared to 245 mg/kg for the quinolone component, fleroxacin. The goal of achieving greater chemical stability was realized; the degradation half-life of **35** at 37 °C was 2–3 days at pH 6.5–7.5 in phosphate buffer, and 21 hours in mouse serum. Against seven of ten infections in the mouse protection test, **35** demonstrated excellent activity (Table 14). It was somewhat less active against *S. pyogenes* and methicillin-resistant *S. aureus* infections and inactive against *P. aeruginosa*. By comparison, the clinical candidate **10** (Table 9) was active against all infections used in this screen.[23] Except for the weaknesses noted for **35**, both compounds showed generally comparable $ED_{50}$ values. The data in Tables 9 and 14 were obtained in separate experiments, and therefore are not tabulated together. Compound **35** was not further developed, since Ro 23-9424 (**10**) appeared to be a more promising candidate. Behavior of **33**, **34**, and **35** toward PBPs from *E. coli* is shown in Table 8. Again, the affinity for PBPs seems to depend largely on the acylamino function. Whether the quinolone is ester-linked, as in **7–10**, or quaternary-linked, as in **33**, binding to PBP 3 for bifunctional compounds with the ATM side chain is the same as that of cefotaxime.

# IV.  MECHANISM OF ACTION

A "dual-action" cephalosporin should exhibit both a mechanism characteristic of a cephalosporin and a mechanism characteristic of the second antibacterial agent (in most cases, a quinolone). Although this statement appears rather facile, its experimental demonstration has proven quite difficult, due at least partly to the limited stability of the cephalosporin-quinolone esters. It is helpful to consider the problem as three basic questions:

1. Do dual-action cephalosporins exhibit both β-lactam- and quinolone-like activities?
2. Do these activities involve the intact molecule or only the biologically active degradation products?
3. If the intact molecule is involved, by what sequence of events are the two kinds of activity expressed?

## A. Do Dual-Action Cephalosporins Exhibit Both β-Lactam- and Quinolone-Like Activities?

Clearly, to justify the name, a dual-action cephalosporin must have both types of activity. The evidence presented by earlier authors for the activity of the second agent is rather stronger than the evidence for β-lactam-like activity. O'Callaghan et al.[13] convincingly showed that active omadine was released from MCO by β-lactamase treatment. On the other hand, the only evidence that MCO also had significant β-lactam activity was that, in three pairs of β-lactamase producer/non-producer strains, MCO was less active against the β-lactamase producer.

Mobashery and Johnston did not claim a dual mechanism of action for their β-chloroalanine-containing cephem esters.[18,19] The release of the dipeptide from the cephalosporin by TEM β-lactamase, cleavage of the dipeptide by leucine aminopeptidase, and inhibition of alanine racemase by the liberated β-chloroalanine were elegantly demonstrated,[18,19] but cephalosporin-like antibacterial activity was not.

Early in the Roche program on cephalosporin 3'-quinolone esters, we noted that such compounds with a 7-phenoxyacetylamino substituent were all more active than the corresponding reference cephalosporin **30** against gram-negative organisms (Table 5) and that the activity added to the spectrum of the reference cephalosporin generally paralleled the activity of the quinolone component.[22] These compounds were generally more active than the quinolone component against gram-positive organisms, reflecting the contribution of the cephalosporin moiety. Moreover, when the in vitro activity of **4** was determined in the presence of β-lactamases, the apparent activity against a quinolone-resistant, gram-positive strain decreased, while the apparent activity against a quinolone-sensitive, gram-negative strain increased.[26] This result suggests that β-lactamase treatment, while inactivating the β-lactam, releases the quinolone in active form.

Throughout our experience with cephalosporin-quinolone esters, we have consistently observed antibacterial activity that reflects both β-lactam and quinolone contributions. However, considerable degradation of these compounds probably occurs during the overnight incubation period required for the MIC determination.

To the extent that this degradation yields biologically active products, the observed antibacterial activity cannot be attributed unequivocally to the intact molecule, leading us to the second basic question.

## B.   Do the Observed Activities Involve the Intact Molecule or Only Biologically Active Degradation Products?

Esters of quinolones are evidently universally inactive,[25,29,38,39] so it is reasonable to assume that the quinolone moiety of an ester-type dual-action cephalosporin must be released as a free acid in order to exert its activity. The question therefore, is whether the quinolone is released as proposed, by a mechanism involving the bacteria (presumably enzymatic), or whether active quinolone is formed primarily by non-enzymatic degradation.

As mentioned above, MIC data are not very useful in answering this question because the time required for the assay is long compared with the stability of the compounds. In general, dual-action cephalosporins are no more active than the more active of the two constituents, so degradation to active products is a plausible hypothesis, at least superficially. For certain strains, however, even MIC data are inconsistent with this hypothesis. Examples of such strains are those shown in Table 11[22] and the ten strains of penicillin-resistant pneumococci shown in Table 15.[15] Recently, Pace et al.[47] have isolated fleroxacin-resistant (MIC = 50 µg/ml) mutants of *E. coli* TE18, a strain over-producing AmpC β-lactamase and resistant to desacetylcefotaxime (MIC = 50 µg/ml). These strains were still susceptible to Ro 23-9424 (MIC = 0.5 µg/ml). Clearly, the activity of the dual-action cephalosporin against these strains cannot be explained by its degradation to less active products.

Several observations based on methods with a more appropriate time scale are also inconsistent with the degradation hypothesis. First, the affinity of Ro 23-9424 for the PBPs of *E. coli* and *S. aureus* is greater than that of desacetylcefotaxime.[29] Second, Ro 23-9424 is a more potent inhibitor of peptidoglycan synthesis in permeabilized *E. coli* cells than desacetylcefotaxime.[25] In fact, all the cephalosporin 3'-quinolone esters tested are more potent inhibitors of peptidoglycan synthesis than the corresponding cephalosporins with conventional 3'-leaving groups.[48] Third, Ro 23-9424 has a stronger bactericidal effect against *E.*

***Table 15.***  In Vitro Activity of Ro 23-9424 Against Ten Strains of
Penicillin-Resistant *Streptococcus pneumoniae*

|  | *MIC$_{50}$ (µg/ml)* | *MIC$_{90}$ (µg/ml)* | *Range* |
|---|---|---|---|
| Ro 23-9424 | 0.12 | 0.25 | ≤0.06–0.5 |
| Desacetylcefotaxime + Fleroxacin | 0.5 | 2 | 0.25–4 |
| Desacetylcefotaxime | 0.5 | 4 | 0.25–8 |
| Cefotaxime | 0.25 | 0.5 | 0.12–2 |
| Fleroxacin | 8 | 8 | 4–16 |

*coli* than an equimolar combination of desacetylcefotaxime and fleroxacin.[25] None of these results can be explained by extracellular hydrolysis of Ro 23-9424. It seems most likely therefore, that these activities are due to the intact molecule, which brings us to the third basic question.

## C. By What Sequence of Events are the Two Kinds of Activity Expressed?

First, Ro 23-9424 apparently crosses the outer membrane of a gram-negative cell as an intact molecule. This is based on the observation of Georgopapadakou et al.[29] that the effect of porin mutations on the growth-inhibiting activity of Ro 23-9424 is similar to the effect on cefotaxime and different from the effect on fleroxacin. This result suggests that Ro 23-9424, like cefotaxime but unlike fleroxacin, is dependent on porins for penetration of the outer membrane of *E. coli.*

The space between the two membranes of the gram-negative cell is known as the periplasmic space. β-Lactamases and PBPs, the two prime candidates for the role of catalyzing release of free quinolone from a dual-action cephalosporin, are localized in this space (see Note 2). In addition, the periplasmic space contains a variety of hydrolytic enzymes that may also play a part.

The original proposal of O'Callaghan et al.[13] was that this role would be played by β-lactamases, as shown in Scheme 1. It has been shown experimentally that hydrolysis by β-lactamases stoichiometrically releases the quinolone moiety from **4** and **10**.[25,26] On the other hand, the in vitro activity of dual-action cephalosporins against gram-negative organisms seems generally to parallel the activity of the quinolone component, regardless of the ability of the organism to produce β-lactamases, regardless of the properties of the enzyme(s) produced, and regardless of the lability of the cephalosporin moiety to β-lactamases. On the basis of these observations, it is easy to dismiss any possible role for β-lactamases in activation of the quinolone moiety.

However, it should be remembered that most, if not all, enterobacteriaceae and pseudomonads produce at least a small amount of chromosomally encoded β-lactamase, such as the AmpC enzyme in *E. coli.* It is conceivable that this small amount of β-lactamase, acting in the periplasm, can release sufficient quinolone to account for the observed quinolone-like activity of dual-action cephalosporins. Thus, the quinolone-like activity would appear to be independent of β-lactamase-related parameters because a completely β-lactamase-deficient control was not observed.

Aztreonam is a potent inhibitor of the AmpC enzyme and should therefore inhibit AmpC-mediated release of quinolones from dual-action cephalosporins. Experimentally, however, aztreonam only slightly affected the inhibition of DNA synthesis by Ro 23-9424 in toluene-permeabilized cells and did not affect the inhibition of thymidine incorporation by Ro 23-9424 in growing cultures of *E. coli* (Christenson et al., manuscript in preparation). It seems likely that, although AmpC β-lactamases may have some role in activating the quinolone moiety, there must also be at least one other mechanism.

As pointed out above, PBPs, like β-lactamases, also open the β-lactam ring according to Scheme 1, forming acyl-enzymes through a homologous serine residue. Dual-action cephalosporins bind strongly to PBPs and are potent inhibitors of peptidoglycan synthesis. Thus, PBPs could also perform the role of activating the quinolone. It may be argued that PBPs, unlike β-lactamases, turn over β-lactams slowly, if at all, so only a very small amount of quinolone can be released by this mechanism. Despite this objection, it can be shown that the release of a single molecule of quinolone by each PBP molecule, within the very small volume of the bacterial cell, would achieve a concentration of quinolone sufficient to have an antibacterial effect. Moreover, PBPs may de-acylate at a much faster rate than has been generally assumed.[50]

A PBP-mediated model would predict that, when a dual-action cephalosporin acts, peptidoglycan synthesis should be inhibited first, followed by inhibition of DNA synthesis. This was found to be the case in toluene-permeabilized cells, but not in growing cells, in which DNA synthesis was inhibited immediately while peptidoglycan synthesis continued at the control rate for about half of one generation time.[25] Moreover, the model would predict that release of free quinolone should be inhibited by pre-blocking of the PBPs with β-lactams. This experiment has been attempted several times with variable results, forcing us to conclude that, although PBPs may be capable of releasing quinolones from dual-action cephalosporins, there must also be at least one other mechanism.

Thus, it seems likely that both β-lactamases and PBPs can activate the quinolone moiety of a dual-action cephalosporin. The relative importance of these two mechanisms, and whether there are other mechanisms as well, will require further investigation.

## V. PHARMACOKINETICS

Initially, it was assumed that ester-type dual-action cephalosporins would be even less stable in serum than they are in buffer, due to the presence of esterases in serum. With the discovery that Ro 23-9424 is *more* stable in serum than in buffer, it seemed likely that the compound would have favorable pharmacokinetic properties.

It is well known that 3'-acetoxy cephalosporins—such as cephalothin cephapirin, and cefotaxime—are de-esterified in vivo to form 3-hydroxymethyl-cephalosporins.[51-53] These metabolites have antibacterial activity, but are less potent than the acetylated cephalosporins. Hydrolysis of the ester bond of a cephalosporin-quinolone ester presumably results in the formation of free quinolone and the respective 3-hydroxymethylcephalosporin, both of which have antibacterial activity. The extent to which a dual-action cephalosporin will exert the proposed mechanism of action in vivo, or act as a combination of active metabolites, is therefore determined by its pharmacokinetic properties.

The pharmacokinetics of **4** and **10** have been studied in animals. Compound **4** in rats has a half-life similar to that of cefotaxime and is fairly stable in human plasma and rat-liver extract,[22] suggesting that it is not highly metabolized.

Pharmacokinetic profiles of Ro 23-9424 (10) were determined in mice, rats, dogs, and baboons after a single intravenous dose of 20 mg/kg.[27] The pharmacokinetic parameters of Ro 23-9424 were similar to analogous data for cefotaxime.[54-58]

Plasma concentrations of fleroxacin after administration of Ro 23-9424 were low (about 1 µg/ml), but persistent in all species. The relatively rapid disappearance of Ro 23-9424 from the plasma without a corresponding increase in fleroxacin concentration suggested a model in which Ro 23-9424 is primarily eliminated by excretion of the intact molecule into the urine or the bile (as is typical of cephalosporins), rather than by hydrolysis to desacetylcefotaxime and fleroxacin.

The validity of this model could be tested by analyzing quantitatively the recovery of the intact molecule and its metabolites in urine and feces, but this has not been attempted due to the instability of the drug in animal excreta under the experimental conditions. High concentrations of intact Ro 23-9424 were found in samples of baboon and mouse urine, however, indicating that urinary excretion of the intact molecule is a major route of elimination of Ro 23-9424, as it is for cefotaxime. Since it is much more practical to quantitate urinary recovery in humans, the crucial test of this model will be part of the initial clinical studies of Ro 23-9424.

It was concluded that, although metabolism is not the major route of elimination of Ro 23-9424, some fleroxacin is formed from Ro 23-9424 in vivo. Since the half-lives of Ro 23-9424 and fleroxacin are substantially different, it appeared possible that plasma levels of fleroxacin might accumulate to unacceptable levels on repeated dosing with Ro 23-9424. A 14-day multiple-dose study in baboons showed only a slight accumulation of fleroxacin, however, with peak fleroxacin concentrations no higher than 4.9 µg/ml.

Thus, although some free fleroxacin is formed in vivo, Ro 23-9424 must be considered as a therapeutic agent in its own right, not simply as a pro-drug, just as cefotaxime is considered to be a therapeutic agent. Nevertheless, the low but persistent concentrations of fleroxacin produced metabolically may have significant therapeutic effects and must also be considered in analyzing the in vivo activity of Ro 23-9424. In fact, free fleroxacin was found in all species long after Ro 23-9424 had disappeared. The persistence of fleroxacin after elimination of Ro 23-9424 is a phenomenon requiring further investigation. Christenson et al.[27] proposed that it results from enteral absorption of fleroxacin formed in the gut from Ro 23-9424 excreted in the bile.

## VI. PERSPECTIVES

The dual-action mechanism, in which bacterial enzyme-mediated opening of the β-lactam results in liberation of a second antibacterial agent, provides a fascinating rationale for drug design, and has led to the synthesis of antibacterials with interesting and unusual properties. It was our strategy to evaluate a number of compounds of various structural types in order to select one or more candidates for clinical development, while concurrently determining whether the compounds have a dual mechanism of action and whether such a dual mechanism is at least plausible in vivo.

The bifunctional cephalosporins incorporating quinolones as ester-linked or quaternary-linked second agents behave in many respects as dual-action cephalosporins would be expected to. Evaluation of these compounds led to the selection of **10** (Ro 23-9424) as a clinical candidate. Research mostly utilizing this compound has established the essential features of the mechanism of action, though some of the details remain elusive. The specific dual-action mechanism (Scheme 1) is difficult to prove unequivocally, and may not be the exclusive mechanism by which these compounds act. However, our investigations have demonstrated that it is indeed realistic to expect a dual mechanism of action to be operative in vivo.

Ro 23-9424 has many fine characteristics as an antibacterial agent, but is less than ideal in certain other ways. Therefore, an optimistic search continues for a "second-generation" dual-action cephalosporin.

## NOTES

1. It is important to distinguish this use of the term "prodrug" from the conventional use of the term. Broadly, a derivative is administered in place of the drug itself in order to impart some desirable property not possessed by the drug itself, or to overcome some undesirable property of the drug. Commonly, a prodrug is considered to be *inactive* until it is transformed to the active compound by a *host* enzyme or metabolic process. In contrast, as will be shown in some detail, dual-action cephalosporins are thought to be *active* as cephalosporins and therefore are prodrugs only for the second active moiety and this second active moiety ideally would be released by a *bacterial* enzyme or process. The term "codrug" proposed by Jones et al.[15] for dual-action cephalosporins may be helpful in making this distinction.

2. PBPs are membrane-bound enzymes, whereas true periplasmic enzymes, such as β-lactamases, are usually soluble. Since the function of PBPs is the extension of the murein sacculus, which is located in the periplasmic space, it is generally assumed that the active sites of PBPs are also located there. The available experimental data seem to confirm this expected orientation.[49] Thus, to the extent that PBPs catalyze the release of free quinolone from dual-action cephalosporins, they do so in the periplasmic space.

## REFERENCES

1. Frere, J.M.; Joris, B. *CRC Crit. Rev. Microbiol.* **1985**, *11*, 299–396.
2. Waxman, D.J.; Strominger, J.L. *J. Biol. Chem.* **1980**, *255*, 3964–3976.
3. Fink, A.L. *Pharm. Res.* **1985**, *2*, 55–61.
4. Knott-Hunziker, V.; Waley, S.G.; Orlek, B.S.; Sammes, P.G. *FEBS Lett.* **1979**, *99*, 59–61.
5. Hamilton-Miller, J.M.T.; Newton, G.G.F.; Abraham, E.P. *Biochem. J.* **1970**, *116*, 371–384.

6. O'Callaghan, C.H.; Kirby, S.M.; Morris, A.; Waller; R.E.; Duncombe, R.E. *J. Bacteriol.* **1972**, *110*, 988–991.

7. Faraci, W.S.; Pratt, R.F. *J. Am. Chem. Soc.* **1984**, *106*, 1489–1490.

8. Russell, A.D.; Fountain, R.H. *J. Bacteriol.* **1971**, *106*, 65–69.

9. Boyd, D.B. *J. Org. Chem.* **1985**, *50*, 886–888.

10. Grabowski, E.J.; Douglas, A.W.; Smith, G.B. *J. Am. Chem. Soc.* **1985**, *107*, 267–268.

11. Page, M.I.; Proctor, P.J. *J. Am. Chem. Soc.* **1984**, *106*, 3820–3825.

12. Sabath, L.D.; Jago, M.; Abraham, E.P. *Biochem. J.* **1965**, *96*, 739–752.

13. O'Callaghanm C.H.; Sykes, R.B.; Staniforth, S.E. *Antimicrob. Agents Chemother.* **1976**, *10*, 245–248.

14. Greenwood, D.; O'Grady, F. *Antimicrob. Agents Chemother.* **1976**, *10*, 249–252.

15. Jones, R.N.; Barry, A.L.; Thornsberry, C. *Antimicrob. Agents Chemother.* **1989**, *33*, 944–950.

16. Uri, J.V.; Actor, P.; Phillips, L.; Weisbach, J.A. *J. Antibiotics* **1978**, *31*, 580–585.

17. Mobashery, S.; Lerner, S.A.; Johnston, M. *J. Am. Chem. Soc.* **1986**, *108*, 1685–1686.

18. Mobashery, S.; Johnston, M. *J. Biol. Chem.* **1986**, *261*, 7879–7887.

19. Mobashery, S.; Johnston, M. *Biochemistry* **1987**, *26*, 5878–5884.

20. Haller, I. *Arzneim.-Forsch.* **1986**, *36*, 226–229.

21. Wolfson, J.S.; Hooper, D.C. *Antimicrob. Agents Chemother.* **1985**, *28*, 581–586.

22. Albrecht, H.A.; Beskid, G.; Chan, K.-K.; Christenson, J.; Cleeland, R.; Deitcher, K.H.; Georgopapadakou, N.H.; Keith, D.D.; Preuss, D.L.; Sepinwall, J.; Specian, Jr., A.C.; Then, R.L..; Weigle, M.; West, K.F.; Yang, R. *J. Med. Chem.* **1990**, *33*, 77–86.

23. Beskid, G.; Siebelist, J.; McGarry, C.M.; Cleeland, R.; Chan, K.K.; Keith, D.D. *Chemotherapy* **1990**, *36*, 109–116.

24. Beskid, G.; Fallat, V.; Lipschitz, E.R.; McGarry, D.H.; Cleeland, R.; Chan, K.; Keith, D.D.; Unowsky, J. *Antimicrob. Agents Chemother.* **1989**, *33*, 1072–1077.

25. Christenson, J.G.; Brocks, V.; Chan, K.K.; Keith., D.D.; Preuss, D.L.; Schaefer, F.F.; Talbot, M.K. 28th Intersci. Conf. Antimicrob. Agents Chemother., 1988a, Abst. No. 444.

26. Christenson, J.; Albrecht, H.; Georgopapadakou, N.; Keith, D.; Preuss, D.; Talbot, M.; Then, R. 28th Intersci. Conf. Antimicrob. Agents Chemother., 1988b, Abst. No. 450.

27. Christenson, J.G.; Chan, K.K.; Cleeland, R.; Dix-Holzknecht, B.; Farrish, H.H.; Patel, I.H.; Specian, A. *Antimicrob. Agents Chemother.* **1990**, *34*, 1895–1900.

28. Christenson, J.G.; Preuss, D.L.; Robertson, T.L. 28th Intersci. Conf. Antimicrob. Agents Chemother., 1988d, Abst. No. 443.

29. Georgopapadakou, N.H.; Bertasso, A.; Chan, K.K.; Chapman, J.S.; Cleeland, R.; Cummings, L.M.; Dix, B.A.; Keith, D.D. *Antimicrob. Agents Chemother.* **1989**, *33*, 1067–1071.

30. Georgopapadakou, N.H.; Albrecht, H.A.; Bertasso, A.; Cummings, L.M.; Keith, D.D.; Russo, D.A. 28th Intersci. Conf. Antimicrob. Agents Chemother., 1988, Abst. No. 442.

31. Koppel, G.A. U.S. Patent 4,042,585, 1976.

32. Kukolja, S.; Chauvette, R.R. Cephalosporin antibiotics prepared by modification at the C-3 position. In:*Chemistry and Biology of β-Lactam Antibiotics*; Morin, R.B.; Gorman, M., Eds.; Academic Press: New York, 1982, Vol. 1, pp. 108–111.

33. Lammert, S.R.; Ellis, A.I.; Chauvette, R.R.; Kukolja, S. *J. Org. Chem.* **1978**, *43*, 1243–1244.

34. Murphy, C.F.; Webber, A. Alteration of the dihydrothiazine ring moiety. In: *Cephalosporins and Penicillins.* E.H. Flynn, Ed.; Academic Press: New York, 1972, pp. 136–138.

35. Bonjouklian, R. U.S. Patent 4,266,048, 1981.

36. Bonjouklian, R.; Phillips, M.L. *Tetrahedron Lett.* **1981**, 22.3915–3918.

37. Yamanaka, H.; Chiba, T.; Kowabata, K.; Takasugi, H.; Masugi, T.; Takaya, T. *J. Antibiot.* **1985**, *38*, 1738–1751.

38. Albrecht, R. *Prog. Drug Res.* **1977**, *21*, 9–104.

39. Kaminsky, D.; Meltzer, R.E. *J. Med. Chem.* **1968**, *11*, 160–163.

40. Gu, J.-W.; Neu, H.C. *Antimicrob. Agents Chemother.* **1990**, *34*, 189–195.

41. Morin, R.B. *25th Intersci. Conf. Antimicrob. Agents Chemother., Session 84* **1985**.

42. Naito, T.; Aburaki, S.; Kamachi, H.; Narita, Y.; Okumura, J.; Kawaguchi, H. *J. Antibiotics* **1986**, *39*, 1092–1077.

43. Albrecht, H.A.; Beskid, G.; Christenson, J.; Durkin, J.; Fallat, V.; Keith, D.D.; Konzelmann, F.M.; Lipschitz, E.R.; McGarry, D.H.; Siebelist, J.; Wei, C.-C.; Weigele, M.; Yang, R. *J. Med. Chem.* **1991**, *34*, 669–675.

44. Ludescher, H.; Ascher, G. Austrian Patent Appl. 8,403,716, 1986.

45. Lunn, W.H.W. U.S. Patent 4,402,955, 1983.

46. Walker, D.G.; Brodfuerhrer, P.R.; Brundidge, S.P.; Shih, K.M.; Sapino, C. *J. Org. Chem.* **1988**, *53*, 983–991.

47. Pace, J.; Bertasso, A.; Georgopapadakou, N.H. *Antimicrob. Agents Chemother.* **1991**, *35*, 910–915.

48. Talbot, M.K.; Schaefer, F.; Christenson, J.G. 28th Intersci. Conf. Antimicrob. Agents Chemother., 1988, Abstract No. 1223.

49. Spratt, B. G.; Bowler, L.D.; Edelman, A.; and Broome-Smith, J.K. Membrane topology of penicillin-binding proteins 1b and 3 of *Escherichia coli* and the production of water-soluble forms of high-molecular-weight penicillin-binding proteins, In: *Antibiotic Inhibition of Bacterial Cell Surface Assembly and Function.* Actor, P.; Daneo-Moore, L.; Higgins, M.L.; Salton, M.R.J.; Shockman, G.D. Eds.; American Society for Microbiology, Washington, D.C., 1988, pp. 292–300.

50. Talbot, M. K.; Schaefer, F.; Brocks, V.; Christenson, J.G. *Antimicrob. Agents Chemother.* **1989**, *33*, 2101–2108.

51. Wick, W.E. In vitro and in vivo laboratory comparison of cephalothin and desacetylcephalothin. In: *Antimicrobial Agents and Chemotherapy-1965.* Hobby, G.L., Ed.; American Society for Microbiology, Washington, D.C., 1965, pp. 870–875.

52. Cabana, B.E.; Van Harken, D.R.; Hottendorf, G.H. *Antimicrob. Agents Chemother.* **1976**, *10*, 307–317.

53. Chamberlain, J.; Coombes, J.D.; Dell, D.; Fromson, J.M.; Ings, R.J.; Macdonald, C.M.; McEwen, J. *J. Antimicrob. Chemother.* **1980**, *6*, (Suppl. A), 69–78.

54. Esmieu, F.; Guibert, J.; Rosenkilde, H.C.; Ho, I.; Le Go, A. *J. Antimicrob. Chemother.* **1980**, *6*, (Suppl. A), 83–92.

55. Marakawa, T.; Sakamoto, H.; Fukada, S.; Nakamoto, S.; Hirose, T.; Itoh, N.; Nishida, M. *Antimicrob. Agents Chemother.* **1980**, *17*, 157–164.

56. Omosu, M.; Togashi, O.; Fujimoto, K. *Chemotherapy (Tokyo)* **1980**, *28*, 1184–1193.

57. Togashi, O.; Maeda, T.; Omosu, M.; Fujimoto, K.; Naito, S. *Pharm. Res.* **1985**, *2*, 124–130.

58. Guerrini, V.H.; English, P.B.; Filippich, L.J.; Schneider, J.; Bourne, D.W. *Vet. Rec.* **1986**, *119*, 81–83.

# SOME THOUGHTS ON ENZYME INHIBITION AND THE QUIESCENT AFFINITY LABEL CONCEPT*

A. Krantz

*Contribution no. 334 from the Institute of Bioorganic Chemistry

Advances in Medicinal Chemistry,
Volume 1, pages 235–261.
Copyright © 1992 by JAI Press Inc.
All rights of reproduction in any form reserved.
ISBN: 1-55938-170-1

# I. RATIONAL APPROACHES TO DRUG DESIGN

## A. The Structural Strategy

Enzyme inhibitors have been of chemical and clinical interest since the early days of modern chemotherapy. Until recently, chemists were guided primarily by considerations of isosterism and sought to mimic substrate structure to develop new synthetic inhibitors.[1] Over the past two decades emphasis has shifted from considerations of the structure of the substrate in the ground state, to the structure of intermediates and transition states produced during the course of enzyme catalysis,[2,3] and very recently, to the three-dimensional structure of the enzyme target.[4-6]

Fields dependent upon novel modes of enzyme inhibition are now being nurtured by two streams of rational strategies: the first, newly emanating from a structural perspective, is rooted in concepts of molecular complementarity; the second issues from a rich mechanistic tradition shaped by the paradigms of organic chemical reactivity. A mature enzyme chemistry is emerging from application of recent technological advances[6] that have provided a descriptive and preparative component that is revolutionizing the field. Thus, cloning and expression techniques have been manipulated to allow the overproduction of specific enzymes in quantities that can be routinely committed to stoichiometric experiments and complex structural determinations. Advances in x-ray crystallography, NMR spectroscopy[6,7] and modern computer technology,[8] enabling chemists to visualize complex macromolecules and their interactions with small molecules, dynamically on a screen, are making enzymology truly a descriptive science on a molecular level.[9]

It is now possible to envisage moving beyond simple organic chemical paradigms, which have governed our thinking about enzymes as chemicals, to a holistic, molecular view of enzymes as finely tuned chemical machines that have evolved in specific ways to catalyze chemical reactions.

To a first approximation, inhibitor design by the structural motif has been reduced to essentially a topographical problem in which the structure of a drug is designed on the basis of its fit to a three-dimensional structure of the enzyme active center. Relevant enzyme functional groups and cofactors are matched with inhibitor functionality, in a complementary fashion, to afford increasingly stable enzyme-inhibitor complexes, relative to the separate components in aqueous media. There are many facets to the structural approach using computer-assisted drug design.[6,9] The screening of small molecule data bases for good fits to specific enzyme binding sites, the identification of opportunities for additional complementary interactions in known enzyme-inhibitor complexes, as well as *de novo* research for new leads, are being pursued vigorously.

Insofar as a few kcal/mol of incremental stabilization, achievable by exploiting simple physico-chemical principles, are enough to tip the balance and turn an unexceptional inhibitor into an extraordinary one, the ability to visualize opportunities for favorable electrostatic and hydrophobic interactions in enzyme-inhibitor complexes is invaluable.

Although admittedly a promising area of research, computer-based structural strategies, even in their infancy, are being touted with unbridled enthusiasm as rational approaches to drug design, in much the same fashion as mechanism-based inhibitors were advanced during the 1970s. A caveat is in order. Multiple binding modes of an inhibitor are often possible,[10] and seemingly small structural changes in an inhibitor can significantly alter binding modes.[11] Thus, even within a closely related series, a great deal of structural information from experiments at the bench may be required for "rational" drug design.

It may also be worth re-emphasizing that no matter how elegantly crafted, potent, or specific an inhibitor may be in vitro, its success in the clinic hinges on issues of pharmacokinetics and bioavailability, as well as toxicity considerations. Currently, rational inhibitor design is a prerequisite for the rational design of drugs which have enzymes as their target, but it is not synonymous with drug design, however important it may be for new lead development. It is precisely at the level of pharmacokinetics and predictions of bioavailability that there is the greatest need for rational design based on systematic studies. Until the fate and locus (loci) of action of new molecules in vivo can be accurately predicted and easily manipulated, a healthy dose of empiricism will still drive the design process.

## B.   Mechanism-Based Inhibition and Other Rational Strategies that Exploit Knowledge of Reaction Mechanisms

The design of mechanism-based inhibitors has been inspired by marrying principles of organic chemical reactivity with knowledge of enzyme mechanisms; this has yielded a rich harvest of novel inhibitors.[12–16] Despite impressively sophisticated design motifs incorporating subtle, often ingenious tactics, a universally acceptable nomenclature for the classification of such inhibitors has not emerged. Rather, the basis to distinguish among the various inhibitors has been somewhat arbitrary and capricious.[16–18] In part, the difficulty of achieving an acceptable classification system, may be a consequence of the distinct perspectives of individual chemists with their emphasis on specific characteristics (kinetic, mechanistic, or structural, etc.) of a complex, multifaceted process, a situation which makes it difficult to completely and satisfactorily categorize enzyme inhibitors. Nevertheless, a nomenclature that conveys mechanistically meaningful information succinctly, focussing on critical elements that epitomize the inhibitory process, is essential for clear thinking, so that trends and novelty can be easily discerned and specific information can be coordinated into general principles for the development of new paradigms.

In the broadest sense, the term mechanism-based inhibitor has been applied to any inhibitor that exploits properties of transition states and reaction intermediates

**Table 1.** Classification of Enzyme Inhibitors

| Type | Binding determinants[a] | Chemically reactive group[b] | Minimal mechanism | Rationale |
|------|------------------------|------------------------------|-------------------|-----------|
| Substrate analog | S | no | $E + S' \rightleftharpoons E{\cdot}S'$ | structural mimicry |
| **Group I.** | | **Tight Binding Enzyme Complements** | | |
| Transition State Analog | t.s. | no | $E + I_{ts} \rightleftharpoons E{\cdot}I_{ts}$ | E binds t.s. tightly |
| Multi-substrate Analog | $S_1 + S_2$ | no | $E + S_1 - S_2 \rightleftharpoons E{\cdot}S_1\text{-}S_2$ | multiple binding effects |
| Ground State Analog | S + R | no | $E + S'\text{-}R \rightleftharpoons E{\cdot}S'\text{-}R$ | multiple binding effects |
| **Group II.** | | **Mechanism-Based Inhibitors** | | |
| Suicide Inhibitor | S | latent | $E + S' \rightleftharpoons E{\cdot}S' \rightleftharpoons E{\cdot}I_r \rightarrow E\text{-}I_r$ | unmasking of latent reactive group |
| Dead-end Inhibitor | S | no | $E + S' \rightleftharpoons E{\cdot}S' \rightarrow E'\text{-}S'$ | S' lacks structural elements for turnover |
| Alternate Substrate Inhibitor | S | no | $E + S'' \rightleftharpoons E{\cdot}S'' \rightleftharpoons E'\text{-}S'' \rightleftharpoons E\text{-}'P \rightleftharpoons E{\cdot}P \rightleftharpoons E + P$ | stabilize E'-S'', E'-P or E•P |
| **Group III.** | | **Affinity Labels** | | |
| Classical Affinity Label | S | yes | $E + S_{rx} \rightleftharpoons E{\cdot}S_{rx} \rightarrow E\text{-}S' + X^-$ | high effective concentration of reactive group at active site |
| Quiescent Affinity Label | S | no | $E + S_x \rightleftharpoons E{\cdot}S_x \rightarrow E\text{-}S' + X^-$ | differential reactivity: chemical vs. enzymatic |

[a]Refers to initial reagent; S = substrate; t.s. = transition state; $S_1 + S_2$ = cosubstrates; R = entity with affinity for non-catalytic site.
[b]S' = pseudosubstrate; $I_{ts}$ = transition state analog; $I_r$ = reactive intermediate; E-$I_r$, E-S', E'-S', E'-S", and E'-P = enzyme-inhibitor adduct; S" = alternate substrate inhibitor; P = product; $S_{rx}$ = reactive affinity label; $S_x$ = quiescent affinity label; $S_1$-$S_2$ = multisubstrate analog; S' – R = ground state analog.

produced during enzyme catalysis. Often the term has been defined in a narrow sense, with specific reference to enzyme suicide inhibitors.[16,17]

Table 1 presents a classification of commonly encountered types of enzyme inhibitors in terms of substrate analogs, tight binding enzyme complements (group I), mechanism-based inhibitors (group II), and affinity labels (group III). Each category will be discussed very briefly in turn, in order to clarify the distinguishing features of mechanism-based inhibitors.

The most straightforward approach to inhibitor design entails mimicking enzyme substrates according to the principles of isosterism. This approach is embodied in the antimetabolite concept and the phenomenon of competitive antagonism, where the structural analogy between the antagonist (or antimetabolite) and the metabolite has been considered to be a basic tenet.

The demonstration by Woods[19] of the competitive nature of the relationship between sulfanilamide and p-amino benzoic acid appeared to open up a direct approach to the rational development of chemotherapeutic agents by structural modification of other essential metabolites.[20] While creative replacements of substrate functionality can sometimes be an effective way of producing new leads and probing the active-site environment of an enzyme,[21] it is essentially an empirical approach whose limitations have been reviewed.[1]

Since enzymes do not generally bind their substrates tightly, mimicking substrate structure would not seem to be a desirable way to optimize inhibitor activity unless more than one substrate is being imitated in a multi-substrate analog. Retrospectively, many inhibitors that have been developed as substrate mimics turn out to be effective, not simply because they bear resemblance to substrate, but serendipitously, for other, more profound reasons (i.e., they are actually transition state analogs, possess latent reactivity, etc.).

Without a good hypothesis or theoretical basis for rationalizing, in structural or mechanistic terms, the potency and specificity of a new lead, the required incremental improvements in activity for drug development are often difficult to obtain. Indeed, Hansch,[22] who has pioneered the use of quantitative structure-activity relationships in medicinal chemistry, has commented with respect to drug development that "designing an effective inhibitor is a bit like standing outside a very large factory making a multitude of parts for a complex instrument and throwing different wrenches through the window in the hope that eventually the right wrench will hit a machine which is making components too fast or too slow and adjust it to the right speed." Fortunately with the application of mechanistic principles to inhibitor design during the past two decades, the odds of designing an effective enzyme inhibitor have markedly increased.

## Tight Binding Enzyme Complements (Group I)

Inhibitors in group I are designed to tightly bind to their enzyme target and therefore their success depends critically on the degree of complementarity between enzyme and inhibitor. For these inhibitors the goal of rational design is to optimize binding by exploiting knowledge (either inferred, or revealed by direct physical evidence of structure) of specific, primarily non-covalent interactions.

A rational basis for inhibitor design follows from the fact that enzymes bind the transition states of the reactions they catalyze far more tightly than they bind their substrates in the corresponding ground states.[23-25] Thus, an important guiding principle has been to design inhibitors that mimic hypothetical transition state structures (or high energy intermediates that lie close to the relevant transition state) (see Note 1[26]).

Examples of *transition state analogs* have been amply documented by Wolfenden[25] and Lienhard.[27] The underlying concept has had considerable impact upon current efforts in drug design through the development of tight binding inhibitors of a number of clinically relevant aspartyl proteinases (carbinol mimics of transient hydrates profoundly inhibit renin and HIV protease[28]), metalloproteinases (phosphinyl mimics of transient hydrates are effective collagenase inhibitors),[29] and of 5α-reductase (amide mimics of a transient enolate are effective inhibitors).[30] A thermodynamic criterion has been proposed for transition state analog inhibition.[31a,b]

Enzyme inhibition by transition state or reactive intermediate analogs can be complicated by the presence of diverse kinetically significant enzyme-inhibitor complexes. The behavior of relevant reversible inhibitors has been classified according to kinetic criteria by Morrison and Walsh.[31c] These authors, as well as Schloss,[31d] have highlighted the important class of inhibitors termed slow-binding inhibitors, which possess slow rates of association but high affinity for their enzyme target. Very often such inhibition takes the form schematically:

$$E + I \rightleftharpoons EI \underset{slow}{\rightleftharpoons} EI^*$$

For enzyme reactions that involve two or more substrates, analogs have been constructed that incorporate the features of more than one substrate. Such *multi-substrate analogs*,[26,31e] combining the affinities of co-substrates $S_1$ and $S_2$, for example, in a single monomolecular species $S_1$–$S_2$, would be expected to have a greater affinity for the enzyme than $S_1$ or $S_2$ alone. For the bi-substrate analog, the binding constant (or dissociation constant) of $S_1$-$S_2$ should be on the order of the product of binding constants (or dissociation constants) of $S_1$ and $S_2$ ($K_{S_1} \times K_{S_2}$). To the extent that $S_1$–$S_2$ mimics the transition state structure, and because of the entropic advantage of combining two molecular entities into one, multi-substrate analogs could conceivably bind several orders of magnitude more tightly to the enzyme than expected from the magnitude of $K_{S_1} \times K_{S_2}$.

The general principle of exploiting the simultaneous interaction of a molecule at two or more binding sites[32] has been applied by Abeles to inhibitors termed *ground state analogs*,[33] which bridge the active site and at least one other binding domain of the target enzyme. Hirulog-1 [(D-Phe)-Pro-Arg-Pro-(Gly)4-Asn-Gly-Asp-Phe-Glu-Glu-Ile-Pro-Glu-Tyr-Leu], which inhibits the thrombin-catalyzed hydrolysis of a tripeptide *p*-nitroanilide substrate, is an example of an inhibitor that interacts simultaneously with at least two distinct binding sites.[34]

This peptide consists of an active-site specificity sequence, (D-Phe)-Pro-Arg-Pro, linked by glycyl residues to a recognition sequence of an anion binding exosite (ABE; a high affinity, $K_d = 10^{-8}$ M, noncatalytic site) of α-thrombin. At micromolar concentrations, individual components of hirulog-1 (the ABE recognition and active-site inhibitory moieties) fail by themselves to inhibit thrombin, but "intact" hirulog-1 inhibits the hydrolytic activity with $K_i = 2.3$ nM.

## Mechanism-Based Inhibitors (Group II)

Whereas thermodynamic principles provide the intellectual underpinnings for strategies in Group I, the rational design of inhibitors in Group II is dependent upon principles of organic chemical reactivity. Group II strategies involve the chemical transformation of the inhibitor by its target enzyme, usually with the establishment of a stable covalent bond between enzyme and inhibitor. Inhibitors in Group II have been referred to by some authors, more or less synonomously, as *mechanism-based* or *enzyme-suicide inhibitors*. Note that the term inactivator is preferred for compounds that covalently and irreversibly inhibit their target enzyme.

The design of suicide inhibitors has attracted widespread interest among chemists with an interest in devising tactics, essentially cunning chemical tricks, designed to stabilize enzyme-inhibitor complexes or form stable adducts. The enzyme suicide designation, however, has come under mounting criticism as a misnomer. Silverman[17] points out that "the main objections are the use of an anthropomorphic term for a chemical phenomenon and the incorrect usage on semantic grounds, of the word suicide." (In light of this latter concern, "enzyme seducing agent" may be a more properly evocative term, since it is certainly the intention of the chemist to disguise the inhibitor in substrate's "clothing," and then undermine the enzyme by leading it astray into a potential energy well from which it cannot easily escape.)

The prototype of the enzyme suicide inhibitor is 3-decynoyl-*N*-cysteamine, which inactivates the enzyme ß-hydroxydecanoyl thioester dehydrase[35,36] according to the mechanistic sequence below.

Im−Enz = Imidazole ring of active−site histidine; R = $(CH_2)_2NHAc$

The actual inactivator is the reactive allenic thioester *A*. This intermediate is produced during the course of catalysis, and combines covalently and irreversibly with the target enzyme in a step that lies outside the normal catalytic sequence for this enzyme. On the basis of this example, the cardinal features of enzyme suicide inhibition are *latent reactivity*, *catalytic processing* of inhibitor to a *reactive intermediate*, and the formation of an *irreversible* adduct.

A mechanism-based inhibitor (or inactivator) is currently described as a "stable compound, which is treated as a substrate, and converted during the normal course of enzyme catalysis into a species, which without prior release from the active site, binds tightly to the enzyme."[16-18] In this expanded definition the reagent has been generalized to embrace virtually any mode of inhibition stemming from a bond-breaking step analogous to that which initiates the normal catalytic sequence.

For example, although peptidyl aldehyde[37,38] and trifluoromethyl ketone inhibitors of specific serine proteinases[11] have been regarded as transition state analogs because of their conversion to tetrahedral oxyanion intermediates (which possess features resembling the rate-determining transition state in normal catalysis), they would be encompassed by the foregoing definition of mechanism-based inhibitor.

$$X = CF_3, H$$

Oxyanion

A more complex example is that of the action of glutamine synthetase (GS) on L-methionine-*S*-sulfoximine[39], cf. phosphinothricins.[40] The enzyme, through the normal course of catalysis, produces a species which binds to it extremely tightly, but apparently non-covalently. Is methionine sulfoximine to be regarded as a suicide inhibitor or as a transition state analog? Or is it best described as an alternate substrate inhibitor that becomes tightly bound to the enzyme?

Since mechanism-based inhibitors are converted to an inhibitory species during catalysis, Group II classifications need not be exclusive of those in Group I. As defined above, mechanism-based inhibitors are not restricted from giving rise to any of the inhibitor types in Group I. This term then covers an incredibly broad and diverse group of stable compounds that qualify as mechanism-based inhibitors merely by the target enzyme's ability to produce the actual inhibitory species during catalysis.

That such a large number of substances have been subsumed under the aegis of the mechanism-based inhibitor concept, embracing many types of active agents, indicates that the classification is in need of further refinement. For example, classification with subtypes such as a *mechanism-based inhibitor/transition state analog* to describe a pseudo-substrate that is catalytically processed into a transition state analog, *not only indicates that the original reagent has been modified during the course of catalysis, but actually stipulates the type of activity directly responsible for the inhibition.*

In line with this proposal the classic example of Bloch[36] (*vide supra*), in which a reactive allene mediates the inactivation, would be described as a *mechanism-based inhibitor/reactive intermediate* in lieu of enzyme suicide inhibitor.

Further, inhibitors which form stable analogs during the normal course of catalysis, but cannot be fully processed to a dissociable product because they lack the necessary structural features, represent yet another distinct category, and have been termed *dead-end inhibitors*.[12] A good example of such an inhibitor is 5-fluorodeoxyuridylate.[41]

In the normal reaction sequence, a ternary covalent intermediate is formed from deoxyuridylate, 5,10-methylene tetrahydrofolate, and a cysteinyl sulfhydryl of the enzyme thymidylate synthetase (TS), that readily undergoes conversion to produce thymidylate. When 5-fluorodeoxyuridylate is processed, the ternary intermediate cannot proceed further along the reaction path. To parallel the normal pathway (loss of a proton), elimination of electron-deficient fluorine would be required, which is too high energy a process.

5−Fluoro−dUMP

Dead-end inhibitors should be distinguished from *alternate substrate inhibitors* in that the latter can, in principle, be fully processed to product by the target enzyme, generally through the normal catalytic sequence. As opposed to dead-end inhibitors, which simply lack the structural features for substrate turnover, alternate substrate inhibitors give rise to stable complexes because the rate of one or more steps in the normal cycle has become extremely slow. This distinction is non-trivial. Unless a step is blocked, high concentrations of an alternate substrate may be required for effective inhibition, with copious amounts of an unwanted product being generated. A challenge to medicinal chemists interested in designing alternate substrate inhibitors is to identify molecules that are rapidly processed to products by their target enzymes, and build structural elements into substrate frameworks that, dramatically and selectively, reduce the magnitude of their off-rates from these enzymes.

2-Amino-4$H$-(3,1)-benzoxazin-4-ones[42,43] are prime examples of alternate substrate inhibitors. The parent 2-aminobenzoxazinone $B$ (R = H, $R_5$ = H) is rapidly isomerized to the 2,4-(1H,3H)-quinazolinedione $Q$ by the serine proteinase, human leukocyte elastase, according to the mechanism below.

The combination of alkyl substitution of the 2-amino group (R > CH$_3$) and $R_5$ profoundly affects the overall rate of deacylation, presumably because of steric effects on O- and N-cyclization, and the hydrolysis pathway. For example, the combination of these effects in 2-(isopropylamino)-5-ethyl benzoxazinone results in an 86-fold increase in $k_{on}$ ($k_1$), and a 770-fold decrease in $k_{off}$ ($k_{-1} + k_2 + k_3$), for an overall 67,000-fold decrease in $K_i$ versus the parent.

The terms "alternate substrate inhibitor" and "dead-end inhibitor" do not indicate which types of active species are being produced and require subtypes (i.e., during the processing of these inhibitors, tight binding inhibitors of group I could, in principle, be produced). Strictly speaking, the term "alternate substrate inhibitor" should be reserved for inhibitors that exhibit measurable turnover..

In summary, catalytic processing by enzymes may serve to convert inhibitors in Group II to:

1. tight binding active-site complements (mechanism-based inhibitor/Group I inhibitor; not formulated in Table 1),
2. reactive intermediates that combine with the enzyme irreversibly in a step

that lies outside normal catalysis (mechanism-based inhibitor/reactive intermediate, or suicide inhibitor),

3. stable analogs that are unable to proceed to product because they lack the requisite functionality for further processing (dead-end inhibitors),
4. stable analogs that have the potential for conversion to product, but are trapped in potential energy wells (alternate substrate inhibitors).

## Affinity Labels (Group III)

A classical affinity label (Table 1, Group III) is an active-site-directed reagent that contains a chemically reactive group.[44-48] According to the generally accepted definition, affinity labels differ from mechanism-based inhibitors in that they covalently label enzyme in a chemical reaction without proceeding through any part of the normal catalytic cycle. In most descriptions of affinity labeling, the desired chemical reaction is usually represented as occurring subsequent to binding, but prior to initiation of the catalytic cycle.

However, other possibilities such as a direct bimolecular reaction without a preliminary binding-step, or the occurrence of a chemical reaction during catalytic processing of the reagent, need to be considered and complicate the classification of these reagents. Covalent modifications of enzymes by affinity labels that are facilitated while the enzyme is undergoing substrate turnover have been described as "syncatalytic modifications".[49]

Classical affinity labels, by virtue of their intrinsic reactivity, are distinguished from mechanism-based inhibitors. But affinity labeling need not be dependent upon chemically reactive active-site directed reagent (see Note 2), but could be driven by the potential hyper-reactivity of the enzyme target (*vide infra*).[50,51] Shaw has developed a number of active-site-directed reagents (**1–3**), which exhibit large rate accelerations in displacement reactions with the cysteine proteinase cathepsin B (EC 3.4.22.1), compared to the corresponding reaction with a model thiol (Table 2).

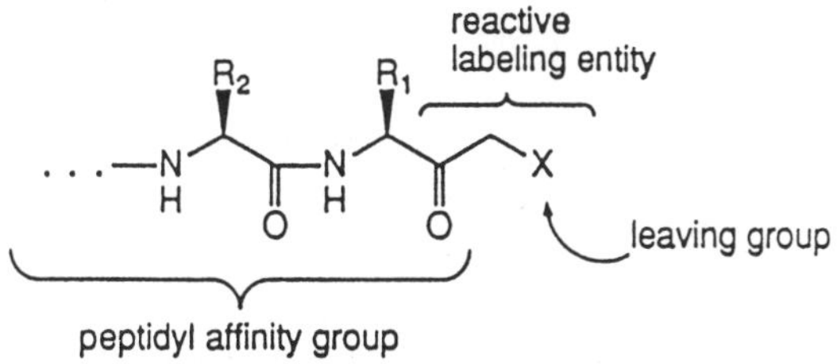

**Figure 1.** General structure of a peptide-based affinity label, containing the leaving group 'X', which is attached to a saturated center vulnerable to nucleophilic attack.

***Table 2.*** Examples of Inhibitors of Cysteine Proteinases

1. Z-Phe-Ala-CH$_2$X  $\xrightarrow{\text{E-SH}}$  Z-Phe-Ala-CH$_2$-S-E      (Ref. 51)

X = F, Cl, Br, R$_1$S$^+$R$_2$, N$_2^+$

2. RCHO  $\xrightarrow{\text{E-SH}}$  R$-\overset{\overset{\displaystyle OH}{|}}{\underset{\underset{\displaystyle H}{|}}{C}}-$S-E      (Ref. 87)

3. RCH=N-NH$\overset{\displaystyle O}{\overset{\|}{\diagup}}$NH$_2$  $\xrightarrow{\text{E-SH}}$  RCH-NH-NH$\overset{\displaystyle O}{\overset{\|}{\diagup}}$NH$_2$      (Ref. 71)
                                                                         S-E

4. Ph-CO-NH-CH$_2$ $-$C$\equiv$N  $\xrightarrow{\text{E-SH}}$  Ph-CO-NH-CH$_2\overset{\diagup NH}{\underset{\diagdown S\text{-}E}{}}$      (Ref. 88–90)

5. Ac-Phe-NH $\diagdown\diagup\overset{X}{\diagup}$  $\xrightarrow{\text{E-SH}}$  Ac-Phe-NH $\diagdown\diagup\overset{X}{\underset{S\text{-}E}{\diagdown}}$      (Ref. 91)

6.  $^-$O$_2$C$\diagdown\diagup\overset{H}{\diagup}\overset{O}{\triangle}\overset{H}{\underset{R'}{\diagdown}}$  $\xrightarrow{\text{E-SH}}$  $^-$O$_2$C$\overset{H}{\underset{S\text{-}E}{\diagup}}\diagdown\diagup\overset{H}{\underset{OH}{\diagdown}}$R'      (Ref. 92)

R ' = -CO-NH-CH(iBu)CO-NH(CH$_2$)$_3$-NHC(=NH)NH$_2$

7. RNHC$_2$H$_4$S-S-C$_2$H$_4$NH$_2$  $\xrightarrow{\text{E-SH}}$  RNHC$_2$H$_4$-S-S-E      (Ref. 93–94)

8. RCO-NH-O-COAr  $\xrightarrow{\text{E-SH}}$  R-CO-NH-S-E      (Ref. 69)

Theoretically, selective affinity labels could be fashioned which are essentially inert chemically at physiological temperatures and pH, if the huge rate accelerations observed for enzyme-catalyzed reactions such as amide hydrolysis,[52,53] could be achieved for displacement reactions between enzyme and certain active-site-directed reagents. The discovery that peptidyl fluoromethyl ketones (4) are potent inactivators of cathepsin B,[54–56] but exhibit merely a small fraction (0.2%) of the chemical reactivity of the corresponding peptidyl chloromethyl ketones toward the model thiol, glutathione, at pH 6.5,[55] is tangible evidence that reagents of modest chemical reactivity can be useful inactivators of this enzyme.

1, $Y = CH_2Cl$         4, $Y = CH_2F$

2, $Y = CHN_2$         5, $Y = CH_2OCOAr$

3, $Y = CHS(CH_3)_2$

## II. ACTION OF PEPTIDYL ACYLOXYMETHYL KETONES ON PAPAIN-TYPE CYSTEINE PROTEINASES

In light of the powerful nucleophilicity of the active-site cysteine residue,[57] affinity labels containing weak nucleofuges seemed to be plausible candidates for effecting selective irreversible inhibition of cathepsin B, a cysteine proteinase of the papain family.[58,59] This enzyme has been implicated in a variety of disease states including proteinuria in glomerular disease,[60] osteoclastic bone resorption,[61-63] tumor metastasis,[64-66] tissue damage in myocardial infarction,[67] and muscle wasting in Duchenne muscular dystrophy.[68]

Accordingly, we describe the use of peptidyl acyloxymethyl ketones (**5**), conceived as affinity labels of low chemical reactivity, that are potent and highly selective inactivators of cathepsin B.[69] These reagents, besides containing a variable structural element in the form of an affinity group, incorporate the novel feature of an aryl carboxylate leaving group that can be systematically varied to provide a rational and exquisitely sensitive means of controlling the potency and chemical reactivity of an affinity label.

### A. Peptidyl Binding Determinants for Cathepsin B Inactivation by Acyloxymethyl Ketones. The Affinity Group

Acyloxymethyl ketones with peptide recognition elements that satisfy the specificity requirements of cathepsin B, span a range of six orders of magnitude in their ability to inactivate this enzyme (Table 3). Our choice of peptidyl recognition elements has been dictated by the high affinity of papain-type cysteine proteinases for aromatic amino acids such as phenylalanine at the P2 position of substrate,[70] and the work of Shaw[51] specifying dipeptidyl affinity groups for cathepsin B. For dipeptidyl acyloxymethyl ketones, a variety of L-α-amino acid residues are well-tolerated at the $S_1$ position of cathepsin B (Table 4). The same dipeptidyl frameworks appropriately linked to chloromethyl-, diazomethyl-, or acyloxymethyl-ketone functions, give rise to inactivators of cathepsin B, although there are differences in the rank order of potency within each series.[71] Very dramatic inhibition is observed with Z-Phe-Lys-CH₂OCO-(2,4,6-Me₃)Ph (**35**), in accord with the affinity of cathepsin B for positively charged amino acids at the $P_1$ position.

Mesitoyloxy compounds with an amino acid of D-configuration (54), or with methyl substitution at the $P_1$ amido nitrogen (i.e., the sarcosine analog 53) are not active against cathepsin B. The peptidyl ketones with β- or γ-amino acids at $P_1$, 51 and 52, respectively, are also inactive.

Furthermore, acyloxymethyl ketones 56 and 57, bearing peptidyl functionality that lack affinity for cathepsin B, (Table 4), are not effective inactivators of this enzyme, despite the presence of a carboxylate leaving group of low $pK_a$ (*vide infra*). The foregoing results and the fact that 50, which possesses only a single amino acid residue, fails to inactivate cathepsin B, clearly demonstrate that a dipeptidyl complement to enzyme $S_1$ and $S_2$ subsites is a critical component of these types of inactivators. The inertness of 50 toward cathepsin B stands in sharp contrast to the action of the corresponding fluoromethyl ketone,[55] which is reported to inactivate both cathepsin B and the serine proteinase chymotrypsin.

It is also noteworthy that the acyloxymethyl ketones, Z-Phe-Lys-CH$_2$OCO-(2,4,6-Me$_3$)Ph (35) (a powerful inactivator of cathepsin B, but showing only $K_i$ = $17 \pm 1$ μM versus trypsin), and Z-Ala-Ala-Pro-Val-CH$_2$OCO-(2,6-(CF$_3$)$_2$)Ph (57) (which bears an affinity group complementary to human leukocyte elastase[72]), do not exhibit significant time-dependent activity toward their serine proteinase "targets." Apparently, serine proteinases lack the nucleophilicity required for facile expulsion of a carboxylate leaving group in an $S_N2$-like displacement. This point underscores the ability of acyloxymethyl ketones as enzyme inactivators to discriminate between serine and cysteine proteinases despite their similar catalytic mechanisms. We have also noted that acyloxymethyl ketone inactivators of cathepsin B do not exhibit time dependent activity against either aspartyl- or metallo-proteinases (D.H. Pliura and Z. Yuan, unpublished results), indicating that such compounds are specific inactivators of cysteine proteinases.

### B. Structural Determinants of the Rate of Cathepsin B Inactivation. The Departing Group

A characteristic feature of cysteine proteinases of the papain family is their nucleophilicity.[73–75] Even against very simple reagents such as iodoacetamide, papain exhibits powerful nucleophilicity at pH 6.0 rivaling the reactivity of the thiolate of glutathione at pH 11.0.[57] Truly spectacular reactivity has been recorded for alkylations involving even sluggish nucleofuges, if the specificity requirements of cysteine proteinases are satisfied. For example, Ala-Phe-Lys-CH$_2$F (2), is a potent inactivator of cathepsin B, reacting essentially $10^8$ times faster with this enzyme than with glutathione at pH 6.4.[76]

The rich lore of organic chemistry provides a logical framework for developing new modalities of cysteine proteinase inhibition based on reactions characteristic of thiol nucleophiles. Potentially reactive moieties linked to affinity groups for

**Table 3.** Rates of Cathepsin B Inactivation by Peptidyl Acyloxymethyl Ketones: Variation of the Leaving Group[a]

| No. | Compound R | k/K ($M^{-1}s^{-1}$) | $pK_a$[b] |
|---|---|---|---|
| | **Z–Phe–Ala–CH₂OCO–R:** | | |
| *benzoyloxymethyl ketones:* | | | |
| 6 | 2,6-(CF₃)₂-Ph | 1,600,000. | 0.58[c] |
| 7 | 2,6-Cl₂-Ph | 690,000. | 1.59 |
| 8 | C₆F₅ | 520,000. | 1.48[d] |
| 9 | 2,6-Me₂-4-COOMe-Ph | 58,000. | 2.67[c] |
| 10 | 2,5-(CF₃)₂-Ph | 38,000[e] | 2.63[c] |
| 11 | 2,6-F₂-Ph | 26,000. | 2.24[f] |
| 12 | 3,5-(CF₃)₂-Ph | 22,000. | 3.18[c] |
| 13 | 3,5-(NO₂)₂-Ph | ca. 19,000.[g] | 2.79 |
| 14 | 2-CF₃-Ph | 17,000.[h] | 2.49[c] |
| 15 | 2,4,6-Me₃-Ph | 14,000.[h] | 3.45 |
| 16 | 2,6-Me₂-Ph | 14,000. | 3.35 |
| 17 | 2,4,6-iPr₃-Ph | 3,800[j] | |
| 18 | 3,4-F₂-Ph | 630.[j] | 3.67[c], 3.79[d] |
| 19 | 4-NO₂-Ph | 610. | 3.43 |
| 20 | 3-CF₃-Ph | 420.[j] | 3.76[c], 3.75[k] |
| 21 | 2,6-(OMe)₂-Ph | 300. | 3.44 |
| 22 | 4-F-Ph | 290.[h] | 4.15 |
| 23 | 4-CN-Ph | 280.[j] | 3.50[k] |
| 24 | 4-CH₃-Ph | 260.[j] | 4.37 |
| 25 | 3,5-(OH)₂-Ph | 140.[j] | 4.04 |
| 26 | Ph | 90.[j] | 4.20 |
| 27 | 4-CF₃-Ph | 80.[j] | 3.67[l] |
| 28 | 3,5-Me₂-Ph | 80.[j] | 4.30 |
| 29 | 4-OMe-Ph | NTD.[i] | 4.50 |
| *alkanoyloxymethyl ketones:* | | | |
| 30 | CH₃ | 140. | 4.76 |
| 31 | CMe₃ | 330.[j] | 5.03 |
| 32 | CH(CH₂CH₃)₂ | 70. | 4.73 |
| 33 | CH₂OCH₃ | 240. | 3.57 |
| | **Z–Phe–Lys–CH₂OCO–R:** | | |
| 34 | 2,6-(CF₃)₂-Ph (HCl salt) | >2,000,000. | 0.58[c] |
| 35 | 2,4,6-Me₃-Ph (TFA salt) | 230,000. | 3.45 |
| 36 | Ph (TFA salt) | 9,200. | 4.20 |
| 37 | 4-OMe-Ph (TFA salt) | 660. | 4.50 |

[a]Conditions: bovine spleen cathepsin B, 100 mM potassium phosphate, 1.25 mM EDTA, 1 mM dithiothreitol, pH 6.0, 25°C, under argon. The parameter "k/K" is the second order rate constant ($k_{inact}/K_{inact}$), as determined from a hyperbolic (saturation) or linear fit, except as noted. Standard errors for k/K ≤ 15%, except as noted.
[b]$pK_a$ (aqueous) of the acyloxy group (RCOOH); values are from Serjeant & Dempsey[95] or Kortum et al.,[96] except as noted.
[c]$pK_a$ determined by UV measurement (or in the case of 9, by HPLC) (this work).
[d]$pK_a$ from Strong et al.[97]
[e]Standard error 20–30%.
[f]$pK_a$ from Strong et al.[97]
[h]Standard error 15–20%.
[i]NTD: No time dependence observed near the compound solubility limit, over a 10–20 min period.
[j]Second order rate constant (k/[I]) determined at one inhibitor concentration near the solubility limit.
[k]$pK_a$ from Ludwig et al.[99]
[l]Calculated $pK_a$.[100]

**Table 4.** Rates of Cathepsin B Inactivation by Peptidyl Acyloxymethyl Ketones: Variation of the Affinity Group[a]

| No. | Compound R' | k/K (M$^{-1}$s$^{-1}$) |
|---|---|---|
| *R'-CH$_2$OCO-(2,6-(CF$_3$)$_2$)Ph:* | | |
| 38 | Z-Phe-Gly | 4,000,000. |
| 39 | Z-Phe-Cys(SBn) | 2,900,000. |
| 40 | Z-Phe-Ser(OBn) | 2,600,000. |
| 34 | Z-Phe-Lys (HCl salt) | >2,000,000. |
| 6 | Z-Phe-Ala | 1,600,000. |
| 41 | Z-Phe-Thr(OBn) | >100,000.[b] |
| 42 | Z-Phe-D-Ala (diastereoisomer of 6) | 7,100.[c] |
| 56 | Z-Pro-Val-CH$_2$OCO-(2,6-(CF$_3$)$_2$)Ph | 30.[c] |
| 57 | Z-Ala-Ala-Pro-Val-CH$_2$OCO-(2,6-(CF$_3$)$_2$)Ph | 200. |
| | | |
| *R'-CH$_2$OCO-(2,4,6-Me$_3$)Ph:* | | |
| 35 | Z-Phe-Lys (TFA salt) | 230,000. |
| 43 | Z-Phe-Thr(OBn) | 30,000. |
| 44 | Z-Tyr(OMe)-Ala | 19,000[d] |
| 45 | Z-Phe-Ser(OBn) | 17,000.[d] |
| 15 | Z-Phe-Ala | 14,000.[d] |
| 46 | Z-Phe-Gly | 9,900. |
| 47 | Z-Phe-Phe | 4,300. |
| 48 | Z-Phe-Cys(SBn) | 4,100. |
| 49 | H-Phe-Ala (HCl salt) | 740. |
| 4 | PhCO | NTD.[e] |
| 50 | Z-Phe | NTD. |
| 51 | Z-Phe-β-Ala | NTD. |
| 52 | Z-Phe-GABA | NTD. |
| 53 | Z-Phe-Sar | NTD. |
| 54 | Z-Phe-D-Ala (diastereoisomer of 15) | NTD. |
| *isomer of 15:* | | |
| 55 | Z-Phe-Ala-OCH$_2$CO-(2,4,6-Me$_3$)Ph | NTD. |

[a]Conditions: bovine spleen cathepsin B, 100 mM potassium phosphate, 1.25 mM EDTA, 1 mM dithiothreitol, pH 6.0, 25°C, under argon. The parameter "k/K" is the second order rate constant ($k_{inact}/K_{inact}$), as determined from a hyperbolic (saturation) or linear fit, except as noted. Standard errors for k/K ≤ 15%, except as noted.
[b]Compound instability was evident.
[c]Second order rate constant (k/[I]) determined at one concentration near the inhibitor solubility limit.
[d]Standard error 15–20%.
[e]NTD: No time dependence observed near the compound solubility limit, over a 10–20 min period.

cysteine proteinases have proven to be effective ways of achieving inhibition (Table 2).

However, most organic chemical reactions are conducted in non-aqueous media and the most valuable reactions synthetically tend to be those that proceed rapidly in high yield under ambient conditions. Because of the stringent specificity requirements for clinically relevant enzyme inhibitors, the most useful chemical reactions would be those that take place infinitesimally slowly (or not at all) at physiological temperatures and pH in aqueous media, but are likely to be vastly accelerated by the target enzyme. The use of $S_N2$ displacement reactions involving inhibitors with variable leaving groups as a means of fine-tuning affinity labels to the practical demands of drug development, would seem to be a suitable approach to controlling reagent reactivity.

Reactivity in $S_N2$ displacement reactions is generally thought to increase with the acidity of the leaving group, within a closely related series of substrates in which the same type of bond is being cleaved.[77] To optimize the potency and selectivity of affinity labels, it would be advantageous to systematically vary the nucleofugality of a leaving group by using substituent effects. The effect of substituents on rates has often been correlated with the acid strengths of benzoic acids.[78–80]

Carboxylates are generally regarded as feeble leaving groups in $S_N2$ reactions unless highly nucleophilic conditions are employed.[81] Larsen and Shaw,[82] who investigated the possibility of employing departing groups other than bromide or chloride to limit side reactions of affinity labeling reagents, could not observe any time-dependent activity against chymotrypsin with Z-PheCH$_2$OAc. It seemed reasonable to us that, for a given dipeptidyl affinity group, a threshold value for enzyme inactivation might be achieved by systematically varying the pK$_a$ of the carboxylate departing group in the presence of a powerfully nucleophilic cysteine proteinase.

Consequently, we have investigated Z-Phe-Ala acyloxymethyl ketones with leaving groups spanning a range of pK$_a$ values from 0.6 to 5.0 (Table 3). The activity of these inhibitors was found to be exquisitely sensitive to the nature of the carboxylate leaving group, as a strong correlation of the logarithm of the second-order inactivation rate with carboxylate pKa was uncovered (Fig. 2, Eq. 1).

$$\log(k/K) = -1.1\ (\pm 0.1) \times pKa + 7.2\ (\pm 0.4);\ r^2 = 0.82;\ n = 26 \qquad (1)$$

There are two obvious ways by which ring substituents could affect the second-order rate constant, $k/K$, and produce the linear correlation shown in Figure 2 (Eq. 1). Substitution of the benzoate framework clearly affects the strength of the RCH$_2$-OCOAr bond, but the sensitivity of the rate to substituent effects depends upon whether this bond is broken in an "early or late" rate-determining transition state. Substituents could also influence the second-order rate by altering the position of equilibrium (or steady state) between free enzyme and inhibitor versus the tetrahedral intermediate **I** (Scheme 1). Electron withdrawal should destabilize the ketone, and thus promote the formation of the tetrahedral intermediate **I**. Such an effect on $K_i$ has been documented by McMurray and Dyckes[83] for the action of α-substituted ketones on trypsin. Whether arylacyl ring substituent effects can be transmitted to a remote keto group and give rise to the large changes observed in Table 3 by virtue of their effect on a dissociation constant, is open to question. Although an ambiguous mechanistic rationale for the effects of ring substituents on the second order rate of inactivation of cathepsin B remains to be established,

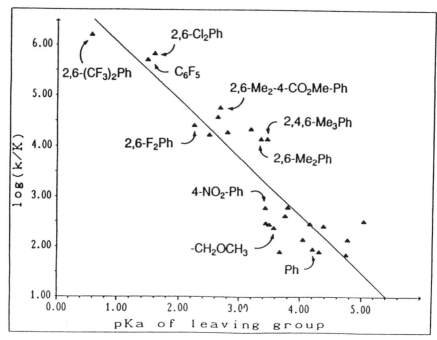

**Figure 2.** Correlation of the rate of cathepsin B inactivation [log(k/K)] by Z–Phe–Ala–CH₂OCOR with pKₐ of the leaving group RCOOH. Data are from Table 3, for all Z–Phe–Ala compounds except **17** and **29**.

the correlation of leaving group pKₐ and rate in Figure 2 provides a useful predictive framework.

For Z-Phe-Ala acyloxymethyl ketones the second-order rate of inactivation falls off to low values when leaving group pKₐ > 4, yet this threshold can be overcome dramatically by using a tighter binding peptidyl moiety. For example, the Z-Phe-Ala-anisate **29** does not exhibit time-dependent activity in our assay, but the anisate linked to the Z-Phe-Lys framework (**37**) inactivates cathepsin B at a rate essentially identical to that observed for the 4-nitrobenzoate in the Z-Phe-Ala series (**19**). The ability to exploit a tighter binding affinity group, to compensate for the effect of down-regulating the nucleofugality of the leaving group, should be a useful design principle for fine-tuning the properties of an inhibitor.

## C. Mechanistic Considerations

We have previously shown that the acyloxymethyl ketone **46** with papain, a model cysteine proteinase, forms an α-thiomethyl ketone adduct.[69] Three mechanisms can be envisaged by which displacement of carboxylate occurs to give

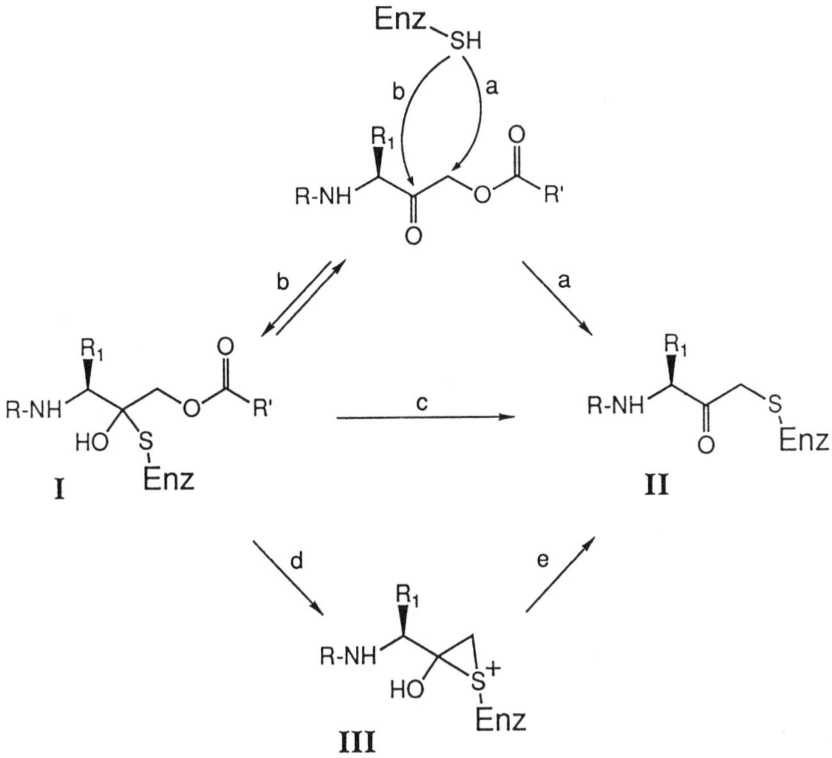

**Scheme 1.** Mechanistic considerations for the inactivation process.

such an adduct **II** (Scheme 1). Path a involves direct displacement of carboxylate in a single step reaction. Path b features a tetrahedral intermediate which might mediate the formation of product by either of two paths (c or d) that differ only in the timing of bond breaking and bond making processes (Scheme 1).

One possibility, involving neighboring group participation by the sulfur atom, features a discrete episulfonium ion intermediate **III** *en route* to product (two step migration). The other is characterized by migration of the sulfur group in a single step concomitant with loss of carboxylate and regeneration of the ketone carbonyl (step c, Scheme 1). The pathway involving the formation of the episulfonium ion intermediate **III** would seem to be the less likely alternative (unless steps d-e are enzyme activated) because of the known chemical stability of simple α-acyloxyl thioethers.[84] Speculation then centers around whether the tetrahedral intermediate **I** lies on the reaction coordinate leading to adduct **II** (steps b-c), or whether direct displacement of carboxylate occurs (step a, Scheme 1).

The latter mechanistic point bears on the classification of acyloxymethyl

$$RCH_2C(O)CH_2X + ESH \xrightarrow{\quad S_N2 \quad} RCH_2C(O)CH_2SE \qquad (2)$$

$$RCH_2C(O)–CH_2X + ESH \rightarrow RCH_2C(OH)(SE)–CH_2X \qquad (3)$$
$$\rightarrow RCH_2C(O)CH_2SE$$

$$RCH_2C(O)–CH_2X + R'S(H) \xrightarrow{\quad S_N2 \quad} RCH_2C(O)CH_2SR' \qquad (4)$$

$$RCH_2C(O)–CH_2X + R'S(H) \rightarrow RCH_2C(OH)(SR')–CH_2X \qquad (5)$$
$$\rightarrow RCH_2C(O)–CH_2SR'$$

**Scheme 2.** Potential enzyme pathways (Eqs. 1 and 2). Potential chemical pathways (Eqs. 4 and 5).

ketones as inhibitors of cysteine proteinases and is not entirely straightforward. For a displacement reaction involving an electrophile $RCH_2COCH_2X$ and the cysteine proteinase ESH, there are at least six distinct cases that can be discerned, depending upon (1) the chemical reactivity of the inactivator at physiological temperatures and pH, (2) the mechanism of the enzymatic displacement reaction, and (3) the mechanism of the displacement reaction between a model cysteine thiol and the inactivator.

Based on Scheme 2 these possibilities are discussed in Cases I–VI below, keeping in mind that catalysis of hydrolytic cleavage of normal substrates by cysteine proteinases, is initiated by formation of a tetrahedral intermediate between enzyme and substrate.[2] In Scheme 2, Equations (2) – (5), $RCH_2COCH_2X$ = inhibitor, ESH = cysteine proteinase and, R'SH = model thiol. For the enzyme reactions [Eqs. (2) and (3)], relevant Michaelis-Menten complexes are not shown. Chemical reactivity [Eqs. (4) and (5)] refers to reactions with model nucleophiles under physiological conditions of temperature and pH.

*Case I.* Enzyme inactivation occurs via an $S_N2$ displacement mechanism (Eq. 2); the inactivator, is chemically reactive via an $S_N2$ displacement mechanism (Eq. 4).

The reagent should be considered a classical, reactive affinity label, since both enzyme and non-enzyme nucleophiles react with a chemically reactive reagent by analogous $S_N2$ mechanisms.

*Case II.* Enzyme inactivation occurs via an $S_N2$ displacement mechanism (Eq. 2); the inactivator is chemically reactive only via the pathway involving a tetrahedral intermediate (Eq. 5).

By virtue of its intrinsic reactivity the inhibitor could be regarded as an affinity label, but the chemical pathway is distinct from the enzyme inactivation pathway.

*Case III.* Enzyme inactivation occurs via formation of a tetrahedral intermediate (Eq. 3); the inactivator is chemically reactive via an analogous pathway (Eq. 5).

In this case latent reactivity is not unmasked. The inactivator is intrinsically reactive. Therefore, the reagent must be regarded as a reactive affinity label (not dissimilar from Case I except for mechanism).

*Case IV.* Enzyme inactivation occurs via formation of a tetrahedral intermediate (Eq. 3); the inhibitor is chemically reactive only via an $S_N2$ displacement mechanism (Eq. 4).

The inactivator would appear to be a reactive affinity label, but a latent reactive mode is unmasked by the enzyme.

*Case V.* Enzyme inactivation occurs via formation of a tetrahedral intermediate (Eq. 3); the inactivator is chemically unreactive at physiological temperatures and pH.

Such reagents would then fulfill the criteria employed for enzyme suicide inhibitors[16,17] in that, (1) *their reactivity is latent*, a consequence of the reactive possibilities inherent in the $\alpha$-substituted hemithioketal intermediate, and (2) *the enzyme inactivation pathway parallels that of normal substrate turnover* by initially producing a tetrahedral intermediate.

*Case VI.* Enzyme inactivation occurs via an $S_N2$ displacement mechanism (Eq. 2). The inactivator is chemically unreactive under physiological conditions of temperature and pH.

Since, in case VI, the enzyme is not inactivated during the normal course of catalysis, the reagent cannot be an enzyme suicide reagent. Neither is it a classical affinity label because it is not intrinsically chemically reactive. Such a reagent should be described as a quiescent affinity label, in that it is capable of reacting facilely only upon binding to its target enzyme, and thus it is "quiescent" in the presence of other bionucleophiles lacking a complementary surface. Binding to the enzyme is tantamount to turning on a switch which renders the inhibitor vulnerable to nucleophilic attack by the enzyme (see Note 3).

This new type of affinity label, like an enzyme substrate, exploits intrinsic binding to effect chemical transformations that would otherwise be excruciatingly slow without some form of catalysis. Whereas suicide substrates utilize binding and normal catalysis to produce reactive intermediates which cause enzyme inactivation, quiescent affinity labels exploit intrinsic binding to lower the overall free energy of activation of an "aberrant" chemical path, leading to facile enzyme inactivation. Perhaps another example that can be encompassed by the latter principle has been reported by Poulter and co-workers,[85] who have described an

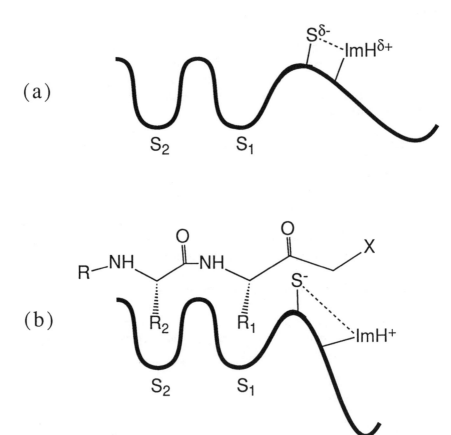

**Figure 3.** Schematic representation of a hypothesis for enhanced active-site thiolate reactivity resulting from tighter inhibitor binding.

allyl fluoride inactivator of the enzyme isopentenyldiphosphate:dimethylallyl diphosphate isomerase.

For cysteine proteinases, in addition to factors normally associated with intrinsic binding,[86] one can envisage a viable model which accounts for enhanced reactivity with tighter binding, assuming that thiolate nucleophilicity is a critical determinant of enzyme reactivity. A reasonable postulate is that tighter binding leads to increased separation of charge in the thiolate-imidazolium ion-pair (Fig. 3). The freer the thiolate, the greater is the nucleophilicity of the enzyme, and the more rapidly displaced is a given leaving group from the inactivator. The implication of this hypothesis is that the stabilization derived from affinity groups which are perfect active-site complements to enzyme active sites, can be used to drive chemical reactions which are unprecedented in the annals of organic chemistry under physiological conditions of temperature and pH, yet distinct from those which the enzyme has evolved to catalyze.

# SUMMARY

Strategies for the design of enzyme inhibitors which draw on chemical principles have had profound impact on the field of medicinal chemistry. Most of these strategies can be classified according to the underlying theoretical basis by which they lead to tight binding inhibitors or stable adducts between enzyme and inhibitor. In contrast to mechanism-based inhibitors, classical affinity labels are chemically reactive, a property which has severely restricted their clinical utility. Comparatively little progress has been made in designing affinity labels with difficultly displaceable groups which might tap the potential reactivity of a specific enzyme target without affecting other macromolecules.

Peptidyl acyloxymethyl ketones are noteworthy because they represent prime examples of chemically stable, highly potent, and specific enzyme inactivators with variable structural elements in both the affinity and departing groups. A broad range of cathepsin B inactivation rates, in some cases exceeding $10^6$ $M^{-1}s^{-1}$, are obtainable by systematically varying these structural elements. Such peptidyl acyloxymethyl ketones are specific inactivators of papain-type proteinases, by virtue of the extraordinary ability of these enzymes to displace weak nucleofuges such as carboxylate leaving groups. This type of reagent extends the boundaries of inhibitor design to a new generation of "affinity labels", but its classification within a mechanistic framework involves subtle considerations of reagent chemical reactivity, and enzyme mechanisms.

# ACKNOWLEDGMENTS

I am grateful to several people who have been responsible for the execution of the work described herein. Dr. Roger A. Smith's efforts are especially noteworthy in the organic chemical domain as are Leslie Copp's contributions to the enzymology. I thank both these individuals and Dr. Diana H. Pliura for valuable suggestions and helpful discussions regarding the manuscript. Peter Coles' support in organic synthesis has also been invaluable. Special thanks are due to Dr. Robin W. Spencer for initial contributions to the biochemistry, to Valerie J. Robinson for NMR investigations and to Dr. Henry W. Pauls for synthetic contributions. I am also grateful for the assistance of students from the University of Waterloo cooperative program (S. Donnelly, A. Erhardt, L. Srait-Gardner, K. Streutker, S. Lukovits, H. Le Blanc, and especially S. B. Heard for his key contributions to this project).

# NOTES

1. Byers has pointed out that an enzyme may in principle stabilize "reaction intermediates" to a greater extent than it stabilizes transition states of a reaction. Accordingly, structural analogs of such reactive intermediates may serve as potent inhibitors.[26]

2. The term "chemically reactive" is ambiguous. To medicinal chemists *reactive* connotes the likelihood of *non-specific* irreversible reactions under physiological conditions during the lifetime of a drug *in vivo*. Thus, molecules containing a chloromethyl ketone functionality that react with glutathione under such conditions are considered to be chemically reactive, and would be expected to react with diverse thiol(ate) functions in a biochemical milieu. On the other hand, certain acyloxymethyl ketones, which contain sluggish leaving groups, and are demonstrably stable in the pH range 1–7 with standard bionucleophiles, would be considered unreactive.

3. For inhibitors that covalently label their target enzyme in a step that lies outside the normal catalytic sequence, the point at which the chemical reactivity is revealed defines the nature of the reagent. It may be useful to regard all such reagents as "affinity labels" insofar as they consist of an affinity group linked to a (potentially) reactive labeling entity. The feature that distinguishes them would be the origin of their reactivity, i.e., whether they are reactive, latent or quiescent in the sense proposed above. Chemically reactive inhibitors such as chloromethyl and bromomethyl ketones would be regarded as *reactive affinity labels* because of their intrinsic reactivity. Suicide inhibitors, because they require catalytic processing would be termed *latent affinity labels*. Reagents which become reactive upon binding to the target enzyme (they are neither chemically reactive nor processed as substrates) would be designated *quiescent affinity labels*. Chemically stable peptidyl acyloxymethyl ketones could conceivably inactivate cathepsin B as latent affinity labels, or as quiescent affinity labels.

# REFERENCES

1. Gale, E.F.; Cundliffe, E.; Reynolds, P.E.; Richmond, M.H.; Waring, M.J. *The Molecular Basis of Antibiotic Action*; John Wiley and Sons: London, 1981, Chapter 1.
2. Page, M.I.; Williams, A., Eds. *Enzyme Mechanisms*; Royal Society of Chemistry: London, 1987.
3. Walsh, C. *Enzymatic Reaction Mechanisms*, W.H. Freeman: San Francisco, 1979.
4. Beddell, C.R. *Chem. Soc. Rev.* **1984**, *13*, 279–319.
5. Hol, W.G. *Angew. Chem.* **1986**, *98*, 765–777.
6. Cohen, N.C.; Blaney, J.M.; Humblet, C.; Gund, P.; Barry, D.C. *J. Med. Chem.* **1990**, *33*, 883–894.
7. Wüthrich, K. *Science* **1989**, *432*, 45–50.
8. Glen, R.C. *Drug News and Perspectives* **1990**, *3*, 332–336.
9. Marshall, G.R. *Annu. Rev. Pharmacol. Toxicol.* **1987**, *27*, 193–213.
10. Meyer, Jr., E.F.; Radhakrishnan, R.; Cole, G.M.; Presta, L.G. *J. Mol. Biol.* **1986**, *189*, 533–539.
11. Brady, K.; Wei, A.; Ringe, D.; Abeles, R.H. *Biochemistry* **1990**, *29*, 7600–7607.
12. Rando, R.R. *Pharm. Rev.* **1984**, *36*, 111–142.
13. Walsh, C. *Tetrahedron* **1982**, *38*, 871–909.
14. Walsh, C. *Annu. Rev. Biochem.* **1984**, *53*, 493–535.
15. Abeles, R.H.; Maycock, A.L. *Acc. Chem. Res.* **1976**, *9*, 313–319.
16. Silverman, R.B. *Mechanism-based Enzyme Inactivation: Chemistry and Enzymology*; CRC Press: Boca Raton, FL, 1988, Vol. 1–2.
17. Silverman, R.B. *Mechanism-based Enzyme Inactivation: Chemistry and Enzymology*; CRC Press: Boca Raton, FL, 1988b, Vol. 1, pp. 3–4.
18. Bey, P. *Recontres Int. Chem. Ther.* **1988**, *16*, 111–122.
19. Woods, D.D. *Brit. J. Exp. Path.* **1940**, *21*, 74–90.

20. Fildes, P. *Lancet* **1940**, *1*, 955–957.
21. Woolley, D.W. *A Study of Antimetabolites*; John Wiley: New York, 1952.
22. Hansch, C. *J. Chem. Ed.* **1974**, *51*, 360–365.
23. Pauling, L. *Nature (London)* **1948**, *161*, 707.
24. Jencks, W.P. *Current Aspects of Biochemical Energetics*; Kaplan, N.O.; Kennedy, E.P., Eds; Academic Press: New York, 1966; pp. 273–294.
25. Wolfenden, R. *Annu. Rev. Biophys. Bioeng.* **1976**, *5*, 271–306.
26. Byers, L.D. *J. Theor. Biol.* **1978**, *74*, 501–512.
27. Lienhard, G.E. *Science* **1973**, *180*, 149–154.
28. Ashorn, P.; McQuade, T.J.; Thaisrivongs, S.; Tomasselli, A.G.; Tarpley, W.G.; Moss, B. *Proc. Natl. Acad. Sci. USA* **1990**, *87*, 7472–7476.
29. Henderson, B.; Docherty, A.J.P.; Beeley, N.R.A. *Drugs of the Future* **1990**, *15*, 495–508.
30. Rasmusson, G.H.; Reynolds, G.F.; Steinberg, N.G.; Walton, E.; Patel, G.F.; Liang, T.; Cascieri, M.A.; Cheung, A.H.; Brooks, J.R.; Berman, C. *J. Med. Chem.* **1986**, *29*, 2298–2315.
31. (a) Bartlett, P.A.; Marlowe, C.K. *Biochemistry* **1983**, *22*, 4618–4624. (b) Bartlett, P.A.; Marlowe, C.K. *Science* **1987**, *235*, 569–571. (c) Morrison, J.F.; Walsh, C.T. *Enzymol. Relat. Areas Mol. Biol.* **1987**, *61*, 201–301. (d) Schloss, J.V. *Acc. Chem. Res.* **1988**, *21*, 348–353. (e) Stark, G.R.; Bartlett, P.A. *Pharmac. Ther.* **1983**, *23*, 45–78.
32. Jencks, W.P. *Proc. Natl. Acad. Sci. USA* **1981**, *78*, 4046–4050.
33. Abeles, R.H. *Drug Dev. Res.* **1987**, *10*, 221–234.
34. Maraganore, J.M.; Bourdon, P.; Jablonski, J.; Ramachandran, K.L.; Fenton II, J.W. *Biochemistry* **1990**, *29*, 7095–7101.
35. Helmkamp, G.; Rando, R.; Brock, D.; Bloch, K. *J. Biol. Chem.* **1968**, *243*, 3229–3231.
36. Endo, K.; Helmkamp, Jr., M.; Bloch, K. *J. Biol. Chem.* **1970**, *245*, 4293–4296.
37. Delbaere, L.T.J.; Brayer, G.D. *J. Mol. Biol.* **1985**, *183*, 89–103.
38. Thompson, R.C.; Bauer, C.-A. *Biochemistry* **1979**, *18*, 1552–1558.
39. Wedler, F.C.; Sugiyama, Y.; Fisher, K.E. *Biochemistry* **1982**, *21*, 2168–2177.
40. Logush, E.W.; Walker, D.M.; McDonald, J.F.; Leo, G.C.; Franz, J.E. *J. Org. Chem.* **1988**, *53*, 4069–4074.
41. Santi, D.V.; Wataya, Y.; Matsuda, A. In: *Enzyme Activated Irreversible Inhibitors*; Seiler, N.; Jung, M.J.; Koch-Weser, J., Eds; Elsevier/North Holland: Amsterdam, 1978; pp. 297–304.
42. Krantz, A.; Spencer, R.W.; Tam, T.F.; Thomas, E.; Copp, L.J. *J. Med. Chem.* **1987**, *30*, 589–591.
43. Krantz, A.; Spencer, R.W.; Tam, T.F.; Liak, T.J.; Copp, L.J.; Thomas, E.M. *J. Med. Chem.* **1990**, *33*, 464–479.
44. Baker, B.R.; Lee, W.W.; Tong, E.; Ross, L.O. *J. Am. Chem. Soc.* **1961**, *83*, 3713–3714.
45. Lawson, W.B.; Schramm, H.J. *J. Am. Chem. Soc.* **1962**, *84*, 2017–2018.
46. Schoellmann, G.; Shaw, E. *Biochem. Biophys. Res. Commun.* **1962**, *7*, 36–40.
47. Wofsy, L.; Metzger, H.; Singer, S.J. *Biochemistry* **1962**, *1*, 1031–1039.
48. Jakoby, W.B.; Wilchek, M., Eds. *Methods Enzymol.* **1977**, *46*.
49. Birchmeier, W.; Christen, P. In: *Methods Enzymol.* **1977**, *46*, 41–48; Jakoby, W.B.; Wilchek, M., Eds.
50. Shaw, E. In: *Enzyme Inhibitors as Drugs;* Sandler, M., Ed. University Park Press: Baltimore, MD, 1980; pp. 25–42.
51. Shaw, E. *Adv. Enzymol.* **1990**, *63*, 271–347.
52. Jencks, W.P. *Catalysis in Chemistry & Enzymology*; McGraw-Hill: New York, 1969; Chapter 1.
53. Kahne, D.; Still, W.C. *J. Am. Chem. Soc.* **1988**, *110*, 7529–7534.
54. Rasnick, D. *Anal. Biochem.* **1985**, *149*, 461–465.
55. Rauber, P.; Angliker, H.; Walker, B.; Shaw, E. *Biochem. J.* **1986**, *238*, 633–640.
56. Shaw, E.; Angliker, H.; Rauber, P.; Walker, B.; Wilkstrom, P. *Biomed. Biochem. Acta* **1986**, *45*, 1397–1403.

57. Halász, P.; Polgár, L. *Eur. J. Biochem.* **1976**, *71*, 563–569.
58. Barrett, A.J.; Kirschke, H. *Methods Enzymol.* **1981**, *80*, 535–561.
59. Katunuma, N.; Kominami, E. In *Current Topics in Cellular Regulation*; Horecker, B.L.; Stadtman, E.R., Eds.; Academic Press: New York, 1983; Vol. 22, pp. 71–101.
60. Baricos, W.H.; O'Connor, S.E.; Cortez, S.L.; Wu, L.-T.; Shah, S.V. *Biochem. Biophys. Res. Commun.* **1988**, *155*, 1318–1323.
61. Delaissé, J.-M.; Eeckhout, Y.; Vaes, G. *Biochem. Biophys. Res. Commun.* **1984**, *125*, 441–447.
62. Kominami, E.; Tsukahara, T.; Bando, Y.; Katunuma, N. *J. Biochem. (Tokyo)* **1985**, *98*, 87–93.
63. Everts, V.; Beersten, W.; Schroder, R. *Calcif. Tissue Int.* **1988**, *43*, 172–178.
64. Poole, A.R.; Tiltman, K.J.; Recklies, A.D.; Stoker, T.A.M. *Nature* **1978**, *273*, 545–547.
65. Sloane, B.F.; Lah, T.T.; Day, N.A.; Rozhin, J.; Bando, Y.; Honn, K.V. In: *Cysteine Proteinases and Their Inhibitors.* Turk, V., Ed.; Walter de Gruyter: New York, 1986; pp. 729–749.
66. Lah, T.T.; Buck, M.R.; Honn, K.V.; Crissman, J.D.; Rao, N.C.; Liotta, L.A.; Sloane, B.F. *Clin. Exp. Metastasis* **1989**, *7*, 461–468.
67. Prous, J.R. (Ed.) *Drugs Future* **1986b**, *11*, 941–943.
68. Prous, J.R. (Ed.) *Drugs Future* **1986a**, *11*, 927–930.
69. Smith, R.A.; Copp, L.J.; Coles, P.J.; Pauls, H.W.; Robinson, V.J.; Spencer, R.W.; Heard, S.B.; Krantz, A. *J. Am. Chem. Soc.* **1988**, *110*, 4429–4431.
70. Schechter, I.; Berger, A. *Biochem. Biophys. Res. Commun.* **1967**, *27*, 157–162.
71. Rich, D.H. In: *Proteinase Inhibitors.* Barrett, A.J.; Salvesen, G., Eds.; Elsevier: New York, 1986; Chapter 4, pp. 153–178.
72. Stein, R.L. *J. Am. Chem. Soc.* **1985**, *107*, 5767–5775.
73. Bender, M.L.; Brubacher, L.J. *J. Am. Chem. Soc.* **1966**, *88*, 5880–5889.
74. Whitaker, J.R.; Perez-Villaseñor, J. *Arch. Biochem. Biophys.* **1968**, *124*, 70–78.
75. Polgár, L. *Mechanisms of Protease Action*; CRC Press: Boca Raton, FL, 1989; Chapter 4, pp. 139–143.
76. Angliker, H.; Wikstrom, P.; Rauber, P.; Shaw, E. *Biochem. J.* **1987**, *241*, 871–875.
77. Stirling, C.J.M. *Acc. Chem. Res.* **1979**, *12*, 198–203.
78. Hammett, L.P.; Pfluger, H.L. *J. Am. Chem. Soc.* **1933**, *55*, 4079–4089.
79. Jaffé, H.H. *Chem. Rev.* **1953**, *53*, 191–261.
80. Lowry, T.H.; Richardson, K.S. *Mechanism and Theory in Organic Chemistry*; 3rd ed.; Harper and Row: New York, 1987.
81. McMurry, J. *Org. Reactions* **1976**, *24*, 187–224.
82. Larsen, D.; Shaw, E. *J. Med. Chem.* **1976**, *19*, 1284–1286.
83. McMurray, J.S.; Dyckes, D.F. *Biochemistry* **1986**, *25*, 2298–2301.
84. Sunko, D.E.; Jursic, B.; Ladika, M. *J. Org. Chem.* **1987**, *52*, 2299–2301.
85. Poulter, C.D.; Muehlbacher, M.; Davis, D.R. *J. Am. Chem. Soc.* **1989**, *111*, 3740–3742.
86. Jencks, W.P. *Adv. Enzymol. Relat. Areas Mol. Biol.* **1975**, *43*, 219–410.
87. Umezawa, H. *Am. Rev. Microbiol.* **1982**, *36*, 75–99.
88. Storer, A.C.; Carey, P.R. *Biochemistry* **1985**, *24*, 6808–6818.
89. Moon, J.B.; Coleman, R.S.; Hanzlich, R.P. *J. Am. Chem. Soc.* **1986**, *108*, 1350–1351.
90. Liang, T.C.; Abeles, R.H. *Arch. Biochem. Biophys.* **1987**, *252*, 626–634.
91. Hanzlik, R.P.; Thompson, S.A. *J. Med. Chem.* **1984**, *27*, 711–712.
92. Hanada, K.; Tamai, M.; Yamagishi; Ohmura, S.; Sawada, J.; Tanaka, I. *Agric. Biol. Chem.* **1978**, *42*, 523–528.
93. Sanner, T.; Pihl, A. *J. Biol. Chem.* **1963**, *259*, 12489–12494.
94. Evans, A.; Shaw, E. *J. Biol. Chem.* **1983**, *258*, 10277–10232.
95. Serjeant, E.P.; Dempsey, B. *Ionisation Constants or Organic Acids in Aqueous Solution*; Pergamon Press: New York, 1979.

96. Kortum, G.; Vogel, W.; Andrussow, K. *Dissociation Constants of Organic Acids in Aqueous Solution*; Butterworths: London, 1961.

97. Strong, L.E.; Brummel, C.L.; Lindower, P. *J. Solution Chem.* **1987**, *16*, 105–124.

98. Strong, L.E.; Van Waes, C.; Doolittle, K.H. *J. Solution Chem.* **1982**, *11*, 237–258.

99. Ludwig, M.; Baron, V.; Kalfus, K.; Pytela, O.; Vecera, M. *Collect. Czech. Chem. Commun.* **1986**, *51*, 2135–2142.

100. Perrin, D.D.; Dempsey, B.; Serjeant, E.P. *pKₐ Prediction for Organic Acids and Bases*; Chapman and Hall: New York, 1981; p. 129.

# INDEX